R. L. Moore

Mathematician and Teacher

©*2005 by*
The Educational Advancement Foundation

Library of Congress Catalog Card Number: 2004113479

Print edition ISBN: 978-0-88385-550-8

Electronic edition ISBN: 978-1-61444-512-8

Current Printing (last digit):
10 9 8 7 6 5 4 3 2

R. L. Moore

Mathematician and Teacher

John Parker

Published and Distributed by
The Mathematical Association of America

SPECTRUM SERIES

Published by

THE MATHEMATICAL ASSOCIATION OF AMERICA

Council on Publications
Roger Nelsen, *Chair*

Spectrum Editorial Board
Gerald L. Alexanderson, *Editor*

Robert Beezer
William Dunham
Michael Filaseta
Erica Flapan
Eleanor Lang Kendrick
Ellen Maycock
Russell L. Merris

Jeffrey L. Nunemacher
Jean Pedersen
J. D. Phillips, Jr.
Marvin Schaefer
Harvey J. Schmidt, Jr.
Sanford Segal
Franklin Sheehan

John E. Wetzel

The Spectrum Series of the Mathematical Association of America was so named to reflect its purpose: to publish a broad range of books including biographies, accessible expositions of old or new mathematical ideas, reprints and revisions of excellent out-of-print books, popular works, and other monographs of high interest that will appeal to a broad range of readers, including students and teachers of mathematics, mathematical amateurs, and researchers.

777 Mathematical Conversation Starters, by John dePillis

All the Math That's Fit to Print, by Keith Devlin

Carl Friedrich Gauss: Titan of Science, by G. Waldo Dunnington, with additional material by Jeremy Gray and Fritz-Egbert Dohse

The Changing Space of Geometry, edited by Chris Pritchard

Circles: A Mathematical View, by Dan Pedoe

Complex Numbers and Geometry, by Liang-shin Hahn

Cryptology, by Albrecht Beutelspacher

Five Hundred Mathematical Challenges, Edward J. Barbeau, Murray S. Klamkin, and William O. J. Moser

From Zero to Infinity, by Constance Reid

The Golden Section, by Hans Walser. Translated from the original German by Peter Hilton, with the assistance of Jean Pedersen.

I Want to Be a Mathematician, by Paul R. Halmos

Journey into Geometries, by Marta Sved

JULIA: a life in mathematics, by Constance Reid

R.L. Moore: Mathematician and Teacher, by John Parker

The Lighter Side of Mathematics: Proceedings of the Eugène Strens Memorial Conference on Recreational Mathematics & Its History, edited by Richard K. Guy and Robert E. Woodrow

Lure of the Integers, by Joe Roberts

Magic Tricks, Card Shuffling, and Dynamic Computer Memories: The Mathematics of the Perfect Shuffle, by S. Brent Morris

The Math Chat Book, by Frank Morgan

Mathematical Adventures for Students and Amateurs, edited by David Hayes and Tatiana Shubin. With the assistance of Gerald L. Alexanderson and Peter Ross

Mathematical Apocrypha, by Steven G. Krantz

Mathematical Carnival, by Martin Gardner

Mathematical Circles Vol I: In Mathematical Circles Quadrants I, II, III, IV, by Howard W. Eves

Mathematical Circles Vol II: Mathematical Circles Revisited and *Mathematical Circles Squared,* by Howard W. Eves

Mathematical Circles Vol III: Mathematical Circles Adieu and *Return to Mathematical Circles,* by Howard W. Eves

Mathematical Circus, by Martin Gardner

Mathematical Cranks, by Underwood Dudley

Mathematical Evolutions, edited by Abe Shenitzer and John Stillwell

Mathematical Fallacies, Flaws, and Flimflam, by Edward J. Barbeau

Mathematical Magic Show, by Martin Gardner

Mathematical Reminiscences, by Howard Eves

Mathematical Treks: From Surreal Numbers to Magic Circles, by Ivars Peterson

Mathematics: Queen and Servant of Science, by E.T. Bell

Memorabilia Mathematica, by Robert Edouard Moritz

Musings of the Masters: An Anthology of Mathematical Reflections, edited by Raymond G. Ayoub

New Mathematical Diversions, by Martin Gardner

Non-Euclidean Geometry, by H. S. M. Coxeter

Numerical Methods That Work, by Forman Acton

Numerology or What Pythagoras Wrought, by Underwood Dudley

Out of the Mouths of Mathematicians, by Rosemary Schmalz

Penrose Tiles to Trapdoor Ciphers ... and the Return of Dr. Matrix, by Martin Gardner

Polyominoes, by George Martin

Power Play, by Edward J. Barbeau

R.L. Moore: Mathematician and Teacher, by John Parker

The Random Walks of George Pólya, by Gerald L. Alexanderson

Remarkable Mathematicians, from Euler to von Neumann, by Ioan James

The Search for E.T. Bell, also known as John Taine, by Constance Reid

Shaping Space, edited by Marjorie Senechal and George Fleck

Sherlock Holmes in Babylon and Other Tales of Mathematical History, edited by Marlow Anderson, Victor Katz, and Robin Wilson

Student Research Projects in Calculus, by Marcus Cohen, Arthur Knoebel, Edward D. Gaughan, Douglas S. Kurtz, and David Pengelley

Symmetry, by Hans Walser. Translated from the original German by Peter Hilton, with the assistance of Jean Pedersen.

The Trisectors, by Underwood Dudley

Twenty Years Before the Blackboard, by Michael Stueben with Diane Sandford

The Words of Mathematics, by Steven Schwartzman

Preface

Robert Lee Moore (1882–1974) was a towering figure in twentieth century mathematics, internationally recognized as founder of his own school of topology, which produced some of the most significant mathematicians in that field. The 50 students he guided to their PhDs can today claim 1,678 doctoral descendants. Many of them are still teaching courses in the style of their mentor, known universally as the Moore Method, which he devised. Its principal edicts virtually prohibit students from using textbooks during the learning process, call for only the briefest of lectures in class and demand no collaboration or conferring between classmates.[1] It is in essence a Socratic method that encourages students to solve problems using their own skills of critical analysis and creativity. Moore summed it up in just eleven words: 'That student is taught the best who is told the least.'[2] A controversial figure, both for his style of teaching and his strong views, Moore was once described as a 'Mr. Chips with Attitude'[3]. The attitude was an integral part of the method of his tuition to generations of students, and it also applied to the unique manner in which he discovered and developed mathematical talent among the young men and women he encountered during his 60-year teaching career. Moore was a born iconoclast, much given to challenging the status quo of academia and the conventional modes of scholarship of his time. With his snowy white hair immaculately combed, his piercing blue eyes always seeking exciting new proofs to complex problems, and

---

[1] Exceptions were Moore's calculus and analytic geometry courses in which textbooks were used for setting problems. His doctoral students were allowed to refer to textbooks mainly to ensure their theses were original.

[2] From *Challenge in the Classroom*, a documentary film on the life and work of R.L. Moore, produced for the Mathematical Association of American in 1966 (now re-issued as *The Moore Method: A Documentary on R.L. Moore*, in the MAA's publications division).

[3] A former Moore student, in conversation with the author.

his well-muscled boxer's physique clad in dark three-piece suits and old-fashioned, hand-made, laced-up black boots, he was a commanding presence on the campus of The University of Texas for 49 years, encouraging and cajoling students in his deeply resonant voice into surpassing their own wildest dreams of mathematical attainment.

Therein lies a vital additional ingredient to his story. It will become evident in the ensuing pages that while the life of this mathematical pioneer is the central theme, there is also an unfolding drama, for such it is, that encompasses what may now be seen as the legacy of the man, and the implications of his work and teaching for mathematics and science in modern times, and for the future. Indeed, the Moore Method has, in recent times, attracted a growing revival of interest, partially because of the activities of former Moore students in places of higher learning throughout the United States of America, Europe, Asia and the Far East. In America, increased emphasis on K–12 education starting in kindergarten through grade 12 following the 1996 review of the National Science Foundation, enshrined in its pamphlet *Shaping the Future*, has contributed to this resurgence.

These chapters, therefore, go beyond a mere biographical study and are intended through closer examination of his personal brand of discovery learning to suggest ideas and possibilities for rekindling mathematical interest and ability among the young. The relevancy of his work on present and future generations in this regard may be discovered by an understanding of the impact of his technique, gleaned from the contributions of those who studied with Moore, while at the same time forging an enlightening and authoritative profile of a man whose influence on the American mathematical community continued over six decades, and remains strong 25 years after his death. That insight, it will be seen, is drawn from some of the most famous mathematical names in America and Europe from the late 1890s onwards, first those who influenced Moore during his own studies, and later those highly respected scholars who had once been his students.

As to his own research, which again was world renowned, he was the first native-born American to become a Visiting Lecturer for the American Mathematical Society, of which he also became president. He published 65 papers and a book which is still referred to 70 years later and which has been the subject of literally hundreds of papers by other mathematicians around the globe. Equally fascinating as a sub-text to his story are the achievements of his students: three of them followed him as president of the American Mathematical Society, three others became vice-president, and another served as secretary of the AMS for many years. Five served as president of the Mathematical Association of America and three, like Moore, became members of the National Academy of Sciences while most of the rest became highly respected and well published researchers and teachers in leading American universities.

Thus, apart from his personal contribution, Moore had a profound influence on American mathematics and the teaching thereof. Given that the presidencies run for two years, his former students were at the helm of one or another of the two major mathematical organizations in the US for a third of the second half of the twentieth century. In 1938 Moore had his photograph taken for his presidency of the AMS and the photographer suggested that he might airbrush from the print a wart on his subject's face. 'Warts and all,' replied Moore. And thus, in this account, I have followed the same guidance. This then is the extraordinary story of R.L. Moore and how he developed the Moore Method, which was bigger than the man (with all his faults and idiosyncrasies), how it equipped its beneficiaries to excel in fields of excellence other than mathematics, and how it has been modified to meet the educational requirements of today. John W. Green, Principal Research Biostatistician at Dupont, says: 'I attribute whatever measure of success I've had as an industrial statistician to the training and experience that I received working under R.L. Moore. There certainly have been other influences in my life, but what I gained and regained with Dr. Moore has had a very profound influence on many of the decisions I've made and how I conduct myself.'[4] There were, however, elements of Moore's teaching that drew criticism and on occasion antagonism from observers and students alike. Some were simply not suited to his style, or the man himself, and psychological bruising was not uncommon. On the one hand, he showed immense patience and dedication, while on the other he could be unflinchingly blunt in rejecting students he either disliked or excluded because he felt they had too much knowledge for his classes, where the entry standard was a virgin mind, untainted by earlier exposure to the work Moore would propose in his courses. There were

---

[4] From *Topology to Statistics: The Influence of R.L. Moore's Training on an Industrial Statistician,* by John W. Green, Principal Research Biostatistician, Dupont, at the 4th Legacy of R.L. Moore Conference 3–5 May 2001.

personal beliefs and attitudes, too, born out of his Southern upbringing, that caused dissent and argument in his classroom, much of it deliberately inspired by himself as part of his pedagogical experiments. In the main, however, Moore is remembered as a teacher of mathematics with honest affection and appreciation for the manner in which he drew from his students latent talent, even from a number who had no intention of becoming mathematicians but eventually went on to become leaders in their fields. An example of that spirit is engendered in a note written to him by one of his students, John Mohat, who saw a passage in a work of fiction that he felt summed up exactly the way Robert Lee Moore, through his style and ideas, fired up their creative talents. The passage came from John Steinbeck's *East of Eden*, which Mohat sent to Moore shortly before Moore retired at the age of 86:

'Sometimes a kind of glory lights up in the mind of a man. It happens to nearly everyone. You can feel it growing or preparing like a fuse burning toward dynamite. It is a feeling in the stomach, a delight of the nerves, of the forearms ... A man may have lived all of his life in the gray, and the land and trees of him dark and somber. The events, even the important ones, may have trooped by faceless and pale. And then— the glory ... Then a man pours outward, a torrent of him and yet he is not diminished. And I guess a man's importance in the world can be measured by the quality and the number of his glories. It is a lonely thing but it relates us to the world. It is the mother of all creativeness, and it sets each man separate from all other men.'

Contents

Preface vii
Acknowledgements xiii
1: Roots and Influences (1882–1897) 1
2: Of Richest Promise (1897–1902) 19
3: On to Chicago (1903) 39
4: A Veritable Hothouse (1903–1905) 57
5: Uneasy Progress (1905–1908) 73
6: A Settling Experience (1908–1916) 91
7: Back to Texas (1916–1920) 109
8: A Rewarding Decade (1920–1930) 125
9: A Change of Direction (1930–1932) 143
10: Politics and Persuasion (1933–1938) 161
11: Moore the Teacher: A New Era (1939–1944) 177
12: Blacklisted! (1943) 193
13: Class of '45 (1945) 207
14: Clash of the Titans (1944–1950) 225
15: His Female Students 241
16: Moore's Calculus (1945–1969) 257
17: Changing Times (1953–1960) 275
18: Axiomatics Continued: (1953–1965) 293
19: The Final Years (1965–1969) 313
Appendix 1: The Moore Genealogy Project 333
Appendix 2: The PhD Students of R.L. Moore 339
Appendix 3: Publications of Robert Lee Moore 359
Appendix 4: Descriptions of Courses 365
Bibliography 367
Photo Credits 373
Index 379

Acknowledgements

The author wishes to express his thanks to the many people who gave of their time and energy to help bring this project to its conclusion, by way of interviews, correspondence, other written or recorded materials and guidance. There is a special thank you to be recorded for the staff of the Archives of American Mathematics at the Center for American History at The University of Texas at Austin and Ralph Elder, Assistant Director for University Archives and Facilities, for their invaluable assistance in tracing the voluminous documents examined for this work. Mention must also be made of the tireless and considerable efforts of Mr. Harry Lucas, Jr. and his team at the Educational Advancement Foundation, Austin, combining their activities with The Legacy of R.L. Moore Project, and Dr. Albert Lewis for the invaluable contribution of his personal knowledge, guidance and encouragement.

Much, however, is owed to one of the most stalwart figures and prime movers of the Project and a tireless contributor to its aims and objects, the late Dr. Ben Fitzpatrick. Harry Lucas recruited Ben Fitzpatrick to the Moore Project in 1996, and he became Lucas's main liaison contact with the mathematics community as foundation work began, and thereafter in all aspects of the project. He was especially successful in instigating and collating a substantial archive of Oral History for The Legacy of R.L. Moore Collection now housed at the Archives of American Mathematics at the Center for American History, The University of Texas at Austin.

Other tape-recordings and transcripts from this archive include those conducted by Douglas Forbes, for his 1971 dissertation: *The Texas System: R.L. Moore Original Edition.* Tape-recorded contributions consulted by the author include those by R.D. Anderson, Steve Armentrout, Joanne Baker, Mrs. B.J. Ball, Lida Barrett, Mary Bing, Ed Burgess, Howard Cook, J.L. Cornette, Jerome Dancis, James Dorroh, W. Eaton, J.W. Green, M.E. Hamstrom, F. Burton Jones, I.W. Lewis, Lee

Mahavier, Jean Mahavier, W.S. Mahavier, Ted Mahavier, E.E. Moise, John Neuberger, James Ochoa, Coke Reed, G.M. Reed, M.E. Rudin, Carol Schumacher, Beauregard Stubblefield, Frank Vandiver, John Worrell, and Gail Young.

1

Roots and Influences
(1882–1897)

Early life and family connections, a generation away from two US presidents and the first president of the Confederacy. Educated privately by a Scot who had served Maximilian in Mexico.

Moore at age ten

The rough and tumble of Texas frontierism in the 1890s may seem a curious and perhaps unexpected point in time to discover the inspirations of a man who was to take his place at the cutting edge of mathematical research in the United States over the ensuing decades. In fact, it offered a setting in which the very genesis of the do-it-yourself principles of what became universally known as the Moore Method emerged and went on to influence his entire career. The defining moment occurred when, at the age of fourteen, Robert Lee Moore decided he would apply for entry to The University of Texas at Austin, founded in the year of his birth, to pursue his fascination with mathematics. To achieve that goal, he had to have calculus but it was not taught at the Waldemar Malcolmson Academy, a private school run by a colorful, enigmatic Scot who had otherwise provided him with a very decent education, and most importantly of all instilled within him a yearning for higher education.

Malcolmson did, however, possess an old calculus book that he lent to his student for private study, although he offered no other tuition. Moore quickly lost patience with the imprecise language of the book and wrote to The University of Texas to ask for a copy of the calculus in use there. He received it by return post and began teaching himself. He recalled in later life he would read a statement of a theorem, while covering the portion of the page that gave the proof. If he failed to prove the theorem after a reasonable length of time, he would uncover the first line of the proof, read that, and then try to prove the theorem without further assistance. If this failed, he would uncover the next line and so on until a satisfactory resolution was achieved. It was not a pleasant experience for him to uncover even a part of the proof and, in those instances that he uncovered much of the proof, he felt he had failed.[1] This was, in essence, the beginning of R.L. Moore's journey of self-discovery that would in turn lead to the creation of a teaching style in which he encouraged his own students to find their own way through the mathematical maze, with a minimal reliance on textbooks, and in many situations, none at all. This précis is, of course, an over-simplification of what eventually transpired and, indeed, the development of the Moore Method, running parallel to Moore's own career, covers far broader concerns than calculus. We will

---

[1] J. Eyles, The Importance of R.L. Moore's Calculus Class, dissertation, 1998, p. 9; from the Center for American History, UT Austin.

delve deeply into the many influences brought to bear on the way in which he taught his students, resulting from his upbringing in those heady days in Texas as well as other subliminal aspects that were surprisingly unknown to himself until much later in his life.

Many of them long pre-dated the moment that Robert Lee Moore joined this world,[2] at a time when America was still in the healing process from the wounds and consequences of the Civil War. With him came an ancestry reflecting that conflict in a quite remarkable way and, when combined with his own future career, we are provided with a classic American tale. It is one that cuts a swathe across defining events in a century of history both in mathematics and the social and political changes that saw the emergence of one nation under the flag, all men supposedly equal under the law, though not necessarily at peace with one another, nor issues of the Civil War fully resolved. Yet, the probability is that Moore himself did not fully appreciate the extent or importance of his own historical connections until quite late in a lifetime immersed in mathematical study and teaching. In fact, he reached adulthood before discovering that he did not come from dyed-in-the-wool Southern stock and it did not especially please him.[3]

This discovery was made during the compilation of a family tree for which over a period of several years he painstakingly researched through distant generations of the American-born families and their European connections and at the end of what, by any measure, was a masterly example of genealogical detective work, he produced a chart of lineage that would embrace two American presidents, the president of the Confederacy and three European royal houses.[4] Moore traced his ancestry through 300 years of American history to one John Moore who was born in England and settled in Sudbury, Mass. where he bought land and built a house in 1642. He married Elizabeth Whale, later moved to Hartford, Connecticut, and died on 6 January 1673. It was from Hartford 185 years later that the family tree, so steeped in New England traditions, developed a southern branch that was to plunge them into the midst of the Civil War and set sibling against sibling.

---

[2] 'No, he never really joined the world' wrote Joe B. Frantz, in his book *The Forty-Acre Follies*, an unofficial history of The University of Texas at Austin where he was a professor of history for 40 years.

[3] Joe B. Frantz, page 113.

[4] See Appendix One: The Moore Genealogy.

R.L. Moore's parents, Charles and Louisa Anne

Moore's paternal grandparents, John Stephens Moore and Caroline Abigail Cowles Moore, were married in 1823 and had six children, five boys and one girl, the first born at Farmington, CT and the others at Glastonbury, CT between 1825 and 1846. John Moore was a physician, practicing in Vermont and Connecticut until the mid-1850s when he informed his family that they were all going on an adventure to the mountainous regions of North Carolina where he wished to study natural drugs and Indian remedies. When the family eventually returned to Connecticut, the eldest son Henry remained in North Carolina and eventually moved to Kentucky. His letters home encouraged R.L. Moore's father, Charles J. Moore, to join him there.

Charles rode straight into the developing bitterness over the abolition issue that reached its point of no return when the Supreme Court's ruling in the Dred Scott case made slavery legal in all territories. Charles arrived in Kentucky in 1858 — the year that Senator William H. Seward of New York referred to differences between North and South as an 'irrepressible conflict'.[5] Such was the feeling in the South by then that

---

[5] William Henry Seward, On the Irrepressible Conflict, delivered at Rochester, New York, October 25, 1858.

Charles had to hide his northern origins and accent, otherwise he would have found neither work nor lodgings. He all but discarded his northern connections as he settled into his new surroundings and became in every respect except birth, a Southerner. When Jefferson Davis and Robert E. Lee ordered the attack on Fort Sumter, Henry and Charles Moore volunteered for service in the Confederacy. Hartford, Connecticut County History would record that the brothers 'gave valiant service to the South'.[6] In the north, meanwhile, two of their brothers enlisted for the Union cause while Henry and Charles in the south went into the Orphan Brigade, 2nd Kentucky Regiment, which came into being on 28 October 1861 at Bowling Green, Kentucky and fought as a unit for the first time at Shiloh. During 1862 the brigade found itself in Vicksburg. Charles Moore was lucky to survive. After just six weeks, the brigade had lost more than 1,300 men, many more from malaria and fever than from battle wounds, with just 542 left standing. The unit was pulled out only to return a few months later, replenished with new recruits in an attempt to save the Confederate garrison from a six-week siege by the Union army. They failed and instead moved to take up a rearguard position with the Army of Tennessee, fought in the Battle of Chattanooga, were in action continuously during 100 days of the Atlanta campaign and suffered 1,860 men killed or wounded. At the end of it, less than fifty men of the original brigade survived without wounds. Both the Moore brothers did indeed survive and with the south desolate and destroyed, joined the rest in attempting to pick up the pieces of their lives. There was no question of them returning North and, in any event, their roots were planted firmly into Southern soil.

The two brothers soon came into contact with another Moore family from Virginia. The upshot of this contact was that the two brothers became engaged to two sisters of that family. Henry married Mary E. Moore and on 10 May 1866 Robert Lee Moore's parents, Charles and Louisa Anne were wed, thus forging the important link on his mother's side of Moore's ancestral chart, one generation away from former US President Zachary Taylor, and to Jefferson Davis who was Taylor's son-in-law. Across the Southern states, the Reconstruction years following the war were marked by a great movement of population, black and white, as they attempted to escape desolation and poverty in a ruined economy. Charles Moore surveyed his options and decided to migrate

---

[6] County Records, City Hall, Hartford, CT.

Robert Lee as a toddler

west from Kentucky with their, then, four children to Dallas, Texas in 1877 to set up a hardware, grain and grocery store just off the town square.[7] It turned out to be a good choice, even if the formative years of the Moore children would be set against the backdrop of what was initially a wild and wondrous place. Dallas was still very much a frontier town, having expanded from John Neely Bryan's trading post for westward bound wagon trains to classic boomtown in barely 25 years. The Confederacy had used it as an administrative center, as did the incoming Union troops and it retained its importance after the war as commercial enterprises blossomed. The addition of a railhead and the influx of farm tool manufacture and the buffalo trade set the population expansion into fast mode. This and the new order of social life after the war also brought its own challenges. With the cotton industry almost in ruins, Dallas saw an influx of families from other less prosperous Southern towns, hoping to rebuild their fortunes. They were to be joined by the first Texas slaves to be freed on 19 June 1865 and others followed, making the trek from distant parts to join the bustle of Dallas

---

[7] On 22 November 1963, shots rang out over the site of the Moores' long-demolished store, as President John F Kennedy was assassinated. The store would have stood on what was by then the famous grassy knoll.

Roots and Influences (1882–1897)

Moore's Feed Store, Dallas, Texas

growth. It is also worth noting that the Ku Klux Klan made its first recorded appearance there as early as 1868.

The first passenger train, the Houston and Texas Central, steamed into Dallas amid great excitement on 16 July 1872 and by the end of that year, the population had doubled. New businesses and buildings began to appear daily. Hotels were being built, water and gas utilities arrived, telegraph lines were installed as the town established itself as a major dispersal center for raw materials, to be shipped to the South and East. Prosperity also attracted outlaws, and some famous names flickered across the Dallas scene. Belle Starr, who had consorted with the James Gang and the Younger Brothers, was already notorious when she arrived in Dallas, first as a dance hall singer and later running a livery stable from which she sold stolen horses and harbored outlaws. Wyatt Earp's pal John 'Doc' Holliday practiced as a dentist there for three years before he shot a man and 'was invited to leave town'.[8] The out-

---

[8] Timeline, Dallas Historical Society (www.dallashistory.org).

law Sam Bass also operated from the Dallas area, and robbed four trains in two months during the spring of 1878 before he was shot dead by Texas Rangers in an ambush near Round Rock. There was refinement, too. Patrons at the Dallas Opera House were able to see James O'Neill in *The Count of Monte Cristo,* while in 1887 customers paid $15 a head to witness Edwin Booth perform *Hamlet.* It was into this cauldron of activity that Charles Jonathan Moore brought his wife and four children, Henry, Jennings, Eleanor and Arthur in 1877, to which were added Robert Lee in 1882,[9] and Caroline Louisa in 1887. Only fragmentary accounts of their lifestyle are available, but it is clear from family correspondence that Charles had, by no means, relaxed the traditional Southern values he had embraced so wholeheartedly before and during the Civil War. His northern roots remained hidden and his children were adults before they learned that they had relatives in Connecticut. In fact, the first recorded visit by a member of the Southern branch of the Moore family to those in the North was in 1907 when Robert Lee stopped off at the home of Charles's younger brother Arthur in Glastonbury while he was at Princeton. There are indications that Charles Moore may have run his home as a not-so-benevolent dictator. Given the surroundings in which he was raising his children, the era and his personal adoption of the Southern code, this was not an unnatural or even unexpected aspect of family life and on Robert Lee's part there appeared to be no lingering animosity in later life. He kept up regular correspondence with his father after he had left home for university and the letters, returned to him after the old family home was sold, displayed nothing but respect and a strong bond of affection. Charles had, after all, ensured that all of his children were given as good a start as he could afford and in Dallas where education had never been a great priority, that meant private schooling. Public school funding was not adequately supported by taxation in Dallas until the early 1900s and a number of private establishments sprung up to fill the void.

Robert Lee, along with two of his brothers, was fortunate indeed in the school chosen for them by their parents. They placed the boys in the hands of a dour Scot named Waldemar Malcolmson who ran an establishment with his wife close to the Moore homestead on Live Oak Street. He called himself a professor, though the grounds for such a title

---

[9] 'A name which fit, except that he lacked that Southern immortal's naturally gentle and generous nature,' Joe B. Frantz, *The Forty-Acre Follies,* page 113.

were somewhat unclear, as was his past. It hardly mattered given the alternatives but, as it turned out, Malcolmson was to be a significant guiding influence on the young Robert Lee and once again, there were interesting historical connections that would add additional weight to the boy's education; not for him, the attentions of a young and inexperienced school teacher, as was often the case in these establishments.

Malcolmson was a well-educated man of the world who spoke fluent French and Spanish. He had traveled to Mexico in 1861 as an officer in a tri-nation army formed under French leadership by Napoleon III to bring to heel the government of Benito Juárez, which owed a huge foreign debt to a consortium of European nations, mainly France, Great Britain and Spain. Napoleon III sent Archduke Maximilian of Austria as the new puppet 'king' of Mexico along with a large army to ensure his safe installation on the Mexican throne. Juárez was driven northward but counter-attacked and Napoleon's forces, crippled by disease that claimed many thousands of lives, evacuated Mexico in March 1867. Maximilian refused to abdicate, and his small remaining force was surrounded, starved, and finally capitulated on May 15, 1867. In spite of petitions from European leaders, and the Pope, Maximilian was executed on a hill outside Querétaro the following month. Malcolmson, who is believed at one stage to have been on the staff of Maximilian, escaped and with many others made a hasty retreat north into Texas, eventually arriving in Dallas with his wife in the early 1870s.

They founded a school under the name of the Dallas Lyceum, although it changed names several times. Robert Lee joined the school register when he reached his eighth birthday. Malcolmson ran a good school, and became noted for the quality of instruction and the variety of subjects. They included arithmetic, algebra, trigonometry, geography, history and penmanship. Seniors were invited at extra cost to select courses in higher mathematics, German, Spanish, French and Latin. Anne Atkins, who attended his school in the 1880s, left this account of Malcolmson:

'He was a rare type and a real scholar, with an odd approach to many things and unorthodox to a degree. He believed that Bacon wrote Shakespeare and was indifferent as to truths in the world around us. But when it came to mathematics, French or Latin, he ranked with the best and prepared another young playmate so well that he entered The University of Texas with the highest grades and became one of the most distinguished members of the State Bar, reading French and Latin with the greatest ease to the end of his life. The professor had an odd way of impressing facts upon us and one thing I have never forgotten: the names of the presidents of the United States up to Cleveland's first administration. This was through learning the final initial of each. From the beginning he gave me, I have gone to a wide development of the French language and literature.'[10]

Other surviving papers from the era in collections maintained by the Dallas Historical Society include an intriguing statement of intent to his clients by Professor Malcolmson that provides a basis for speculating that it might well have sown the seed for the unique style of teaching eventually adopted by Robert Lee Moore: 'The method of instruction adopted is concise and lucid, not burdening the minds of young scholars with useless matter, but omitting nothing that is necessary to a thorough, practical course of instruction. The Lyceum aims to be second to no institution of learning in Dallas and a student will be able to receive instruction in all the branches taught at schools of the highest grade without leaving the building.' And so, Robert Lee Moore began his studies with Malcolmson and remained largely under his tuition until it was time to move on to university. It was a good grounding, and one in which the teacher encouraged his pupils into home study in areas in which they showed an aptitude. In Moore's case, his fascination with mathematics was evident before he entered his teens and a collection of grade reports among his papers relating to his time at Malcolmson's school reflect an unbroken line of As for mathematics. On that basis, Malcolmson had already singled out Moore as a potential candidate for The University of Texas at Austin, which had opened its doors in 1883, and the teacher was anxious from his own standpoint to have students passing through his establishment to successful higher education.

He had already made contact with a number of the teachers and departmental heads at Austin, including George Bruce Halsted, a brilliant, if eccentric, mathematician who accepted a professorship at Austin in 1884. We will meet Halsted later, looming large in our subject's early career, although he was already in the foreground of Moore's immediate future as he dedicated himself to securing a place at Austin. Encouraged by Malcolmson, he began earnest preparations to gain entry to the university. He wrote Dr. Halsted and obtained the calculus and other books for mathematical study in preparation for his anticipated

---

[10] Collections and other documents, Dallas Historical Society.

arrival at Austin and, as an additional brainteaser, he learned shorthand, which he used to make copious notes in the margins of books.

Knowing full well he would also need Latin for a university place, he wrote in his newly commenced diary, written on pages of an Ivory Soap order pad[11] doubtless picked up from his father's store, that he intended to begin studying Latin at the very latest in January 1898. He was considered sufficiently proficient by the summer of that year and duly applied for enrollment to The University of Texas while still only fifteen years old. With his application went a letter of recommendation from Waldemar Malcolmson dated 19 July 1898: '...the hardest [working] student I have ever had, nothing pleases him but continuous study, he can read Spanish at sight having studied with me for seven years. In mathematics, he has used Professor Halsted's books as well as others on the same subject and has gone through his Calculus. He takes special pleasure in the study of mathematics in many cases solving, without aid, the most difficult propositions in an original manner.... [T]his boy Moore is without exception the student who has shown the greatest application to his studies. It in fact amounts to a passion with him.'[12]

The University of Texas was as old as Robert Lee Moore, building work having begun in 1881 after political machinations had delayed funding. The Texas Constitution of 1866 directed the legislature to put the university in operation at an early date. In 1871 the legislature established the Texas Agriculture and Mechanical College, but the university was postponed. The Texas Constitution of 1876 specified that the legislature, as soon as practicable, was to establish, organize, and provide for the maintenance and support of a 'university of the first class' to be located by vote of the people and styled The University of Texas, for promotion of the study of literature and the arts and sciences. An agricultural and mechanical branch was mandated. The same article of the constitution made A&M a branch of the university and also ordered the legislature to establish and maintain a college or branch university for instruction of black youth, though no tax was to be levied and no money

---

[11] Commencement date, June 1897, partly illegible and some notes in shorthand and Latin; R.L. Moore Papers in the Archives of American Mathematics at the Center for American History, The University of Texas at Austin (referred to as the R.L. Moore Papers in the AAM).

[12] Waldemar Malcolmson to Dr. Garrison (The University of Texas), 19 July 1898, from the R.L. Moore Papers in the AAM.

appropriated out of the general revenue for such a school or for buildings of The University of Texas.

This prohibition prevented establishment of a branch of the university for African-Americans, although Austin was selected for its site in 1882. The campus of the main university originally consisted of the forty-acre tract on College Hill set aside when Austin became the state capital, and it opened in 1883 with a strong staff, predominantly weighted in teachers from the South, who included three former colonels in the Confederate army and one professor who had been chief of staff to General Stonewall Jackson. The arrival of Professor Halsted, a Princeton man, the following year rather put the cat among the pigeons. While evidently in sympathy with Southern ideals, his eccentricity did not do him any favors among those stern ex-colonels. By the time R.L. Moore made his application, the university was well established and only the previous year had taken steps to raise entry standards and to strengthen its catalog. Beneficial from Moore's point of view was the fact that the School of Mathematics had established a new instructorship to enable the teaching force to divide the freshman class into six sections instead of two, with two instructors of mathematics fully committed. The university catalog pointed out that smaller sections would enable students to get more individual instruction from his professor; enthusiasm and interest would be aroused and 'a desire created to pursue the subject in its higher branches'.[13] In R.L. Moore's case that is exactly what happened. He took his entry examination in May 1898 and on 11 July, recorded in his diary 'received news today from the University; examinations papers have been pronounced satisfactory'.[14] He enrolled in UT to begin an association that would last more than seventy years, entering a school with almost four hundred students, and he never once wavered in his intention to become a mathematician. He arrived in Austin on 6 September, staying initially at the Driskill Hotel until he found lodgings. He had been given an advice sheet from the university that informed him:

'Some students do their own cooking and housework, and are thus enabled to live at an expense not exceeding $5 a month. They serve as waiters in boarding houses, or do other work in private families, which relieves them of expense of board. Regular board, with furnished room,

---

[13] *The University Record* (The University of Texas, Austin, Texas, 1898) Vol. 1, No. 1, December 1898, pp. 32–33.

[14] Ivory Soap pad diary.

Young Master Moore

can be obtained near the University, at prices varying from $12.50 to $20 a month. A large number of students pay the former price. In University Hall board, furnished room, lights and fuel may be obtained for $15 a month. Two large student clubs have, during the present session, further reduced the price of board and lodging... [to] an average expense of $11.25 monthly for each member.' Traylor adds: 'Many of the students support themselves by doing work in private families, milking cows, making fires, cooking, tending the horse; others waited on the tables in boarding houses or attended to the rooms; others taught, acted as clerks, stenographers, typists, accountants, or surveyors.'[15]

---

[15] Traylor, pp. 13–14.

So, only the prices changed down the decades. As to the academic standards that confronted him in 1898, Austin had put together a strong catalogue and had brought together a broad based and experienced staff. As already noted, admission standards had already been raised as demand for university education continued to rise. For a BA course, for example, candidates were to take entrance examinations in English, History and Mathematics and in Latin or Greek. They would be tested on their knowledge of the elements of English Grammar and English Composition, and 'especially upon their ability to write paragraphs of idiomatic English properly spelled and punctuated.... subjects for such paragraphs will be taken from the books named below, whose subject matter the candidates must be familiar with: [Joseph Addison's fictitious] Sir Roger de Coverly Papers in *The Spectator*,[16] Dryden's *Palamon and Arcite*, Cooper's *The Last of the Mohicans*, and Burke's *Speech on Conciliation with America*.'[17]

In Latin, candidates were to become proficient in grammar, 'with special stress upon inflections and the syntax of the simple sentence; the translation of elementary English prose into Latin; in *Viri Romae*, any books of Caesar, the four lives of Nepos that bear upon Roman history (Hamilcar, Hannibal, Cato, Atticus), any four orations of Cicero and the first book of Virgil's *Aeneid* with the scansion of the dactylic hexameter. Greek would be examined on inflections and syntax; in any three books of Xenophon's *Anabasis*; in the translation of easy Greek at sight; in the translation of elementary English prose into Greek. Knowledge of accent [was] required.'[18]

The courses offered at that time in the School of Pure Mathematics included: 1] Spherics, Solid Geometry, Algebra, Plane and Spherical Trigonometry, with Applications to Surveying and Navigation. 2] Conic Sections, Analytical Geometry. 3] Calculus for Physics and Engineering. 4] Differential and Integral Calculus. 5] Integral Calculus, Differential

---

[16] Addison's contributions to *The Spectator* are said to have perfected the essay as a literary form. His prose style was the model for pure and elegant English until the end of the 18th century; his comments on manners and morals were widely influential in forming the middle-class ideal of a dispassionate, tolerant, Christian world citizen. His fictitious Sir Roger De Coverly Papers, according to William Makepeace Thackeray, give a full 'expression of the life of the time; of the manners, of the movement, the dress, the pleasures, the laughter, and the ridicules of society..

[17] *The University Record*, Vol. 1, No. 2, April 1899, pp. 102–103.

[18] Ibid.

Calculus, and Differential Equations, for Physics, Engineering, and Economics. 6] History of Elementary Mathematics. 7] Advanced Integral Calculus: Definite Integrals, Differential Equations, Functions of a Complex Variable. 8] Modern Geometry, Metric Geometry, Recent Geometry. 9] Geometry of Position. 10] Theory of Equations, Theory of Functions. 11] History of Mathematics. 12] Noneuclidean Geometry. 13] Hypergeometric Functions. 14] Algebra of Logic.[19]

On his makeshift Ivory Soap pad diary, Moore kept up a running commentary of his life at the University. The entries, once again, pointed to a young man intensely interested in study, the day's classroom activity, work done, comments of teachers and so on, while reflecting little social activity apart from church and long walks. Many of the entries were timed late at night, Moore having spent the evening studying. His first two sets of grades provided him with As in his mathematics and Latin courses, a B in physiology and hygiene, a C in English and Ds and Es in the remainder. He also took an examination in Spanish with the freshmen and passed, thus to get a credit for Freshman Spanish. Later, he also obtained a credit for Freshman German. But it was mathematics that had gripped his imagination and Halsted soon recognized this fact. Moore records a conversation he had with his teacher in which he was admonished for talking too fast. The diary relates a conversation entry for 12 January 1899 when Halsted was talking of recommending two of his students for a teaching fellowship at the University that year. Halsted had clearly penciled him in as a future candidate for that role:

'Mr. Moore, if you are going to make mathematics as a profession, you should pay more attention to the pedagogical and learn to talk slow.'

'How old are you, Mr. Moore?' [asked Halsted.] 'Sixteen,' I replied.

'It is a great advantage to be young,' Halsted continued. 'You are gifted in mathematics.... I do not think there is anyone here more so.... But there are other things to be considered ... and you being young have time to see to them. Demand in mathematics is always greater than the supply.'

Moore added a further note in his diary of a later conversation in which Halsted told him, 'If you go on, your career is assured.' In February, Halsted confirmed he intended at some point in the future to put Moore's name forward when he promised, 'If everything turns out

---

[19] *The University Record,* Vol. 1, No. 2, April 1899, pp. 105–109, 111.

all right in the mathematical departments, I don't know but what I will recommend you as a student assistant.'

This additional praise came after his student's encouraging start to a course on noneuclidean geometry, which began in the New Year of 1899. Thus, Halsted was already guiding the young Moore along parallel lines to his own interests as is indicated in Moore's diary entries for May, 1899 when he makes reference to studying Bolyai. The work was a translation of Bolyai's original by none other than Halsted himself, published under the title of *The Science Absolute of Space Independent of the Truth or Fallacy or Euclid's Axiom XII*. Bolyai is mentioned increasingly in Moore's notes and it is possible, even at that early stage of Moore's career, that Halsted had developed a somewhat romantic notion that here, in this new University of Texas, from the unlikely place of Dallas, he had discovered a new János Bolyai. The ingredients were not dissimilar and Halsted knew the story well. Born in Transylvania, Bolyai had mastered the calculus at the age of thirteen under the guidance of his father, Farkas. He became an accomplished linguist and between 1820 and 1823 he prepared a treatise on a complete system of noneuclidean geometry, but before the work was published, he discovered that Carl Gauss had anticipated much of his work. Although Gauss had never published his work in this area, this was a severe blow to Bolyai. However, when Bolyai's work was published in 1832 as an Appendix to an essay by his father, Gauss wrote to a friend saying, 'I regard this young geometer Bolyai as a genius of the first order.'[20]

To Bolyai's father he wrote, 'To praise it would amount to praising myself. For the entire content of the work ... coincides almost exactly with my own meditations which have occupied my mind for the past thirty or thirty-five years.'[21]

That Halsted had Bolyai in his mind as he began to take a deeper interest in Moore, noting his characteristics, intensity and ability, was to some extent borne out by a letter Halsted wrote to him a few years later with glowing largesse: 'I have always held that your work is the most profound, unexpected and epoch-making ever contributed to mathemat-

---

[20] *The University Record,* Vol. 1, No. 2, April 1899, pp. 105–109, 111.

[21] Information on Bolyai can be found at www.gap.dcs.st-and.ac.uk/~history/mathematicians/bolyai.html.

Roots and Influences (1882–1897)

Moore and unknown person in cap and gown

ics by an American.... Like John Bolyai, you had from the very first *one* who appreciated you.'[22]

That kind of support for Moore's ability had been growing in the mind of his teacher from the early days, and his praise was carefully administered at the beginning so as not to spoil his protégé. Moore seems to have understood.

Having been invited to Halsted's house where a reception was being given for a member of the staff, Moore indicated in his diary that he had no wish to involve himself in this social activity: 'I tried to excuse myself on ground of examinations ... telling him something like "I will try to come".... But I won't go, very sorry that I can't well do it for I would hate to offend such a true friend as I think Dr. Halsted is to me and whom I like so much and have so much respect for.'[23]

---

[22] Halsted to Moore, 12 December 1916, R.L. Moore Papers in the AAM.
[23] R.L. Moore's entry in his Ivory Soap pad diary, 8 June 1899, R.L. Moore Papers in the AAM.

2

Of Richest Promise
(1897–1902)

Professor G.B. Halsted had introduced his protégé to the work of Bolyai, Lobachevsky and Hilbert, and Moore, at 18, makes his name producing a 'delightfully elegant and simple proof' demonstrating the redundancy of one of Hilbert's geometry axioms, which draws the attention of E.H. Moore.

*1898 Christmas holidays
R.L. with his sister,
Caroline Louisa Moore*

*R*obert Lee Moore's confidence in and respect for George Bruce Halsted never wavered and late in life, after years of being able to reflect and consider, he insisted there was no other person he would rather have studied under in his early Texas days. Some mathematicians, and a number of authors of papers on Moore, have found this curious, given the personalities he encountered in the early stages of his mathematical adventure, notably his doctoral mentor Professor E.H. Moore. Despite accumulating many accolades for his work, there was always a question mark against Halsted among the mathematical community, one that became more pronounced as the years passed, and his eccentricities began to form a greater part of conversation than his prodigious output.

It was to Moore's personal dismay that he would find himself the unfortunate catalyst to growing animosity towards Halsted at The University of Texas, which escalated into open warfare with the hierarchy and finally a parting of the ways. The bitterness that Halsted harbored for the rest of his life was regularly reflected in his letters to Moore, not surprisingly in that it marked the beginning of the decline of his own career. The endnote eventually came from Halsted himself in a mordant comment he wrote for a Princeton biographical questionnaire: 'I am working as an electrician as there is nothing in cultivating vacant lots.'[1] As we will see, Moore would be blameless in this affair and fortunately it did not erupt until he had acquired three years of solid tuition and guidance from Halsted, followed by many more years of regular exchanges by correspondence until, as Halsted admitted, the student eventually became the teacher. Moore was therefore able to garner all that Halsted could offer him and benefit from an impressive mathematical legacy that was as star-studded as Moore's personal genealogy chart. Nor would he ever forget that it was Halsted who, as they turned into the 1900s, provided linkage to the past, the present and the future of the American mathematical research community in which Moore was to play such a major part. In doing so, he became a pivotal figure in the expansion of this professional family tree that expanded far and wide across the United States of America to embrace some of the most gifted mathematicians of the twentieth century.

---

[1] L.E. Dickson, Biography, Dr. George Bruce Halsted (1853–1922), *Amer. Math. Monthly* 1 (1894), 337–340; and Halsted entry by H. S. Tropp in the *Dictionary of Scientific Biography*, New York, 1970.

Moore would undoubtedly have figured among them anyhow, but Halsted's own background and experiences are important because they reflect developments that brought outstanding additions to Moore's mathematical pedigree. Halsted had grown up with the finest. The son of Oliver Spenser and Adela Meeker Halsted, of Newark, New Jersey, he was the seventh child in a family of ten children. His brother, father, an uncle, a grandfather, a great uncle and a great-grandfather were all Princeton graduates and the Halsted family had given funds for an early astronomical observatory. Halsted was awarded his B.A. there in 1875 and his M.A. three years later. In 1876, however, he became the first doctoral student of Professor James Joseph Sylvester on his appointment to the chair of mathematics at The Johns Hopkins University. Recognized as one of the founding fathers of the American school of mathematical research, he and Halsted struck up an immediate rapport. Unfortunately it did not last.

Halsted had virtually begged to be taken on by Sylvester[2] who had a history that was, and still is, well known across the boundaries of international mathematics. After schooling at the Royal Institution in Liverpool and St. John's College, Cambridge, he was refused graduation because, at the time, it was necessary to swear a religious oath to the Church of England. Being Jewish, he could not take that oath so could not graduate and had difficulty in finding work. At the age of 27, he won the chair at Thomas Jefferson's nonsectarian University of Virginia at Charlottesville, although his appointment lasted only three months. Halsted loved to tell one of the many stories that add spice to the mathematical history of America:

'As Sylvester would not sign the 39 articles of the Established Church, he was not allowed to take his degree, nor to stand for a fellowship to which his rank in the tripos entitled him. Sylvester always felt this religious disbarment bitterly. His denunciation of the narrowness, bigotry, and intense selfishness exhibited in these creed tests was a wonderful piece of oratory in his celebrated address at Johns Hopkins. No one who saw will ever forget the emotion and astonishment exhibited by James Russell Lowell while listening to this unexpected climax.

---

[2] Karen Hunger Parshall, "America's First School of Mathematical Research: James Joseph Sylvester at The Johns Hopkins University 1876–1883," *Arch. Hist. Exact Sci.* 38 (1988), 153–196. See also Parshall, *James Joseph Sylvester Life and Work in Letters*, Oxford University Press, 1998, and bibliography.

James Joseph Sylvester

Thus barred from Cambridge, he accepted a call to America from the University of Virginia. In Sylvester's class was a pair of brothers, excruciatingly pompous. When Sylvester pointed out one day the blunders made in a recitation by the younger of the pair, this individual felt his honor and family pride aggrieved, and sent word to Professor Sylvester that he must apologize or be chastised. Sylvester bought a sword cane, which he was carrying when waylaid by the brothers, the younger armed with a heavy bludgeon.

'The younger brother stepped up in front of Professor Sylvester and demanded an instant and humble apology. Almost immediately he struck at Sylvester, knocking off his hat, and then delivered with his heavy bludgeon a crushing blow directly upon Sylvester's bare head. Sylvester drew his sword cane and lunged straight at him, striking him just over the heart. With a despairing howl, the student fell back into his brother's arms screaming out, "I am killed! He has killed me!" Sylvester was urged away from the spot by [an onlooker], and without waiting to collect his books, left for New York, and took ship back to England. Meantime a surgeon was summoned to the student, who was lividly pale, bathed in cold sweat, in complete collapse, seemingly dying, whispering his last prayers. The surgeon tore open his vest, cut open his shirt and at once declared him not in the least injured. The fine point of the sword cane had struck a rib fair, and caught against it, not penetrating. When assured that the wound was not much more than a mosquito-bite, the dying man arose, adjusted his shirt, buttoned his vest, and walked off, though still trembling from the nervous shock.'[3]

Back in London, Sylvester worked for a time as an actuary for an insurance company and tutored part-time in mathematics. One of his private pupils was Florence Nightingale. In 1846 he became a law student at the Inner Temple, and in 1850 was admitted to the bar. It was while working as a lawyer that Sylvester met Arthur Cayley and began the long partnership in which they created the theory of algebraic forms. From 1855 to 1870 Sylvester was professor of mathematics at the Royal Military Academy, Woolwich. From that position, he built up an international reputation. His style has been described as flamboyant and his output powerfully imaginative. He wrote in haste and was continually coining new technical terms, few of which actually survived. Recognition eventually superceded religious bigotry and he became the second president of the London Mathematical Society and the first recipient of the Gold Medal that the Society awarded to honor its first president, Augustus DeMorgan.

His retirement from the military academy at the early age of fifty-five came at an opportue time. The organization of The Johns Hopkins University in Baltimore was already under way and Sylvester was widely tipped for the chair of mathematics. There were detractors, including many of America's own mathematicians. At Harvard, mathematician Benjamin Peirce, then receiving attention for his own innovative work on linear associative algebras, stepped in with support. He had known Sylvester since his first visit to America and wrote to university president Daniel Gilman:

'If you inquire about him you will hear his genius universally recognized but his power of teaching will probably be said to be quite deficient… Among your pupils sooner or later, there must be one who has a genius for geometry. He will be Sylvester's special pupil the one pupil who will derive from the master knowledge and enthusiasm and that one pupil will give more reputation to your institution than ten thousand who will complain of the obscurity of Sylvester, and for whom you will provide another class of teachers.'[4] The upshot was, of course, that

---

[3] Quoting Halsted, writing in *Amer. Math. Monthly* (1894), 294–298, quote from 296.

[4] From the Gilman Papers, Special Collections Division, Milton S. Eisenhower Library, The Johns Hopkins University, as quoted in Karen Hunger Parshall and David E. Rowe, *The Emergence of the American Mathematical Research Community 1876–1900: J.J. Sylvester, Felix Klein and E.H. Moore*, American Mathematical Society, 1994, pp. 73–74.

Sylvester was appointed and as Parshall and Rowe tell us, his arrival on American shores, 'did more for American mathematics than resuscitate Peirce's work. It signaled the beginnings, however modest, of the entry of the United States into the international mathematical arena, a development which would take place within the context of neither the college nor the federal government nor the general scientific society but within that of the emergent American university.'[5] As for the one genius of geometry that Peirce spoke of, was this to be George Bruce Halsted? Or perhaps we should rephrase that and say it could have been him, but for some peculiar twists of fate. Halsted was in no doubt where he wanted to be in 1876 as Johns Hopkins was being set up. He had heard that the university was to offer a limited number of graduate fellowships and in what was to become a familiar trait, he drew up letters plastered with unashamed egotism about his background and sent them to the powers-that-be in Baltimore, including one to president Daniel Gilman. In it he spoke of 'such overmastering anxiety to be a partaker in your rich feast of learning that I cannot wait a single day, but would this very instant lay before you my humble petition'.[6]

Regardless of the over-gilded CV, Halsted, then 23, was chosen along with Thomas Craig, a graduate of civil engineering from Lafayette College, Pennsylvania, to become the first two candidates to receive the fellowship award. The Johns Hopkins mathematics department opened for business with seven undergraduates and eight graduates and in that fall of 1876 Halsted took on the role of research assistant to Sylvester who began the semester in a manner that one of his students described as 'fiery and passionate'.

By the second academic year, however, Halsted was showing a clear divergence from his professor's continuing theme of invariant theory and veered increasingly towards geometrical research, notably non-euclidean geometry that was to take up of much of his time in the years ahead. He began his investigations in 1877 and published a bibliography in two parts in 1878 and 1879, entitled *Bibliography of Hyperspace and Non-Euclidean Geometry*. This, he reported in his fellowship application for that year, had been a substantial task, completed on his own account without aid, discovering what had been written in virtually

---

[5] Parshall and Rowe, p. 49.
[6] Letter, Halsted to D.C. Gilman, 10 April, 1876, Gilman Papers, quoted in Parshall and Rowe, p. 78.

G.B. Halsted

every language. It was, he said, a major contribution towards ending the 'woeful ignorance' that existed in America on the subject. Halsted's tendency towards outspokenness also came to the fore in 1878 when he was preparing a lecture on Modern Logic. He had planned to refer to the contributions of Charles Peirce,[7] son of the mathematician Benjamin Peirce, but then declined to do so, claiming that Sylvester thought the work to be pretentious. Sylvester made it known to Peirce that this was a gross misrepresentation of his views and Peirce who, according to his own biographer Joseph Brent, had his 'own characteristic attraction to controversy', resumed cordial relations with both. Even so, Halsted left Johns Hopkins in the fall to return to Princeton where he completed his dissertation on *Basis for a Dual Logic* and was awarded his PhD at Johns Hopkins *in absentia*. Back at Princeton, Halsted became an instructor in the mathematics department and it appeared he had his mind set on advancing his career with that institution.

In 1950, Katherine S. Eisenhart, wife of Luther Pfahler Eisenhart, Dean of the Princeton Graduate School, compiled a selection of notes on earlier Princeton faculty members. In her references to Halsted, she mentions a 'curious little booklet' he published in 1882, entitled *Some*

---

[7] Peirce, of course, went on to become the principal founder of America's pragmatic school of philosophy.

Testimonials and Credentials. The pages contained a summary of his career, along with letters from a number of eminent mathematicians and scientists carefully chosen to bolster the Halsted image. Among them was one from Henry Burchard Fine who had been a student of Halsted at Princeton, and was by then a teacher in mathematics. He wrote that Halsted had steered him into a mathematical career.[8] If the intended purpose was to gain a bid for a professorship, it did not succeed. Mrs. Eisenhart identified the reason why Princeton shunned his self-promotional efforts: 'He was a man of ability in his chosen field, but certain eccentricities prevented him from attaining the success either as a teacher or writer which his powers seemed, in his youth, to promise.'[9] The following year therefore found Halsted attending an interview for the chair at The University of Texas, which had opened its doors just twelve months earlier. His credentials were impressive. The appointment was confirmed and he took his place among the ex-colonels of the Confederacy in 1884 and remained until 1902. He re-organized the mathematical curricula, popularized a series of faculty lectures and devised various methods to try to put The University of Texas on the map. He also pursued a fairly broad base of mathematical teaching and research with some vigor, particularly in the area of noneuclidean geometry. He would state fiercely that after two thousand years geometry was almost as Euclid had left it, circa 300 BC. He maintained that Cartesian coordinates, algebracized geometric notions and projective geometry were adaptations or refinements of euclidean geometry.

As we have seen from the early recollections of R.L. Moore, he began to focus particularly on the work of the Hungarian Bolyai in his outlining of a 'hyperbolic' geometry in 1833. Whereas Euclid assumed that in the two-dimensional plane, given a point not on a given line, one and only one line can be drawn parallel to the given line, this new hyperbolic geometry asserted that infinitely many lines could be drawn parallel to the given line. A surface known as the pseudo-sphere is a realization of this hyperbolic geometry. Halsted translated Bolyai's supplement into English in 1890 and Lobachevsky's papers the following

---

[8] Fine later developed into a distinguished administrator at Princeton and into a gifted expositor. Princeton named a building Fine Hall, as a home for the mathematics department.

[9] Katherine S. Eisenhart, pp. 31–33 in a typescript prepared in 1950, deposited in the Princeton University Library.

year. Professor Robert Greenwood, who arrived later at the mathematics department at Austin, was able to study his work and formed a precise view: 'In a very real sense Halsted introduced and facilitated the study of noneuclidean geometry in [the United States], and also to some extent in other English speaking areas. Halsted visited eastern Europe and became acquainted with several of Lobachevsky's followers in Kazan.... In Austin, Halsted authored some geometry textbooks (in English), and one of these was translated into French. As U.T. Austin grew, two lines of mathematical curricula were developed. Thomas Ulvan Taylor was added to the staff in 1888 as an assistant professor of mathematics. Taylor taught the courses which later came to be designated as "applied" mathematics and Halsted taught those courses which later became "pure" mathematics.'[10]

Halsted also began the tradition of attracting quality students and achieved success with a number of students prior to the arrival of Robert Lee Moore. Some were to loom large and imposing in the young man's future. They included Milton Brockett Porter from Sherman, Texas, who was a student at The University of Texas from 1889–1892. He went on to Harvard where he received his M.A. in 1895 and his PhD in 1897. He returned to Texas as an instructor in pure mathematics in 1897 and accepted a similar post at Yale in 1898, subsequently being promoted to an Assistant Professorship, which he held until returning to The University of Texas (to replace Halsted) as Head of Pure Mathematics in 1903. A gifted mathematician of high ideals and character, he remained at The University of Texas for the next forty-two years, and then as Emeritus Professor from 1945–1960. Another was Harry Yandell Benedict who came to Austin in 1889, graduating in 1892. During 1891–1892 he was a Fellow in Pure Mathematics, and in 1892–1893 a teacher in Pure and Applied Mathematics at the University. He moved to the University of Virginia and during 1893–1895, he was an assistant in astronomy at that university. He resigned that post to study at Harvard, graduating from there with a PhD 'in mathematics, especially astronomy',[11] in 1898, having held two scholarships while at Harvard. Benedict, affectionately known to all as

---

[10] R.E. Greenwood, *History of the Various Departments of Mathematics at the University of Texas at Austin: 1883–1983*, p. 7. Unpublished manuscript, R.E. Greenwood Papers at the Archives of American Mathematics.

[11] *The University Record*, Vol. 1, No. 2, April 1899, p. 223, as quoted in Traylor, p. 27.

Dean Benny, spent one year at Vanderbilt University, took entire charge of the department of mathematics at that institution and was subsequently elected president of The University of Texas, 1927–1937.

Leonard Eugene Dickson was another student who credited early contact with Halsted for the successful start to his career, which he pursued in an outstanding manner. Dickson was from Cleburne, Texas, where his father was a successful merchant. He graduated in 1893 and was awarded a teaching fellowship at The University of Texas in pure mathematics. He completed the course for his Master of Arts degree under Halsted's supervision. He then applied for doctoral fellowships at Harvard and Chicago and received an offer from both, electing to join Professor Eliakim Hastings Moore who, like Sylvester, was in the process of founding a mathematics department of substance at the University of Chicago. Dickson received his PhD in 1896 for a dissertation entitled *The Analytic Representation of Substitutions on a Power of a Prime Number of Letters with a Discussion of the Linear Group*.[12] Dickson went to Europe for a year, studying with Sophus Lie at Leipzig and with Jordan, Appel, Picard, Hermite and Painlevé at Paris. He returned to become an assistant professor of mathematics at the University of California during 1897–1899, resigning that post to become an associate professor of mathematics at The University of Texas where he remained for only a year before being recruited by Professor E.H. Moore at the University of Chicago, there to remain pursuing an outstanding career until 1939.[13] The staffing line-up in pure mathematics at The University of Texas as Robert Lee Moore began his own work in earnest was therefore impressive: Professor George Bruce Halsted, Associate Professor L.E. Dickson, instructors H.Y. Benedict and T.M. Putnam, a promising young mathematician from the University of California who had studied with Dickson. R.L. Moore was exposed to all of these brilliant minds at various times and as can be seen from his early diary notes, dating from the end of 1898, Halsted was leading the teenager whom he described as a 'young man of marvelous genius, of richest promise'[14] down the path of his own specialist interests. In the summer of 1900 Halsted, while blowing his

---

[12] His work there was to be a precursor to the arrival in Chicago of Halsted's current (and last) protégé, R.L.Moore, which will be discussed in Chapter Four.

[13] A.A. Albert, Leonard Eugene Dickson (1874–1954), *Bull. Amer. Math. Soc.* 61 (1955), 331–346.

[14] George Bruce Halsted, *Science Magazine,* Friday, Oct. 24, 1902, p. 645, as quoted in Traylor, p. 36.

own trumpet for achievements at The University of Texas, singled out Moore, still only eighteen, as one of the stars of the future in that field.

He went public with this view in the June summary[15] of the Pure Mathematics Department at Austin, which he described as a 'bloom period'. He noted that for sixteen years, his department had pursued a decidedly geometric character, believing that this was the 'most remunerative as it is the most charming part of all mathematics'. There were courses in Modern Synthetic Geometry, Recent Geometry of the Triangle and Circle (the Lemoine-Brocard Geometry), Geometry of Position (Projective Geometry), Noneuclidean Geometry, and these were further strengthened by one in Group Theory by Dickson. Halsted then revealed that Dickson had accepted the position of an Assistant Professorship at the University of Chicago, pointing out that the head of department, Professor E.H. Moore, had finally included in his program 'the magic name non-Euclidean geometry'. In announcing Dickson's departure, Halsted added in the same summary that R.L. Moore was one of two students that year showing 'that the splendid quality of Texas youth is of undiminished vigor'. Indeed, the academic year of 1900–1901 was to bring significant results for Moore, and instant fame within the mathematical world. Before going on towards completion of this mathematical backdrop to R.L. Moore's own progression, it would be remiss not to mention the other side of the Halsted coin, that is the folklore surrounding him during his time at Austin. His eccentricities, wrote T.U. Taylor, were widely discussed: 'Of all the rare and odd professors that have been on the faculty of The University of Texas, I think George Bruce Halsted will rank as number one ... for about sixteen years his sayings and doings in the classroom and in public lectures were the talk of the campus and the town.'[16] Halsted married Miss Maggie Swearingen in June 1886. They had three sons, Arthur, Halcyon, and Harbeck. The last two sons received M.D. degrees; the eldest (born in 1887) became an electrical engineer. Professor Halsted built a house for his family on the northeast corner of Twenty-fourth and Guadalupe Streets in Austin, which became the object of humorous comment. It was erected to his own design high above the ground with

---

[15] *The University Record*, Vol. 1, No. 3, June 1900, pp. 165–166, as quoted in Traylor, pp. 30–31.
[16] Among various recollections of Halsted contained in T.U. Taylor's book, *Fifty Years on Forty Acres*, published by Alec Book Company, Austin, 1938.

brick columns tall enough for a man to walk under without bumping his head. One theory popular at the time in the university chambers of gossip was that Halsted feared that a break in the dam on the Colorado River would flood Austin and wash away his home, whereas with his design, the water would merely flow underneath the building.

Another theory was that he built his house high so that he could stable his Shetland ponies, pigs and chickens under his home and would not need a barn or other shelter for them. Halsted and his Shetland ponies became a familiar sight around Austin. He had constructed a cart made from a single-plank mounted on two axles and connected by shafts to the animals. Taylor recalled: '[He] drove over town at a rattling pace, irrespective of traffic laws or right or left side of the street. One day in the faculty meeting, one of the professors happened to refer to the police court and ended with the remark that he had never been before a police court. Immediately, Dr. Halsted interjected: "I have the advantage over you. I have been there several times."'

Students, of course, had a great deal of fun amongst themselves. The 1895 edition of their annual yearbook, *Cactus*, included a mention that Professor Halsted won the Roman Chariot Race for that year! Some were able to record another incident that occurred after he'd made a trip to Mexico and found a plant whose juice, he decided, would revive youth in old age. 'He brought a boiled-down syrup from it to class and made each of his students, who weren't exactly suffering from the onset of senility, drink some of the potion. Unfortunately, we have no statistics on how the students held up sixty years later,' wrote Joe Frantz.

Similarly, his appearance at national meetings of mathematical and scientific bodies, at which he was in regular attendance, often raised a few eyebrows, not to mention some telling looks down noses. He was also well known for speeches outside the mathematical sphere, which he apparently delivered as a performance, like a hammy actor in a touring repertory company reciting Shakespeare not very well. An example survives. He gave the address on the eve of George Washington's birthday in 1899 to the Charles Broadway Rouss Camp of Sons of Confederate Veterans. It took the form of a collective deification of George Washington, Robert E. Lee and the Texas Ranger Sul Ross. The prose was purple and the applause loud, as he spoke in these terms:

'What words can fitly paint the mighty, the peerless paladins of the South. But for purity of contour, perfection of form, grace of power, elegance of mode, adorable, heart-winning Southern dignity, the world inevitably singles out Robert E. Lee...'

He had the speech printed as a pamphlet by Ben C. Jones and Co, Austin and presented a copy to Robert Lee Moore.[17] Notwithstanding these idiosyncrasies, Halsted worked hard to draw attention to the work being done at Austin and to promote scientific and mathematical discussion of all kinds, and especially about his own research. At a time when rail links made it difficult and expensive for potential lecturers and speakers of note to attend, he participated in the affairs of the Texas Academy of Science to encourage regular contact and discussion between educators and scientists. He was its president in 1894 and his retiring address, *Original Research and Creative Authorship, the Essence of University Teaching,* was reprinted in *Science*, the journal of the American Association for the Advancement of Science. He was also the mathematics section chairman and vice-president of the AAAS.[18] The mixture of peculiarities with the underlying importance and prodigious nature of his study, research and involvement in the mathematics and scientific communities at large combined to establish Halsted as a character who, sadly, was to be reviled as much as he was revered. Whether Halsted was a good teacher was also a matter of debate. Some said he was not, because of his tendency to talk about everything under the sun in class, other than the topic at hand, such as his travels and experiences and opinions. He was also impatient, insulting or chastising toward those he considered laggards, which, among seniors, occasionally led to heated exchanges and on one famous occasion, fisticuffs. If judged on results and publications, however, Halsted had an excellent record although again, it may be said that he paid particular attention to those who might develop into star performers. His colleague, T.U. Taylor, was in no doubt: 'The one outstanding thing that must be said to [his] credit was that he inspired men to study and research and in this respect he made a genuine contribution to American scholarship in mathematics.'

R.L. Moore certainly fell into that category and at his youthful age was quite clearly in awe of his teacher. His diary entries largely concerned his work, apart from attempting to talk a certain Miss Joynes into

---

[17] G.B. Halsted, *Washington, The Ideal of the South; Resurgent in Lee and Ross: Address Before the Charles Broadway Rouss Group of Sons of Confederate Veterans,* February 21, 1899, R.L. Moore Papers in the AAM.

[18] His retiring vice-presidential address in 1902, The Message of Noneuclidean Geometry was published in the *Proceedings of AAAS* 53 (1903–1904), 349–371.

going to church with him or fighting with his roommate over his persistence in washing his nose in the bowl and opening the blinds when Moore was still in his night gown. Moore had to slap him, and grazed his head.[19] The time of 11:15 pm appeared regularly on his jottings, indicating that he used it as a break point when he finished his studies, and took time out to make a few notes in his diary before retiring. He had piled on the courses and the credits to such an extent that he gained his B.S. and M.A. degrees simultaneously in 1901, having been given an option on a B.S. or B. Lit. and choosing the former. He had already taken classes as a student assistant and for the academic year of 1901–1902 was granted a teaching fellowship, much to the amusement and admiration of some of his colleagues. As one of them recalled in the students' magazine, *The Alcalde*, in 1917:

'It is reasonable to assume that as a boy he attended some sort of school. It is alleged that he never studied algebra or geometry but was born with knowledge of them. The writer of this article has no recollection of learning the abc's and he does not believe that R.L. Moore can remember when he did not know geometry. However that may be…during the year of 1901–02 … he taught … a course in analytical geometry administered from Puckle's *Conic Sections*.… Though Moore was at the time only some nineteen years old, noticeably younger than the average students in his class, he handled the subject after the manner of a veteran and his subjects (pupils) even more so. Never, from the first week did the class doubt his knowledge of Conic Sections or his intention to run his class. He often criticized the text, a thing at that time new to most of us. One morning when class assembled, there appeared on the board the following: "Quiz For Math 2: 1] Find the equation of the boundary of Moore's intellect. 2] Find the points of intersection of this curve and Puckle's *Conic Sections*." It was during this year that he distinguished himself by proving the redundancy of Hilbert's set of geometric assumptions. This was the real beginning of his work.'[20]

Moore was obviously enjoying himself, although he seldom showed great emotion in the entries in that same Ivory Soap pad diary that he began on the first day he arrived on the Austin campus. He simply noted that he 'like[d] the business [of teaching] *very* well'.[21] In another entry

---

[19] Ivory Soap pad diary, 8 June 1899, R.L. Moore Papers in the AAM.
[20] *The Alcalde* 5 (March 1917), 427.
[21] Ivory Soap pad diary, 6 December 1901, emphasis in original, R.L. Moore Papers in the AAM.

for the same day, he mentioned what was to develop into the most important discovery of his short career. 'My, what a fine thing that Hilbert's geometry is. How fine, how pure, and free from ideas of space. Now at last it seems it has been rendered possible for one to write out in this year, 1901, a geometry in which every proposition is arrived at from certain explicit assumptions with the use of no other assumptions whatever except "the laws of pure logic." How pure!! How unexpected. No, I don't believe I would have thought last session this year I would be studying so fine a thing as this "geometry" of Hilbert.' Given an inherent talent for mathematics such as R.L. Moore early evinced, the direction in which it would take him was clearly determined by the kinds of mathematical problems on which his mentors and colleagues were concentrating, as well as his own tastes. R.L. Moore had arrived at the point of eureka, although he hadn't quite realized it. Raymond L.Wilder, future president of the American Mathematical Society and one of Moore's own early doctoral students, would record in reflections of that era:[22]

'Geometry was finally being put on a satisfactory axiomatic basis and Halsted's greatest interest at the time was in geometry, particularly [then] in Hilbert's recently published *Grundlagen der Geometrie* (1899). The outstanding characteristic of this work was its attempt to found the geometry of the plane and three-space on a rigorous axiomatic basis. Not only did Halsted apparently acquaint R.L. Moore with Hilbert's work, but Moore was induced to check one of the axioms (Axiom II 4) for independence.'

In fact, Halsted had been in touch with the renowned German mathematician, David Hilbert, who, in 1895, was appointed to the chair of mathematics at the University of Göttingen, where he continued to teach for the remainder of his career. After the publication of the *Grundlagen der Geometrie* (which became *The Foundations of Geometry*, in English in 1902), much of the book's popularity was due to a list of 23 research problems he set out at the International Mathematical Congress in Paris in 1900. His famous speech, 'The Problems of Mathematics', was translated by Halsted who described it as brilliant. It was an in-depth appraisal of mathematics of that period and offered a preview of the mathematical problems that he believed

---

[22] R.L. Wilder, The mathematical work of R.L. Moore: Its background, nature, and influence, *Arch. Hist. Exact Sci.* 26 (1982), 73–97, quote from p. 74.

would arise in the twentieth century. Halsted, meanwhile, was hurriedly preparing a Supplementary Report for *Science* on Hilbert's set of axioms. Hilbert had called one group of his assumptions the Axioms of Arrangements but Halsted renamed them the Betweenness Assumptions.

Of these assumptions, the fourth one stated:

Any four points, A, B, C, D, of a straight line can always be so arranged that B lies between A and C and also between A and D, and furthermore C lies between A and D and also between B and D.

In his translation of Hilbert's statements, to arrive at the above, Halsted had come to an interpretation of 'angeordnet'. He discussed this with R. L. Moore, posing the question as to whether that fourth assumption might be demonstrated from the other assumptions. Based on Halsted's translation, we have:

If A, B, C are points of a straight (line), and B lies between A and C, then B also lies between C and A. If A and C are points of a straight, then there is always at least one point B, which lies between A and C, and at least one point D, such that C lies between A and D. Of any three points situated on a straight there is always one and only one which lies between the other two.

DEFINITION. The system of two points A and B, which lie upon a straight a, we call a *sect*, and designate it with AB or BA. The points between A and B are said to be points of the sect AB or also situated *within* the sect AB; all remaining points of the straight are said to be situated *without* the sect AB. The points A, B are called *endpoints* of the sect AB.

Let A, B, C be three points not co-straight and a a straight in the plane ABC striking none of the points A, B, C: if then the straight a goes through a point within the sect AB, it must always go either through a point of the sect BC or through a point of the sect AC.

At the time, Halsted was preparing to write a geometry textbook, using Hilbert's axioms as a basis for it and had written to Hilbert, asking whether he 'recognized any desirability for change'. Hilbert's answer to Halsted was received 14 April 1902 and read, in part, 'Instead of II 4 (the fourth assumption), I believe it suffices simply to say: If B lies between A and C and C between A and D, then lies also B between A and D; and then to prove my old II 4 as theorem.'

Halsted suggested that Moore might attempt to fill in the proof. His student went away to work on the task and before the day was out he was able to demonstrate Hilbert's new axiom, eliminating II 4 and

David Hilbert

reducing The Betweenness Assumptions from five to four. Moore would recount this discovery many times in the years ahead. He was so excited that he dashed over to the campus late in the evening. Looking up toward the old main building, he saw the light shining from Halsted's window and hurried to his room. Halsted wrote down Moore's oral arguments and sent the paper to the *American Mathematical Monthly*. He also wrote Professor E.H. Moore, at the University of Chicago, to draw attention to his namesake's achievement, knowing full well that E.H. Moore had also been working on Hilbert's conclusions. In fact, Professor Moore had already established the redundancy himself and had written a paper on it, published in the *Transactions of the American Mathematical Society*[23] the previous January. However, the young Moore's proof was decidedly shorter and more elegant. In fact, said the professor in a note to Halsted, it was 'delightfully simple' and he took the trouble to write to R.L. Moore to explain exactly what had happened to stave off any disappointment:

'Apparently he [Dr. Halsted] has not called your attention to the fact that the redundancy was pointed out by me and proved in my paper, which I am sending under separate cover, on the projective axioms of

---

[23] On the Projective Axioms of Geometry, *Trans. Amer. Math. Soc.* 3 (1902), 142–168.

geometry, published in the January number of the *Transactions*. In accordance with correspondence with him [Halsted], it was in connection with this paper of mine that he wrote to Hilbert and received Hilbert's response which led to your work on the subject. You will see that it was my desire to survey the whole system of projective axioms, and to exhibit a new system, and, in that connection to show that Hilbert's axioms I 4 and II 4 were in his system redundant, and moreover, to furnish a satisfactory account of the roles of the axioms I 3, 4, 5 which had been held by Schur[24] to be redundant. As to the axiom II 4, you will see that, by considerations of the other linear axioms alone, and so in particular without the use of II 5, or of my axiom 4, I prove on page 151 that the axiom II 4 is a result of the statement 21 which statement is the statement of your theorem I. Thus to complete the proof of the redundancy of II 4, in Hilbert's system, I should today make use of your proof of theorem I. The proof that I give, in that it involves my triangle transversal axiom 4, is necessarily much longer. I have supposed that you might be interested in understanding how your paper impresses me, and remain with considerable interest in the progress of your mathematical career.'[25]

Halsted was not impressed by E.H. Moore's letter. He would later write to R.L. Moore:[26] 'Neither Hilbert nor anyone else acknowledged that E.H. Moore's obscure and bungling adumbration proved anything before I published your beautiful proof.'

Despite the touch of rancor in Halsted's part, Professor Moore had obviously made a note of the young Moore's ability, which was Halsted's intention in the first place, and it augured well for the future. At The University of Texas, however, Dr. Halsted did not come out of these exchanges as well as his student, and already an aura of disquiet had descended over Halsted's tenure there. There had been many disagreements between him and the Board of Regents, especially over his outspoken views, which, as a senior university figure, might be taken as the official University of Texas line. The animosity had been building between them for some time and now it was to affect R.L. Moore. Halsted had proposed him for an instructorship vacancy in the universi-

---

[24] See Schur's statement in Chapter Four.
[25] E.H. Moore published a copy of the letter to R.L. Moore in The Betweenness Assumptions, *Amer. Math. Monthly* 9 (1902), 142–158.
[26] Halsted to Moore, March 19, 1904; R.L. Moore Papers in the AAM.

ty teaching staff, which was to occur for the 1902–1903 year. Moore had recorded this in his diary, with an uncommon show of excitement in May 1902. The euphoria was to be short-lived.

The following month, Halsted was informed that his candidate, R.L. Moore, had been rejected in favor of Mary E. Decherd who had graduated from the university with a B.S. in 1892 and had since been teaching at Austin High School. The daughter of an Austin family whose four children all received degrees from The University of Texas, she had also completed some graduate work for her Masters at the University of Chicago. There were, as Joe Frantz pointed out, Decherds all over the Austin area. They were people of some prominence with local civic connections, although 'none compared with R.L. Moore for raw promise'. Halsted was furious at this news, and promptly fired off an article published in *Science*[27] in which he pointedly criticized the university for its decision. He pulled no punches and the piece resounded through the halls of The University of Texas hierarchy:

'If the keenest, brightest, most gifted of the young people reject the scientific career, then fellowships serve only a dull, stale, tired clique of incompetents. Even after the possession of the rare and precious gift of scientific genius has been clearly, competitively proven ... the exquisite bud in its tender incipiency may be cruelly frosted.... This young man of marvelous genius, of richest promise, I recommended for continuance in the department he adorned. He was displaced in favor of a local schoolmarm. Then I raised the money necessary to pay him, only five hundred dollars and offered it to the President here. He would not accept it.... The bane of the state university is that its regents are the appointees of a politician. If he were even limited by the rule that half of them must be academic graduates, there would be some safety against the prostitution of a university, the broadest of human institutions, to politics and sectionalism, the meanest provincialism.'

Through no fault of his own, R. L. Moore found himself at the center of a huge row and given Halsted's notoriety, onlookers throughout the mathematics community watched with interest. Halsted confirmed by letter to Moore on 8 September that he had raised the $500 himself to fund his employment: 'He rejected it. He would not have you even if you cost him nothing. Of course, I made a fuss about it. He sent a letter

---

[27] Published in The Carnegie Institution, *Science* 16 (24 October 1902), 642–652, quote from p. 645.

after me, saying that after the present session, "my services would not be required."... So you see there is no hope for you here.'[28]

Halsted's departure suddenly became more imminent. He published another article, mentioning R.L. Moore, in the December issue of the prestigious *Education Review*, which was available in the third week of November for those awaiting the next installment of this rather public brawl. In a paper entitled 'The Teaching of Geometry', Halsted was highly critical of published aids to teaching which, he said, were in a 'hopeless muddle'. America awaited a *Geometry for Beginners* written by someone familiar with the new, penetrating and critical research in the principles of geometry. Meanwhile 'every conscientious teacher must better from his own studies the book he is still forced to use. He has no right to dump into a helpless scholar what he knows to be trash'.

University administrators acted swiftly, and in a later letter to Moore, Halsted wrote: 'My professorship has been declared vacant.... My salary is stopped and the blow has affected me considerably. It will, I fear, hurt my chances, which were good, from getting a place at another university.'[29] In another note, his consuming anger was evident: 'I like all my predecessors, am left after all the honors I have conferred on the University, to starve insofar as [they] can accomplish it. If I succeed in getting a place, I will want you at once and the new president will not be the malicious fool to refuse to take you, even "gratis".'[30]

The Board of Regents issued a statement in the most pointed terms, confirming that with immediate effect, Halsted was no longer connected with The University of Texas. And so Halsted's departure was confirmed and R.L. Moore was caught in the turbulence.

---

[28] Halsted to Moore, 8 September 1902, R.L. Moore Papers in the AAM.
[29] Halsted to Moore, 19 January 1903, R.L. Moore Papers in the AAM.
[30] Halsted to Moore, 8 September 1902, R.L. Moore Papers in the AAM.

3

On to Chicago (1903)

Arrival at the University of Chicago, then a hotbed of mathematical development and progress under E.H. Moore; scene setting and influences including such luminaries as John Dewey, Bolza, Maschke, Klein and Weierstrass.

'Bobby' Moore

That Robert Lee Moore did not receive the Austin instructorship temporarily sent his youthful career into the sidings. Neither did The University of Texas cater for doctoral studies at that time and the plan that he should continue his studies with Halsted, with the clear hope of following L. E. Dickson to Chicago, could not be fulfilled. Halsted, meanwhile, was still desperately trying to find a position that would enable him to take Moore with him but that possibility very quickly diminished to the point of despair, as the professor made clear in another letter to Moore about their joint predicament:

'As yet I have not secured any place for next year. Of course, I would rather have you with me than anyone else in the world but it is very doubtful if I will have any choice of assistant the first year.... I feel sure you can get good places henceforth. If you can afford it, why not go to Germany and take a PhD under Hilbert? Write to me often.... Yours Always, Sincerely, G.B. Halsted'[1]

Moore, however, could not afford to go to Germany and in a couple of his diary entries, he mentioned 'having to go to papa' again. In early November 1902, further exchanges between himself and Halsted came to naught, with the latter now suggesting that he should write directly to Professor E.H. Moore and Dr. Dickson about the possibility of taking his studies towards a doctorate at Chicago. '[Y]our future is assured,' Halsted told him, 'I wish I could say as much for my own. The strain of my position here [is] now terrible.'[2]

R.L. Moore's situation was no more encouraging either and although he wrote to Professor Moore in Chicago, there was no fellowship available. In the fall of 1902, he accepted a teaching post at Marshall High School, in the north east of Texas, 260 miles from Austin but closer to his Dallas home. He hoped it would be a stopgap appointment, teaching students almost as old as himself and certainly a good deal worldlier after the cocooned life of study he had led thus far. The gloom was apparent in his first diary entry in the Ivory Soap book, on 9 December 1902: 'Tonight, I was walking along and passed by one of the Yates boys (George, I believe). After I had passed on a certain distance, someone (I think it was he) called out: "Look at that bull-headed school teacher. Hasn't he got a cap, that bull-headed teacher?" Am I going to ask him if he was that person who called out if I see him on the street

---

[1] Halsted to Moore 22 October 1902; R.L. Moore Papers in the AAM.
[2] Halsted to Moore, 13 November 1902; R.L. Moore Papers in the AAM.

by himself some time and is a fight going to follow? Well, that perhaps will be "seen" in the future.'

This was to be his first and last diary entry during his time at Marshall. It was not a happy place for him, as was evident from the continuing correspondence with Halsted, whose own dilemma stumbled from bad to worse. At least the exchanges between them provoked a fairly regular flow of mathematical challenges and discussion, which was a good deal more than the young teacher could expect at his new place of employment. Although some have suggested the months spent there would allow him to 'grow' as a teacher, it was unlikely he gained much from the experience and he certainly did not enjoy it. He stood in front of a classroom of long rows with girls on one side of the room and boys on the other. Some of the tough young farm lads delighted in playing him up. One of his former pupils, Inez Hughes, recalled that Moore turned up at school one day with his shoes tied with cord laces. After lunch he discovered a pair of shoelaces on his desk with verse attached:

> An eagle flew from North to South
> With Bobby Moore in his mouth,
> But when he saw he had a fool,
> He dropped him in the Marshall School.[3]

Moore never learned the identity of the author of that piece of doggerel and after some early attempts by students to disrupt lessons, he chose the course of all-inclusive discipline, keeping the entire class after school if any one of them had misbehaved. He was considered short-tempered by his students, which some saw merely as a challenge to taunt him further. After a while, things settled down and, as a rule, the students came to like him. 'We found we could get on his good side by asking a lot of questions, so we did.'[4]

Meanwhile, Moore's correspondence with Dr. Halsted during this unhappy period took an interesting turn. It occurred, coincidentally, as a debate had opened up between Halsted and John Dewey who, like Halsted, had gained his doctorate from Johns Hopkins University, his in philosophy in 1884. Both had long ago attended the lectures on the logic of Charles Peirce that resulted in the rift between Halsted and Sylvester. In response to Halsted's paper published in the *Education Review* in December 1902 in which he chastized teachers for dumping 'into a

---

[3] Quoted by Traylor, in *Creative Teaching*, p. 44.
[4] Ibid.

helpless scholar what he knows to be trash', Dewey published a follow-up paper. It was entitled 'The Psychological and the Logical in Teaching Geometry' and it discussed, among other things, the difference between truth and the whole truth. It was a wide-ranging treatise in which he proclaimed that while, for example, Halsted's favorite starter for beginners, the definition of a straight line, was a matter for mathematics alone, there was a need for making clear that the content of a book or lesson for a given grade of pupils was not just a matter for mathematics. It was a psychological matter as well.[5]

The professor found himself compelled to offer some guidance on matters psychological, logical and spiritual over a period of several exchanges on the issue of using unproven statements as an initial premise for reasoning. In the first, which appeared to be the result of Moore questioning religious assumptions, he stated: 'We *must* postulate God and we must study and follow Jesus, and strive to get in our minds the Holy Spirit. Your tendency, I fear, is too critical. Do not go on in any doubting. *Will* to believe. Go to Church but only to select out the good and what you can approve. Cultivate love for all men, even those you find it very hard to like. Begin to do definitely helpful acts every day towards someone.'[6] The theme of this correspondence continued for some weeks, during which time Halsted had finally found a new position, as a professor of mathematics at St. John's College, Annapolis, Maryland, from where he wrote to R.L. Moore:

'[Y]ou should read William James *The Will to Believe* and also his later Gifford Lectures. I supposed you had intellectually passed the point where you supposed there was any tenable meaning in the phrase you use "simply in search of the truth". You might be in search of what postulates created or accepted by you would make the best universe for life but to suppose there is an objective truth independent of our will is absurd. As Tennyson says, "For this is truth to me, and that to thee."

'Every successful and happy community has been and is held together by a postulated God. I know you can postulate a Godless universe but such "truth" will make you isolated and unhappy and is in no sense truer than the truth of your happy neighbor, the [C]hristian who sees that after he has postulated God then Christ is divine, as containing and revealing

---

[5] Albert C. Lewis: *Reform and Traditions in Mathematics: The Example of R.L. Moore*, 15 April 1999.
[6] Halsted to Moore, 5 February 1903; R.L. Moore Papers in the AAM.

the proper nature and will of God. You are in a very primitive state of the so-called *scientific* mind, which does not see that your creative postulating belief makes that which you can call true. "Will to believe" is "will to *make true*." Accept the plan of creating your universe first expounded by Jesus. You are a good example of the consequences of creating a non-Christian universe. That you do not feel inclined to bow your head when the others do, even if only out of sympathy, is to me a sort of self-deception that begins alienation from fellow men and so begins dreadful unhappiness.

'Changes come to the mass of mankind gradually and are minute variations of their created universe. The wayward are *eliminated*. By all means go to church and bow your head. Your illustration from geometry turns against you. The latest understanding of the parallel-postulate is that it is a *definition*, and whoever believes it is right. All we need to see is that another space could be postulated (Bolyai's or Riemann's) just as I see that a [G]odless universe can be postulated. There is no "our space" except the space we make and in the making of it, we make it by definition Euclidean or non-Euclidean. I could very well tell you "will to believe" Euclid's parallel-postulate. If it made such a difference for you to believe God, I would urge you *will* to believe it.'[7]

The discussion by correspondence continued and again although it is impossible to be specific about the questions, the answers in Halsted's response[8] provide the clues as to the thoughts and requirements of Moore's search for 'the truth':

'The universe is not *entirely* created by our *individual* self since we are imitative animals and accept the language used about us. Yet your supposedly cogent illustration is badly chosen. You ask "Suppose one should will to put his hand in a fire and also at the same time to believe he felt no pain. Could he create such a state of affairs?" I answer, Yes, certainly he could. In fact, that very state of affairs has often been created.... Of course, if society, ancestors and contemporaries agree to believe certain things, then certain others logically follow. They agree to think the essences of things together into a personal God. Though you have the power to believe otherwise, and to create a scheme antagonistic, yet as one of the things which follow from social life is dependence on fellows, then also follows unhappiness from the making of sympa-

---

[7] Halsted to Moore, 26 February 1903; R.L. Moore Papers in the AAM.
[8] Halsted to Moore, 31 March 1903; R.L. Moore Papers in the AAM.

thy more difficult, the self-isolation of an antagonistic scheme. We believe that society eliminates the wayward. I postulate that you at least partially understand me. That seems a universal assumption on which to base social life.

'The creation of a Godless universe has usually had as a concomitant unhappiness. This sentence I may safely assume to convey to you a meaning. I believe, and so do you, that if an individual creates in a certain way he thereby creates, as a concomitant, unhappiness.... Since we both believe this, we may put it among the assumptions. It gives meaning to my sentence about "Every successful and happy community has been held together by a postulated God." The fact undeniable that "many and important things true for one man are not true for another", has only relevancy insofar as it enables us to conceive that there might be a man who could believe the opposite of this statement. It remains for you, understanding it, to make it part of your universe or to try to make the opposite belief part of your equipment for action. I cannot coerce your creation. I can only suggest alternatives and say which I believe leads to happiness. Remember we have in the language the words perversity and pervert.'

Moore was clearly at some sort of crossroads spiritually and mathematically but for the time being this was set aside. There was, included in that last note, also a reference to the possibility that Moore might soon hear good news from Chicago. In fact, Halsted knew it was highly likely. He had already received a letter from Oskar Bolza, one of E.H. Moore's partners in the mathematics mission being undertaken at the University of Chicago. Bolza revealed that R.L. Moore had applied for a fellowship at the university and Bolza was now seeking a reference regarding his mathematical abilities and attainments. Overjoyed at the news, Halsted dashed off a note to his protégé: 'I have just written a tremendously strong letter to Professor Bolza in your favor... and I think you may surely count on receiving the appointment.'[9]

That prophecy proved correct and Moore, in a further uncommon display of moderated emotion in his Ivory Soap diary: 'Gee whiz! I have become a Fellow in Mathematics in the University of Chicago.' When confirmation arrived, he wrote to his father and of course to Halsted, who replied: 'I greatly rejoice.'[10]

---

[9] Halsted to Moore, 16 March 1903; R.L. Moore Papers in the AAM.
[10] Halsted to Moore, 14 April 1903; R.L. Moore Papers in the AAM.

He and Moore both, and they had good reason. The mathematics department of the University of Chicago had, in its short history, attracted international attention and acclaim as a place of significant and exciting developments in an otherwise bland and largely under-performing arena of American scientific and mathematical research. Chicago was well on the road to establishing itself as one of the outstanding universities in the United States offering true graduate education in mathematics. There were only two others at that time, Johns Hopkins, whose development we have already examined, and Clark University, founded two years before Chicago. These three shone out like beacons at a time when, compared with their counterparts in Europe, mathematics courses taught at the majority of American universities were, according to one view, 'paltry, the spirit of research almost nonexistent and the quality of the faculty vastly inferior'.[11]

Chicago's first president, William Rainey Harper, set about generating great innovations in every aspect of higher education and the process was still in its youth when R.L. Moore was finally able to extricate himself from Marshall High School to arrive in Chicago in September 1903. He was in for a culture shock. This city was the modern, northern equivalent of the Wild West, and more. It was bigger, much bigger, brasher and far more violent than any place in Texas. Chicago had experienced an incredible rate of growth, expanding in seventy years from a small trading post on the swampy river mouth near the southwestern tip of Lake Michigan to a city of over one million by 1900, and rising. Its people were drawn largely from the northeastern states and across Europe. The strategic inland water location provided the powerhouse that drove its climb towards becoming one of the most renowned and wealthiest commercial centers in the world. The boomtown rose with the Civil War. Just as Dallas was a major supply center for the Confederacy, Chicago had been a major logistical base for the Union, although downstate in southern Illinois there had been sympathy toward the Confederate cause. The city had survived the Great Fire of 1871, when four square miles of property, including the business district, was burned to the ground, killing 250 people and rendering 90,000 homeless. The cost of the damage was $200,000,000 and the rebuilding program saw the birth of steel-framed skyscrapers, with the completion

---

[11] Quoting David E. Zitarelli, in Towering Figures in American Mathematics 1890–1950, *Amer. Math. Monthly* 108 (2001), 606.

of the Home Insurance Building in 1885, followed, during the next decade, by twenty-one buildings ranging from twelve to sixteen stories.

That was fine, but great festering slums were being created just as fast, with the population swelling during the second great wave of European immigration. Now, Russian Jews, Italians, Poles, Serbs, Croatians, Bohemians and others from Eastern Europe joined the earlier immigrants predominantly from Scandinavia, Germany, Britain and Ireland. The 1900 census showed that more than three-quarters of Chicago's population was made up of the foreign-born and their children, and it now surpassed Philadelphia as America's second most populous city.

By 1903 when R.L. Moore arrived, it was a principal center of not just the economy of the United States, but of social insurgency, immigration, housing and education. It also boasted the most elegant brothels in the country, which entertained royalty from abroad and millionaires from the newly sprawling suburbs. Scholars were attracted by this new seat of learning, while writers and authors arrived for inspiration from its brawling streets. 'I have struck a city,' wrote Rudyard Kipling. 'A real city, and they call it Chicago. The other places do not count.... Having seen it, I urgently desire never to see it again. It is inhabited by savages.'[12]

In the middle of this sprawling, steaming morass of human flotsam and jetsam sitting uncomfortably alongside boiling commercialism arose the University of Chicago, founded in 1891 by the American Baptist Education Society and Standard Oil magnate John D. Rockefeller. He later described the University as 'the best investment I ever made' although its president, Dr. William Rainey Harper, was its true creator. He envisaged a university that went well beyond the liberal arts colleges of the 1890s, where classical education ruled. He wanted important additions such as the graduate research pioneered in Europe, managed by a faculty of dedicated teachers and researchers who were to follow specific guidelines that precluded stocking the students' minds with knowledge of what had already been accomplished. He demanded new horizons, new research and new lines of investigations. The mission, he insisted, could not be allowed to fail and in forcing into existence a hotbed of new thinking, Harper provided unparalleled opportunities for the linchpins in a new generation of American

---

[12] *American Notes,* 1889.

mathematicians and scholars. Frederick Rudolph, professor of history at Williams College, has written: 'No episode was more important in shaping the outlook and expectations of American higher education during those years than the founding of the University of Chicago, one of those events in American history that brought into focus the spirit of an age.'[13]

The whole focus of the Chicago enterprise emerged through the foresight of Dr. Harper who was not only a great barnstorming manager of university affairs but an exceptional role model for all who gathered around him. Among them were some brilliant personalities to add to the chart of R.L. Moore's mathematical ancestry and, once again, a detour from the main thrust of our story is necessary to place them in the chronology of his exposure to a rich seam of research and development in progress immediately prior to, and after, his arrival at Chicago.

Harper himself was not a mathematician. He had pursued Hebraic studies at Muskingum College, New Concord, Ohio, from which he graduated in 1870 at the age of fifteen. He received his PhD from Yale at the age of nineteen for studies in the Indo-Iranian and Semitic languages and moved into academy teaching before being awarded a professorship in Hebrew at the Baptist Union Theological Seminary in Chicago. He authored a number of textbooks[14] and aids for the teaching of Hebrew. He returned to Yale in 1886 as a professor in Semitic languages and in 1889 was simultaneously appointed Woolsey professor of biblical literature. Even as the ground plans for the University of Chicago were approved, Harper was among those short-listed to become president. He held back until the last, and was finally persuaded by Rockefeller himself by being given virtual carte blanche to organize a liberal and comprehensive outlook for the new university where students would be given the freedom to do research. He announced his intention to divide the traditional collegiate program into two parts, devoting the first two years to general education and the last two years to higher study. He would keep the university open all year round and other innovations included the introduction of correspondence courses.

---

[13] Frederick Rudolph, *The American College and University: A History*, 1962.

[14] Among his more important books are *Religion and the Higher Life* (1904); *A Critical and Exegetical Commentary on Amos and Hosea* (1905); *The Prophetic Element in the Old Testament* (1905) and *The Trend in Higher Education* (1905). Harper remained at Chicago as president and head of the department of Semitic languages until his death in 1906.

More important were the people he drew in, carefully selecting those who would head up the teams to attract strong faculty and students alike in a research-orientated environment. They would include such luminaries as John Dewey whose influence on the American education field, for good or ill, was to reverberate through the twentieth century. He left Michigan in 1894 to become professor of philosophy and chairman of the Department of Philosophy, Psychology, and Pedagogy at the University of Chicago. Dewey's achievements there had brought him national fame, which came to fruition at the time of R.L. Moore's arrival.

The emphasis on new thinking and the pedagogical development became a priority in Harper's plan, and none more so than the Department of Mathematics which had been highlighted as a crucial area in Chicago's objectives, which in turn had great bearing on young careers, like that of R.L. Moore. Harper wanted to appoint a man to bring European thinking and personalities into the new institution and there were a number of established American or European mathematicians he could have selected to head the new department. In fact, he had taken advice and a short list of six well-known names had been drawn up for him. But, as with many of his decisions, Harper confounded those around him by choosing a virtual unknown who was not even on the list. He settled on Eliakim Hastings Moore, a relatively untried assistant professor with no great publishing history, then residing not far away at Northwestern University in Evanston, Illinois.

Moore, son of a Methodist minister from Athens, Ohio, was barely 28 years old when Harper first approached him and had published only four papers, apart from his doctoral dissertation, which had caused hardly a ripple of interest. Harper had spotted something others had missed, the possibility of genius inside the excitable persona of E.H. Moore. It was undoubtedly a gamble, but it paid off handsomely. Harper had completed a thorough research job on Moore's background, which was solid enough. It included experience among some of Germany's finest mathematicians, although it was not what might be termed a spectacular CV. He also saw that the assistant professor was undoubtedly hemmed in by his present surroundings at Northwestern, whose work was not especially adventurous, although a good deal better than some. E. H. Moore had studied at Yale and worked one summer with Ormond Stone, director of the Cincinnati Observatory, whose mathematical output was substantial. His own mentor was Hubert Anson Newton, pro-

Eliakim Hastings Moore

fessor of mathematics at Yale who inspired and guided Moore into research. Newton was professor of mathematics at Yale from 1855 until his death in 1896, although in the latter stages of his career he turned towards astronomy.

He was not noted for his own research work, but in recognizing the limited facilities for the training of future mathematicians in the United States, encouraged his most talented mathematics students to undertake research by self-discovery and especially to take account of the work being done abroad. In fact, he was so impressed by E.H. Moore that after he had earned his PhD at Yale in 1885, Newton personally financed a year's study in Germany. There, armed with an introduction from Newton, Moore met and worked with great mathematicians of the day, including Schwartz, Klein and Weierstrass. He returned in confident mood, took work first as an instructor in the Academy at Northwestern University, then as a teacher at Yale University for the next two years before returning to Northwestern in 1889 as assistant professor. He was still there when Harper began to cast his net for his own staff and it must surely have been Hubert Newton, an old friend and former colleague at Yale, who tipped him off about the genius of E.H. Moore.

Harper made the approach, but Moore did not rush to accept. He played a canny game. There were apparently several exchanges, especially over salary, before Moore finally accepted the offer of a professorship and the appointment as acting head of mathematics at the University of Chicago, starting in the fall of 1892. Four years later, recognizing the remarkable job Moore was accomplishing, Harper made him permanent head, a position he held until 1931. From the outset, Harper and Moore were in complete agreement over the need to attract universally brilliant scholars to the University, and especially recruit faculty members from Germany. The first important addition was Oskar Bolza who in 1892 was engaged in a rather bitter dispute at the fiscally challenged Clark University.

Nine of the eleven members of the permanentfaculty had put in their resignations en masse, and although they eventually withdrew them the situation remained volatile. Clark had been cutting back heavily on equipment and research, in which the university had boasted a leadership role in America. Financially, however, the chickens were coming home to roost. Harper immediately targeted a number of the unhappy members of Clark faculty and in what became known as Harper's Raid[15] secured the services of no less than fifteen of them, including instructors and fellows. Fed up with the lack of resources at Clark, Bolza himself had decided to return to Germany.

E.H. Moore had other ideas. Bolza was the one man he especially wanted from Clark because he had the experience and the connections that Moore himself had sampled in Germany. Bolza had studied mathematics under Christoffel and Reye at Strasbourg, under Hermann Schwarz at Göttingen and Karl Weierstrass at Berlin. He was present for the lecture course on the calculus of variations given by Weierstrass in 1879, which was a defining moment in Bolza's own search for mathematical direction. Finally, and most importantly from E.H. Moore's standpoint, he achieved his doctorate from the University of Göttingen in 1885 supervised by Felix Klein, best known for his work in non-euclidean geometry, the connections between geometry and group the-

---

[15] The Clark crisis, outlined in *American Higher Education: A Documentary History*, Wilson Smith and Richard Hofstadter (eds.), University of Chicago Press, 1961, Vol. 2, pp. 759–761.

ory, and for results in function theory. Plagued by ill-health and depression, Klein had by then curtailed his own research work but had set out to inspire others with the foundation at Göttingen of what became a model for mathematical research centers the world over. It was an experience that E.H. Moore, who had witnessed it himself first hand in 1886, wanted to tap into.

Nor was that the end of the story as far as the Klein connection was concerned. Bolza's closest friend and another of Klein's students was Heinrich Maschke who had migrated to the United States in 1888. Having acquired Bolza, E.H. Moore now approached Maschke, who worked for the Western Electrical Instrument Company, Newark, New Jersey. He was glad of the opportunity to join his old friend Bolza and the two went to Moore's department as professors of mathematics. So was created one of the most influential mathematical research teams in the United States whose work as a forceful and inspirational triumvirate was to become the guiding spirit for so many brilliant young mathematicians over the next decade and beyond, including R.L. Moore. As these arrangements were being completed, the Chicagoans were presented with a heaven-sent opportunity to make a very firm stamp of their authority on the American mathematical scene when the Windy City was chosen as the venue for the 1893 World Fair.

To coincide with that event, E.H. Moore's team began work on organizing the Chicago Mathematical Congress, a weeklong extravaganza of lectures and papers from America's own mathematical hierarchy, including one from Dr. G.B. Halsted,[16] as well as from abroad. On the first day, a paper by David Hilbert entitled 'Invariantentheorie' was the star attraction, and the Congress concluded with a lecture by Felix Klein.[17] Never had such an influential collection of views been assembled under one roof in the United States and it was in the vein of what might be termed calm assertiveness that the mathematics department of the University of Chicago under Moore, Bolza and Maschke grew in stature and performance. It was a team totally dedicated to mathematical excellence and Archibald's description hit the right note:

---

[16] Halsted's Congress lecture was entitled 'Some Salient Points in the History of Non-Euclidean and Hyper-Spaces.'

[17] The title was *Concerning the Development of the Theory of Groups during the Last Twenty Years*. Klein then went straight on to deliver the Evanston Colloquium Lectures on Mathematics, from 28 August to 9 September 1893.

'These three men supplemented one another remarkably. Moore was a fiery enthusiast, brilliant, and keenly interested in the popular mathematical research movements of his day. Bolza, a product of the meticulous German school of analysis led by Weierstrass, was an able and widely read research scholar. Maschke was more deliberate than the other two, sagacious, brilliant in research, and a most delightful lecturer on geometry.'[18]

Into this hotbed of exploration and discovery stepped R.L. Moore, arriving in the fall of 1903 when the triumvirate was at its peak. It was also a measure of the Texas Moore's ability — and Halsted's very strong recommendation of it — that he was offered a fellowship at such an important time in the Chicago experiment. It was, after all, not the norm to take a young man teaching high school in some out-of-the-way Texas town for such a position, but more importantly at this juncture was the fact that all eyes were on the Chicagoans. They were being watched by the whole mathematics community, a scrutiny that became rather more focused when Professor Moore became president of the American Mathematical Society in 1901 and even more so when he received a good deal of publicity during his term of office for suggesting the use of what he described as 'laboratory' teaching techniques at Chicago. To put it bluntly, his department had gained a reputation but now had to prove the worth of his pedagogical experiments. Moore chose his 'stars' carefully. Only the best would do and indeed, as will be seen, he experienced difficulty in teaching those who were slower on the uptake. Of his early PhD students L.E. Dickson was outstanding. As already noted, Dickson arrived from Texas having gained his B.S. and M.S. under G.B. Halsted and was described by E.H. Moore as 'the most thoroughly prepared student I have ever had'. He joined Moore's department in 1900 for what was the beginning of a long and fruitful career, remaining at Chicago until his retirement in 1941.

In 1901 his book, *Linear groups with an exposition of the Galois field theory,* was published as an expanded version of his 1896 doctoral thesis. It is described by Karen H. Parshall as 'a unified, complete, and general theory of the classical linear groups — not merely over the prime field $GF(p)$ as Jordan had done, but over the general finite field $GF(pn)$, and he did this against the backdrop of a well-developed theo-

---

[18] *Semicentennial History of the American Mathematical Society, 1888–1938*, New York, 1938.

L.E. Dickson

ry of these underlying fields.... [H]is book represented the first systematic treatment of finite fields in the mathematical literature'.[19]

Dickson went on to publish 17 books and many articles and papers. He was awarded numerous honors and degrees from rival universities, became president of the American Mathematical Society (1917–1918) and supervised more than 60 PhD students. It has been said that mathematicians at this level might consider it fortunate indeed to have one such thoroughly brilliant student in even a decade but E.H. Moore brought on three others who were to arrive in Chicago at his behest at a particularly relevant time. The first was Oswald Veblen, widely considered to be E.H. Moore's first true prodigy, followed quickly by R.L. Moore and George David Birkhoff. Before examining their arrival and friendship in more detail, it is timely to inject a recollection of E.H. Moore's style and attitudes as he drew in graduate students of such high caliber and who themselves were to carry forward this great flair and influence on the creativity of the American mathematical school. For

---

[19] K.H. Parshall, A study in group theory: Leonard Eugene Dickson's *Linear groups*, *Math. Intelligencer* 13 (1991), 7–11.

this we can accurately rely on the personal descriptions of L.E. Dickson himself, in a biographical account he jointly authored with another outstanding Chicago student of the earliest era, Gilbert Bliss, who began his doctoral studies working on the calculus of variations. He gained his PhD in 1900, producing a dissertation, *The Geodesic Lines on the Anchor Ring,* which was supervised by Bolza. Bliss joined Dickson on the Chicago faculty in 1908, following the untimely death of Maschke in that year and, like his comrade, remained until retirement. Their joint summary of experiences with E.H. Moore may be seen to hold the seeds of inspiration for the style of instruction eventually developed by R.L. Moore. In the biographical memoir, Dickson and Bliss identified E.H. Moore's success as an educator as being due to his profound interest in mathematics and his faculty for inspiring his colleagues, and especially the strongest graduate students, with some of his own enthusiasm. He pursued educational experiments fearlessly and in every one he undertook, as at every other stage of his leadership in his department, he had one permanent characteristic: 'He believed in the exercise of individuality in classroom instruction, and he gave his colleagues unlimited freedom in the development of their classroom methods. He expected and insisted on success, and he was always sympathetically interested in a new proposal or procedure, but so far as is known to us he never prescribed a textbook.... Two of the characteristic qualities of his research were accuracy and generality. He was a master of mathematical logic, and his originality in making one or more theories appear as special instances of a new and more general one was remarkable.'[20]

His classes were of indeterminate length because he would bring before his students topics that he found of absorbing interest and he, and they, would investigate those ideas together until the topic was exhausted. Formal class schedules, time of day nor mealtime had any bearing on his pursuit of that topic. His style was not to everyone's taste. Dickson and Bliss recalled that students not so far advanced found difficulty in keeping up with him. At times, because of his total absorption in the subject at hand, he would be unaware offense had been given by something he had said, and would be extremely gentle in his expressions of regret when it was called to his attention that someone's feelings had been hurt by his impatience. Weak students often shunned his

---

[20] *Biographical Memoir of Eliakim Hastings Moore (1862–1932),* by G.A. Bliss and L.E. Dickson, National Academy Biographical Memoirs, Vol. XVII, 1936, p. 89.

classes because of the demand placed on them. Strong students and those of quickest mind were attracted to him. He became a teacher who taught the teachers of mathematics. His teaching skills were the strongest when dealing with graduate students, although with students not so far advanced he was less accessible. His own mind was so quick and his concentration so thorough that it was difficult for him to await the comprehension and the slower development of understanding among those who were less experienced. However, he did edit an arithmetic book for use in elementary schools in 1897, and in 1903–1904 and following years, he influenced radically the methods of undergraduate instruction in mathematics at the University of Chicago. He gave courses in beginning calculus himself, casting aside textbooks, and concentrating instead on the fundamentals of the topic and their graphical interpretation.

He also put forward a proposal for laboratory courses, meeting two hours each day, which E.H. Moore had outlined in his presidential retirement address before the American Mathematical Society, the principal theme of which was a call for reforms across the American educational system in the teaching of science and mathematics:[21]

'This program of reform calls for the development of a thoroughgoing laboratory system of instruction in mathematics and physics, a principal purpose being as far as possible to develop on the part of every student the true spirit of research and an appreciation, practical as well as theoretic, of the fundamental methods of science.... To provide for the needs of laboratory instruction, there should be regularly assigned to the subject two periods, counting as one period in the curriculum.... This pedagogic principle of concentration is undoubtedly sound. One must, however, learn how to apply it wisely.'

The laboratory methods apparently intrigued R.L. Moore and he mentioned it to G.B. Halsted in a letter to him on 2 November 1903. Halsted was similarly intrigued, though sceptical: 'What have you found out about them? Do you find anything admirable or desirable in it? Or is it merely a fad?'[22] R.L. Moore reserved judgment and appears to have made no further reference to it in correspondence with Halsted, perhaps because it was apparent from his tone that Halsted believed it

---

[21] Retirement address, *On the Foundations of Mathematics*, delivered before the AMS, at its ninth annual meeting on 29 December 1902; *Bull. Amer. Math. Soc.* 9 (1902–3), 402–424.

[22] Halsted to Moore, 15 November 1903; R.L. Moore Papers in the AAM.

to be a ridiculous notion. There were certainly pros and cons that created debate among those at the hub and, as with many such new plans, the amount of work required of the instructor was considerable. As Dickson and Bliss pointed out in their memoir, the two-hour period that was the feature of the experiment had to be abandoned because of the very practical difficulty in finding hours on schedules which would not interfere with the offerings of other departments. In the end, it proved an unworkable burden upon the curriculum schedules. Perhaps the most notable edict was also included in E.H. Moore's presidential retirement address:

'The teacher should lead up to an important theorem gradually in such a way that the precise meaning of the statement in question, and further, the practical truth of the theorem is fully appreciated and the importance of the theorem is understood and indeed the desire for the formal proof of the proposition is awakened before the formal proof itself is developed. Indeed, in most cases, much of the proof should be secured by the research work of the students themselves.'

E.H. Moore's grand plan, with its inherent flaws such as the pressures it brought to bear on the curriculum, eventually led to the demise of the laboratory experiment. Perhaps the overriding aspect that came out of these discussions in the Chicago Mathematics Club was E.H. Moore's insistence on great care being taken in using precise language. This reinforced a trait already evident in R.L. Moore's earliest work, and even in his letters to Halsted and his own notebooks, where he precedes even the most mundane statements with his 'Suppose...' Those who came to his classes in years hence quickly discovered that precision and clarity in written and spoken statements were vital ingredients of any Moore Method experience.

4

A Veritable Hothouse
(1903–1905)

Under the general guidance of E.H. Moore, Bolza and Maschke, R.L. Moore takes his place at Chicago with nervous enthusiasm. Befriended by Veblen and in the company of Dickson, Birkhoff (G.D.), Bliss and Co., E.H. Moore's finest group of young mathematicians begin their journey to the cutting edge of American mathematics. Letters and papers provide R.L. Moore's running commentary.

Mister Robert Lee Moore

*A*n aura of excitement, competitiveness and newness developed at Chicago that was in many ways unique. There was so much going on that the sheer drama of the moment — for such it surely was — combined effectively with the inspirational qualities of instructors to lead them into territory previously uncharted in American methods of tuition and in the subject matter upon which they focused. E.H. Moore's own courses might, in retrospect, be looked upon as elitist in that they ought to have carried a health warning for the faint-hearted or slower students. It might also be added that an additional ingredient that gave rise to the (R.L.) Moore Method which is ultimately at the heart of these chapters, must surely have come to him in his days at Chicago (1903–1905) when he was exposed to that dynamic scenario created by E.H. Moore, Bolza and Maschke, each guiding 'students along lines of investigation consonant with his own evolving research interests'.[1]

The latter probably had a greater influence than the system in injecting into the psyches of the aspirants the qualities and ambitions that produced a fair number of mathematical stars who went on to substantial endeavor as E.H. Moore and his team skillfully identified areas of modern work, often aligned to European studies, and pushed forward the boundaries of research in the United States. In doing so, Professor Moore set up a program at Chicago that by 1900 put it ahead of the half dozen or so rival universities with graduate programs in mathematics and thus attracted a continuing flow of talent into his own program. By 1915, the University of Chicago had produced 63 mathematics doctoral students, well ahead of any other American establishment.[2]

As history now shows, the sparkling collection of students who gathered at Chicago in those early years included a number who were brilliant enough to take their place at the cutting edge of America's mathematical development in the first half of the twentieth century, which saw crucial advances in group theory, algebra, differential geometry, topology and logic. But, of course, that was Moore's prime objective in his aborted plan for the 'laboratory' periods for which he directed that students and instructors should combine their efforts in the acquisition of

---

[1] Parshall and Rowe, *The Emergence of the American Mathematical Research Community*, p. 393.
[2] Analysis and studies by Dell Dumbaugh Fenster and Karen Hunger Parshall in *The History of Modern Mathematics*; Boston, Academic Press, Inc., 1994, pp. 179–227.

Heinrich Maschke

skills and knowledge. He made that quite clear in his 1902 AMS presidential address, "On the Foundations of Mathematics":

'Some hold that absolutely individual instruction is the ideal and a laboratory method has sometimes been used for the purpose of attaining this ideal.... The instructor utilizes all the experience and insight of the whole body of students. He arranges it so that the students consider that they are studying the subject itself, and not the words, either printed or oral, of any authority on the subject ... I am convinced that the laboratory method ... is the best method of instruction for students in general and for students expecting to specialize in pure mathematics.'[3]

Consequently, students and instructors at Chicago were inspirational to each other, a fact no better demonstrated than in the work of five, in particular, who went on to contribute significantly to the American mathematical world for the ensuing four decades: Leonard Dickson, Gilbert Bliss, Oswald Veblen, Robert Lee Moore, and G.D. Birkhoff. Of particular interest for this account was the relationship that developed between R.L. Moore and the brilliant young doctoral student Oswald Veblen, later to achieve great fame at Princeton. The two struck up an immediate friendship, both on a personal level and in a very close and fervent relationship in their early work. The central interest in this

---

[3] Retiring presidential address to the AMS Ninth Annual Meeting, *Bull. Amer. Math. Soc.* 9 (1902–3), 402–424, quote from pp. 419–420.

coming together of two young minds was R.L. Moore's astute approach to geometry and axiomatics that emerged under Halsted, and which had brought him to the attention of E.H. Moore in the first place.

Veblen, likewise, was engaged in similar research and both came into Professor Moore's sphere at a time when he himself was pursuing the foundations of geometry and analysis. It will be recalled from early references in Chapter Two that Hilbert had published his *Grundlagen der Geometrie* in which he pursued the work of the German mathematician Moritz Pasch and the Italians, Giuseppe Peano and Giuseppe Veronese, and others. Hilbert stated that his aim was to discover a complete set of independent axioms and 'to deduce from these the most important geometrical theorems in such a manner as to bring out as clearly as possible the significance of the different groups of axioms and the scope of the conclusions to be derived from the individual axioms'.[4] Hilbert proposed twenty such axioms and he analyzed their significance in *Grundlagen der Geometrie*, thus putting geometry in a formal axiomatic setting. His work became a major talking point, highlighting the axiomatic approach to mathematics, which he followed up with a famous presentation at the Paris congress in 1900. He opened with one of his most quoted statements:

'Who of us would not be glad to lift the veil behind which the future lies hidden; to cast a glance at the next advances of our science and at the secrets of its development during future centuries? What particular goals will there be toward which the leading mathematical spirits of coming generations will strive? What new methods and new facts in the wide and rich field of mathematical thought will the new centuries disclose?'[5]

Hilbert then posed 23 problems and challenged mathematicians the world over to resolve fundamental issues. He went on to stress the undeniable importance of definite problems for the progress of mathematical science in general: 'We hear within ourselves the constant cry: There is the problem, seek the solution. You can find it through pure thought.'[6]

---

[4] D. Hilbert, *The Foundations of Geometry*, Open Court Publishing, La Salle, Illinois, 1965, p. 1.

[5] Hilbert's famous speech *The Problems of Mathematics* was delivered to the Second International Congress of Mathematicians in 1900 and published in *Bull.Amer. Math. Soc.* 8 (1902), 437–479.

[6] Hilbert references: M. Toepell, On the origins of David Hilbert's Grundlagen der Geometrie, *Arch. Hist. of Exact Sci.*, 35 (4) (1986), 329–344. H. Weyl, David Hilbert and his mathematical work, *Bull. Amer. Math. Soc.* 50 (1944), 612–654. H. Weyl, David Hilbert. 1862–1943, *Obituary Notices of Fellows of the Royal Society of London* 4 (1944), 547–553.

With these words ringing in the ears of like-minded teachers, and their students, urged on by reviews of Hilbert's book and papers by such personalities as Poincaré, Halsted and Hedrick, the whole topic was ready meat for seminar discussion and Oswald Veblen was drawn into it as soon as he arrived at Chicago in the fall of 1901. Veblen's background included schooling in Iowa, a B.A. from the University of Iowa in 1898 and then working there for a year as a laboratory assistant. He spent a year at Harvard University before leaving to undertake graduate research at Chicago and arrived at the very moment when Professor Moore was introducing a seminar on the foundations of geometry and analysis in which his principal reference for his lectures was Hilbert's *Grundlagen der Geometrie*.

The course was especially topical in view of a paper published that year contesting Hilbert's work and which, for E.H. Moore, introduced an up-to-the-minute level of discussion. The paper, by Friedrich Schur,[7] claimed redundancies in Hilbert's two sets of axioms. As he discussed the contested elements in his class, Moore there and then discovered that while redundancies existed, they related to one axiom of connection and one of order. As already noted, Moore published the new mathematical result ahead of R.L. Moore's discovery of the same redundancy that led to the Chicago professor praising Halsted's protégé for a 'delightfully simple proof'.

Veblen was so inspired by his exposure to his professor's method and subject matter that he chose to pursue the issue in his studies.[8] His abilities and sustained interest drew the close attention of E.H. Moore, Bolza and Maschke and they engaged the 21-year-old's receptive brain in an intensive program of instruction that laid the basis for the important work he was later to achieve in the fields of foundations of geometry, projective geometry, differential invariants and spinors.[9] Such was the tenor of Veblen's work that he progressed rapidly towards the decision to focus his doctoral dissertation on euclidean geometry, entitled *A System for Axioms for Geometry*. E.H. Moore was so impressed by his student's achievement in this direction that he made a point of mention-

---

[7] Friedrich Schur, Über die Grundlagen der Geometrie, *Math. Ann.* 55 (1901), 265–292.
[8] Parshall and Rowe, *The Emergence of the American Mathematical Research Community*, p. 383.
[9] R.C. Archibald, *A Semicentennial History of the American Mathematical Society 1888–1938*, New York, 1938.

Oskar Bolza

ing it at the beginning of his presidential address to the American Mathematical Society in 1902:

'In his dissertation on Euclidean geometry, Mr. Veblen, following the example of Pasch and Peano, takes as undefined symbols "point" and "between" or "point" and "segment." In terms of these two symbols alone he expresses a set of independent fundamental postulates of Euclidean geometry, in the first place developing the projective geometry, and then as to congruence relating himself to the point of view of Klein in his *Erlangen Programm* [sic], whereby the group of movements of Euclidean geometry enters as a certain subgroup of the group of collineations of projective geometry.'

In Veblen's system then there were only two primitive notions, point and order (the points *a*, *b*, *c* occur in the order *abc*).[10] When Veblen completed his thesis in 1903 he became an associate in mathematics at the University of Chicago and immediately became active in advising other young mathematicians, and the following year the friendship established with R.L. Moore on his arrival into this maelstrom of activity was apparent from the version of the dissertation revised for publi-

---

[10] *Oswald Veblen (1880–1960), A Biographical Memoir*, Saunders Mac Lane, National Academy of Sciences, Washington, DC, 1936; passim.

cation. It not only mentioned R.L. Moore's work at Texas but carried a footnote: 'I wish to express deep gratitude to Professor E.H. Moore who has advised me constantly and valuably in the preparation of this paper and also to Messrs N.J. Lennes and R.L. Moore who have critically read parts of the manuscript.'[11] It was a proud moment for the young Moore, then poised to receive his own fellowship, to be mentioned in such a way.

The rapport between R.L. Moore and Oswald Veblen began with this common interest in the foundations of axiomatic geometry, in which they were immersed in the ongoing theme of investigations, and developed into a lively mathematical friendship that was to continue over many years. This relationship may be seen as one of the defining elements of their respective careers. They were clearly fascinated, excited and undoubtedly filled with a youthful zest that turned into passion. It would form the central plank of the work of both men for the first decade or more of the twentieth century, by which time they had themselves moved to new pastures, though they remained constantly in touch. Quite apart from the thrills attained from their chosen lines of research, the enthusiasm with which encouragement was heaped upon them in this hothouse environment of study was an experience that identifiably underscored their future progress, both in their work and development as researchers and teachers.[12] A further contributor to this arena of research during R.L. Moore's early days at Chicago was

---

[11] As quoted in Traylor, p. 59.

[12] Veblen remained at Chicago until 1905 when he moved to Princeton University to begin a long and illustrious career which kept him at the university until 1932. He established Princeton as one of the leading centers in the world for topology research, his first of many works on the subject being published just before he left Chicago. His continuing interest in the foundations of geometry led to his work on the axiom systems of projective geometry and together with John Wesley Young he published *Projective Geometry* (1910–18) which they introduced with the words: 'Since it is more natural to derive the geometrical disciplines associated with the names of Euclid, Descartes, Lobachevsky, etc. from projective geometry than to derive projective geometry from one of them, it is natural to take the foundations of projective geometry as the foundations of all geometry.' Veblen's 1922 work, *Analysis Situs* was seen as the first systematic coverage of the basic ideas of topology and contributed much to the development of modern topology. Soon after Einstein's theory of general relativity appeared Veblen turned his attention to differential geometry. This work led to important applications in relativity theory, and much of his work also found application in atomic physics. He followed E.H. Moore as an active member of the American Mathematical Society, serving as president in 1923–1924 and was honored by other mathematical societies around the world.

Gilbert Bliss, who brought to the table the added dimension of having become one of the first American mathematicians to benefit from further study in Europe. He entered the University of Chicago in 1893, receiving his B.S. four years later, and then began graduate studies in mathematical astronomy. In 1898, inspired by a lecture from Bolza, he began his doctoral studies working on the calculus of variations, and after receiving his doctorate in 1900, spent the following two years at the University of Minnesota before leaving for Göttingen. There, he was exposed to the work of several of the most famous European mathematicians of that era including Hilbert and Klein. He was thus well equipped with up-to-date knowledge when he returned to the United States and the University of Chicago in 1903 at the time of R.L. Moore's arrival. Bliss remained for a year before moving to the University of Missouri, but he, Veblen and R.L. Moore were all to meet up again at Princeton.[13] The fifth influential member of the early 1900's group at Chicago was G.D. Birkhoff, first as an undergraduate before moving to Harvard, returning to enter the Chicago graduate program in 1905 for his doctorate. By then, he had completed graduate work at Harvard and jointly authored with H.S. Vandiver a paper which contained a number-theoretic result that turned out to be pivotal in Wedderburn's proof of his eponymous theorem. The coincidence of that, apart from the fact that Birkhoff[14] also became one of the most

---

[13] Bliss returned to Chicago on the death of Maschke and remained until his retirement. Throughout his professional life, he took a major role in mathematics in the United States, and especially in the American Mathematical Society: Colloquium Lecturer, 1909, Vice President 1911 and President 1921. He received many awards for his work and most famously worked on ballistics during World War I, designing new firing tables for artillery. In 1918 he joined Veblen in the Range Firing Section at the Aberdeen Proving Ground, a military weapons testing site established in 1917 in Harford County, Maryland. There he very effectively applied methods from the calculus of variations to solve problems relating to correcting missile trajectories for the effects of wind, changes in air density, rotation of the Earth and other perturbations.

[14] A leading American mathematician of the early twentieth century, who formulated the ergodic theorem, which transformed the Maxwell-Boltzmann ergodic hypothesis of the kinetic theory of gases (to which exceptions are known) into a rigorous principle through use of the Lebesgue measure theory. Birkhoff taught at the University of Wisconsin, Madison (1907–1909); Princeton University (1909–1912); and Harvard University (1912–1944). He was a stimulating lecturer and director of research. In the mid-twentieth century, many of the leading American mathematicians had either written their doctoral dissertations under his direction or had done postdoctoral research with him. Birkhoff served as president of the American Mathematical Society (1924–1926), as dean of the

highly respected mathematicians of the first half of the century, was that his co-author, Vandiver, was to become a life-long (if uneasy) associate of R.L. Moore after they both landed in Austin, Texas.

Evidence that R.L. Moore, as the youngest in the Chicago group, received guidance and stimulation from all those around him can be gleaned from the ongoing entries in his Ivory Soap pad diary. Generally, the diary (where it is possible to decipher the oft-illegible scrawl of a young man in a hurry or, late at night, very tired) reads like the naïve jottings of a very unworldly student who has missed his youth while his face was buried in his work. There is virtually no mention of any social activity, sports, girlfriends or other interests that might otherwise amuse and engage someone of his age.

The diary concerns itself mainly with reflections and comments about his work, intentions and notes of anything faintly praiseworthy of himself, or advice as to how he might proceed. His entries include mentions of Leonard Dickson remarking, for example, that those were 'pretty theorems this morning, Mr. Moore, those were all right'.[15] G.A. Bliss, meanwhile, gave him advice on presentation in preparation for talks to the university's Mathematics Club, formed in the 1890s by E.H. Moore and which was recognized as another area of group inspiration. He records advice from Bliss: 'You know, most people don't understand a paper, anyway, but if you present things in a certain way you will get a reputation of being a person who can't present things well and you are too good a man to have that said of you. We want to let people see what kind of material we have here. Don't you see?'[16]

There were numerous diary entries noting encouragement from E.H. Moore himself. He was seemingly keen to see Moore establishing himself in mathematics on a broader front than university study, and suggested quite early on that he should join the American Mathematical Society as soon as he could afford it: 'I think you are in mathematics for good, are you not?'[17]

R.L. Moore replied: 'I hope so.'

Harvard faculty of arts and sciences (1935–1939), and as president of the American Association for the Advancement of Science (1936–1937). His works include *Relativity and Modern Physics* (1923), *Dynamical Systems* (1928), *Aesthetic Measure* (1933), and *Basic Geometry* (1941; with Ralph Beatley).

[15] Ivory Soap pad diary, 31 March 1904; R.L. Moore Papers in the AAM.
[16] Ibid., 6 March 1904.
[17] Ibid., March 1904.

He also noted Professor Moore's numerous references to him taking up summer teaching posts at the university. This, along with the professor's teaching effort and style, clearly made a lasting impression, not least in the way that the younger Moore was able to observe his precision, clarity, and logical correctness in his presentations. In fact, the manner in which he delivered his presentations was thought by some to show that E.H. Moore had a speech impediment. But as Dickson and Bliss pointed out in their memoir, 'This arose from his speaking habit which was designed to allow accurate description of what he wished to state. It would seem at times that he would begin to say a word, testing it almost to see if it would convey what he desired and, determining that it would not, switch in mid-breath to another word.'

This precision was applied not merely to mathematical statements, but all aspects of everyday conversation, so that those who conversed with him were forever on their mettle to state exactly what they meant. R.L. Moore himself recalled[18] that one day when late for an appointment, he began by explaining that his alarm clock did not go off at the appointed time. Then, remembering his professor's requirement for accurate statement, corrected himself and added that it may have gone off but he had not heard it. Stumbling over that excuse, he corrected himself again by saying that of course it was possible that the alarm had gone off and that he had heard it and switched it off only then to have fallen asleep again. At this point, R.L. Moore was relieved of the need for further explanation. It was another lesson learned in the need for precise and clear statements, in both mathematical presentations and general conversation. Equally, the manner in which E.H. Moore approached his teaching of mathematical concepts had distinct influence on R.L. Moore. According to Traylor,[19] much of the content in R.L. Moore's course "Introduction to the Foundations of Analysis" which he offered at The University of Texas years later was seen by him first with E.H. Moore or Oswald Veblen.

Veblen, in his dissertation, gave a system of axioms for geometry in which the so-called Heine-Borel property appeared as Axiom XI: If there exists an infinitude of points, there exists a certain pair of points A, C such that if σ (σ denotes a set or class of elements, any one of which is denoted alone or with an index or subscript) is any set of seg-

---

[18] In conversation with Traylor, *Creative Teaching*, p. 49.
[19] Traylor, *Creative Teaching*, pp. 45–63.

ments of the line AC, having the property that each point which is A, C or a point of the segment AC is a point of a segment σ, then there is a finite subset $\sigma_1, \sigma_2, \ldots, \sigma_n$ with the same property. He called Axiom XI his Continuity Axiom, noting that Schoenflies had called that property the Heine-Borel theorem. Borel is given credit for first stating the theorem in 1895, as a theorem of analysis, but Heine in 1871 used that notion in the proof of a theorem of uniform continuity. Veblen credited fellow student N.J. Lennes with the idea of the equivalence of his Axiom XI with the Dedekind cut axiom while R.L. Moore also made further contributions.

Veblen stated as axioms: [Axiom II:] If points A, B, C are in the order ABC, they are in the order CBA. [Axiom III:] If points A, B, C are in the order ABC, they are not in the order BCA. [Axiom IV:] If points A, B, C are in the order ABC, then A is distinct from C. He then refers to R. L. Moore as having suggested that 'If A is a point, B is a point, C is a point would be a more rigorous terminology for the hypotheses of II, III, IV, inasmuch as we do not wish to imply that any two of the points are distinct'.

R.L. Moore was therefore closely in touch with Veblen's work as he honed his dissertation to perfection and conversely, it was Veblen, in his capacity as R.L. Moore's supervisor in his doctoral studies — under the overall direction of E.H. Moore — who suggested topics for investigation toward his own thesis. The choice eventually became *Sets of metrical hypotheses for geometry* in which he gave a set of assumptions concerning point, order, and congruence that, together with other order assumptions, a continuity assumption, and a weak parallel assumption, were sufficient for the establishment of ordinary euclidean geometry.

Veblen meanwhile published a paper in 1905 entitled 'Theory of plane curves in non-metrical analysis situs' in which he argued a proof of the fundamental theorem proposed by Jordan: A simple closed curve lying wholly in a plane decomposes the plane into an inside and an outside region. The setting for this theorem was taken by Veblen to be a space satisfying axioms I–VIII, XI of his dissertation, stating explicitly that nothing is assumed 'about analytic geometry, the parallel axiom, congruence relations, nor the existence of points outside a plane'. Again, discussion over these issues had taken place between himself and R.L. Moore and in fact there was input from another unexpected source, G.B. Halsted.

Throughout his time at Chicago, and almost two decades afterward, R.L. Moore had maintained a long and detailed correspondence with his original mentor, G.B. Halsted as he moved through the last and somewhat unchallenging posts of his illustrious, if bumpy, career. In 1903 he took up a position at St. John's College, Annapolis, Maryland, where he stayed for just a few months before moving to Kenyon College, Gambier, Ohio (1903–1906). He remained prodigious in his output, writing papers and articles, notably translating Poincaré's 1902 review of Hilbert into English. He also pursued his interest in mathematical education and often took up his pen to criticize the careless way that mathematics was presented in the textbooks. In 1902, he also began work on his book, *Rational Geometry,* eventually published in July 1904.

As will be seen from extracts from the correspondence below,[20] R.L. Moore made a significant contribution to this work and, as an extraordinary sequence of letters demonstrates, the pupil eventually began to match, and then surpass, the teacher in his modern mathematical concepts. In all the correspondence, the formality of past contact was maintained, every one beginning 'Dear Mr. Moore', and usually ending 'Yours always, G.B. Halsted'. The selection of edited extracts published below is, of course, a one-sided affair since the letters received by Halsted from Moore were not available. However, it is not difficult to imagine the full exchanges, or indeed to discover the hope held by Halsted that at some point in the future, he and his pupil might be reunited professionally.

November 15, 1903: 'Your letter of November 13 has just arrived. You certainly had a hard night of it to keep you up to 2:15. Be careful. You cannot do the best mathematical thinking that way.'

November 25, 1903: 'Your letter with the proofs and your notes came duly, and I find you have lost none of your keenness and subtlety... I especially credit you for your beautiful proofs, of which I have made an appendix, but I wish also to mention my indebtedness to you in my very brief preface where I shall mention no one else but Hilbert. Do you care to suggest the phraseology?... Please give a little meditation to the "Construction problem" and let me hear from you, as I do not see my way clear how to change it and introduce the distinctions...'

January 16, 1904: 'I have been so severely sick that I had two physicians and the case was diagnosed as tuberculosis and I was ordered to

---

[20] Halsted to Moore, various dates, R.L. Moore Papers in the AAM.

give up work and leave for a milder climate at once; but I would not, and told the doctor I was going to get well, and so I am now free from fever and much better, but I will hardly undertake now to change the chapter on Constructions to fit the very searching discussion which you wrote me. Now that I am back at work, I hope to be able to send you some new proof sheets.'

January 24, 1904: 'I have just received your letter of Jan. 19 and I think your five postulates for Constructions therein contained are the best yet. But as the insertion of them now in the book would necessitate the entire re-making of every printed page thereafter ... I doubt if I will attempt any sweeping changes in the text. I send you by this mail so much of the text as I now have, and in its present form. You will notice great changes and will perhaps see other changes which should still be made. Among so many changes surely some will have escaped me. If you will kindly indicate these, I may still have time to insert them.... My health has much improved and I am inclined to think the physician was hasty in his diagnosis.'

March 19, 1904: 'I have just received your note of March 16. The proof sheets you sent came safely. What I say in the Appendix is: "The above proofs are due to my pupil R.L. Moore, to whom I have been exceptionally indebted throughout the making of the whole of this book." If you would prefer some other form of acknowledgement kindly let me know. I prefer that yours should be the only name, because I owe so much more to your criticism than to all others. Neither Hilbert nor anyone else acknowledged that E.H. Moore's obscure and bungling adumbration proved anything before I published your beautiful proof. [Postscript:] What suggestions have you as to 'The Value of Non-Euclidean Geometry to the Teacher' [a paper Halsted had prepared for delivery in April]?'

April 13, 1904: 'Your note written April 7 but posted April 11th has just reached me. There was a tremendous meeting at Columbus and the paper on 'The Value of Non-Euclidean Geometry to the Teacher' was enthusiastically received. I am glad that you have been awarded the fellowship for 1904–5. I have myself to teach Elementary Mechanics ... I wish you would give me the benefit of your investigations in the choice of a suitable book ... and I hope you will hit upon something satisfactory. I send you now for the first time the whole of the [proofs for] *Rational Geometry*. To the footnote ... I have added that I have been <u>exceptionally</u> [double underlined] <u>indebted</u> to you throughout the making of all the book. No other name but yours is to be mentioned as in any way helping me.'

During the summer months there was considerable correspondence between them in regard to Moore's proof of $AB = BA$ for a line segment joining A and B, and by the fall, Halsted was reporting his meetings with some old friends.

October 22, 1904: 'I was very glad to hear from you again ... I was with Poincaré almost constantly at St. Louis and am now just finishing translating his great address there. I have the original MS now before me. No copy exists. When my translation is published of course you will get a copy. I will send you the first. Poincaré now appreciates your proof. He has seen it in my book, of which he thinks very highly.'

December 12, 1904: 'I was exceedingly interested in your speaking on the Non-Euclidean and using my Bolyai. Have you notes written out that you could send me?.... The *Rational Geometry* is to have the unexampled honor of translation into French. Meantime ... M. Dehn[21] has made an onslaught upon it from which I hope to get in good time for the French translation. I send you herewith my letter to him, as I wish to ask you also what you think on all these points. Why is the angle-sum-excess in a circle-arc-triangle not proportional to the area? Explain all this to me. Please return the letter when you answer the questions. I enclose postage. Please be thinking how we should modify the book in the French translation. You know I have grown to rely implicitly on your judgment.'

December 15, 1904: 'Please tell Dr. Dickson that I give you full authority to change, add to or take from my article.[22]... What you say about Hilbert is true, but your axioms are better than his and amount to cutting out a part of his assumptions. As to Dehn, I saw no reason to attach importance to the treatment of <u>equivalence</u> for tetrahedra, and have entirely omitted it.... Perhaps a few words might be inserted in the book to make this standpoint plain. Please let me know if there is anything radically untenable in these positions.'

The correspondence between pupil and professor diminished somewhat as Moore began the most important era of his mathematical life,

---

[21] Max Dehn, one of Hilbert's former students who eventually emigrated to the United States in the 1930s. It is also worth noting that Dehn features in a recent work by Benjamin Yandell, *The Honor's Class: Hilbert's Problems and Their Solvers*, 2002.

[22] For publication in the *Transactions of the American Mathematical Society* of which Professor Dickson was now editor.

working towards his doctorate. The remaining letters from Halsted contain a brief commentary of the build up to that event.

May 23, 1905: 'The article, my *Hilbert's Foundations of Logic and Arithmetic* is already in print and will appear in the forthcoming number of the *Monist*... I wish I could have you with me next year, but I have not been able to make arrangements. I have been asked to be Director of a new University for Mexico. If it goes through[23] I want you as one of the 3 Professors of Mathematics.... Will you undertake to write a *Rational Algebra* with me? If we go to Mexico it will be issued in English and Spanish and be the official textbook. If not, it will supercede all American algebras for use where the new rigor is growing into appreciation. I will of course adopt it here. *Rational Algebra: A textbook for colleges based on Hilbert's Axiomatic Founding of the Number System and Utilizing the New Graphic Methods. By Dr. G.B. Halsted and Dr. R.L. Moore*'[24]

July 18, 1905: 'My Dear Dr. Moore, Congratulations on your Professorship.'

---

[23] It did not.
[24] The joint project was never activated.

5

Uneasy Progress (1905–1908)

R.L. Moore is credited in Veblen's dissertation for assistance given and Veblen supervised Moore's PhD, secured in the fall of 1905. The work set the scene for much of Moore's early research which was, like Veblen's, devoted to the axiomatic foundations of geometry. His dissertation was also closely related to Veblen's, whose axioms I and III–X it utilized. His first appointment, however, did not call upon these talents...

E. H. Moore and
R.L. Moore
in Chicago

*R*obert Lee Moore was still only 22 years old when he received his PhD from The University of Chicago and there has always been some confusion as to just who acted as his principal adviser during his two years of doctoral studies. In 1938, the issue first raised itself publicly. In that year, Raymond C. Archibald named Veblen[1] as R.L. Moore's doctoral adviser in his *Semicentennial History of the American Mathematical Society*. Eyebrows were raised at the time because, by then, E.H. Moore had been dead six years and both Veblen and R.L. Moore had risen to the status of being among the top five most influential mathematicians in the United States. Moore himself was president of the American Mathematical Society in that important semicentennial year, a role that Veblen had already fulfilled. Moore was also known across the land as the first American to be honored with the American Mathematical Society's sponsorship as Visiting Lecturer,[2] which had taken him to universities throughout the country.

Both were universally respected and Veblen, especially, had achieved international recognition in that decade as a founding professor of the Institute for Advanced Study established at Princeton. He was largely responsible for the selection of the Institute's early mathematics faculty that contained such names as Einstein, Alexander, Morse, von Neumann and Weyl and 'his effect on mathematics, transcending the Princeton Community and the country as a whole, [would] be felt for decades to come.... his assistance was decisive for the careers of dozens of men'.[3]

It was with some interest, therefore, that readers of Archibald's book would discover that Veblen had more than a friendly interest in the launching of R.L. Moore's mathematical career. As already noted, Veblen himself was a student of E.H. Moore at the time of R.L. Moore's arrival at Chicago, and after gaining his own PhD he stayed on as an associate in mathematics. Reverting back to that time, a collection of mathematical papers discovered in the R.L. Moore files[4] (and read by the present author) clearly indicate that there was input from E.H.

---

[1] In *Semicentennial History of the American Mathematical Society, 1888–1938*, New York, 1938, p. 209.

[2] In 1931–1932.

[3] Deane Montgomery, in Oswald Veblen, a biographical memoir, *Bull. Amer. Math. Soc.* 69 (1963), 26–36. Montgomery is quoting a statement by the faculty and trustees of the Institute for Advanced Study.

[4] R.L. Moore Papers in the AAM.

Moore during his doctoral studies. In the same collection, there is a single sheet headed *Vita,* written in R.L. Moore's own hand. It was the draft for inclusion in his published version of his dissertation and certainly provided room for ambiguity. Indeed, R.L. Moore set the scene for confusion when he wrote: 'In the University of Chicago, I was a Fellow in 1903–1904 and 1904–1905 and an assistant in mathematics in the summer of 1904. Let me here express my thanks to my instructors Professors Bolza, Dickson, Laves, Maschke, Moulton, Moore, and Drs. Bliss and Lunn and in particular to Dr. O. Veblen who has given me so much and continued assistance and many suggestions in conjunction with the preparation of this thesis.' There was a further sentence that had been crossed out of the draft, but perfectly legible, which said: 'In fact, part of the work was done in conjunction with his [Veblen's] collaboration.'

Twenty-three years later, he went further and allowed Veblen unequivocal credit. On 12 March 1938 when Archibald was finalizing the manuscript for his *Semicentennial History*, he wrote to Moore at Texas: 'My dear Moore ... One interesting item has just come to me which I wish to have checked by yourself. The statement was made that although you really got your doctor's degree at Chicago, it was done under Veblen and not under E.H. Moore. Do you agree with this? Just a postcard reply is all that is necessary.' R.L. Moore allowed the Archibald reference to stand. In his dissertation entitled *Sets of Metrical Hypotheses for Geometry*, Moore gave axioms for euclidean geometry using as primitive notions *point, order* and *congruence*. As R.L. Wilder points out,[5] much of Moore's early work was, like Veblen's, devoted to the axiomatic foundations of geometry. His dissertation was also closely related to Veblen's, whose axioms I and III–X it used. Alternative sets of axioms were considered, some of them being systems in which ordinary ruler and compass constructions are possible, as well as a set for Bolyai-Lobachevskian geometry. In showing that every circle is a Jordan curve, he had to use a definition given by Veblen in 1905[6] in terms of order and continuity conditions.

---

[5] In The Mathematical Work of R.L. Moore: Its Background, Nature and Influence, *Arch. Hist. Exact Sci.,* 26, (1982) 73–97. R.L. Wilder, a president of both the AMS and MAA, was Moore's fourth doctoral student.

[6] O. Veblen, Theory on plane curves in non-metrical analysis situs, *Trans. Amer. Math. Soc.* 6 (1905), 83–98.

The Veblen influence continued for some time, elements of it evident in Moore's work for many years afterward. Their close working relationship under the auspices of E.H. Moore, Bolza and Maschke ended in the summer of 1905, however, when Veblen went to Princeton to launch what was to become an outstanding career solely from that base while R.L. Moore himself took up relatively low-grade work in terms of both pay and stature, at the University of Tennessee.

The appointment in the fall of 1905 began what turned out to be a particularly unhappy tenure. Departing from the excitement and friendship of Chicago appears to have had an adverse psychological effect on a young man who had had little practice in the art of general communication with life outside of the university campus. Even his letters home, evident from those retained in his papers which he wrote his father[7], were generally concerned with his work, often in complex detail. Whether his father understood was another matter.

There is scant evidence during his time at Chicago of any extracurricular activities or interest, other than walking, and only a fleeting reference to the opposite sex in his diary. Even that was with the critical eye of a man whose time in the north had not tempered his Southern-based opinions one bit. He noted with obvious disdain that during his last days at Chicago, he had joined a party that went riding in a sleigh and a female student actually sat with the driver! He wrote that a greater freedom existed among Northern people, especially in regard to 'relations between the sexes (in public! etc etc)' than in the south.[8] It is also to the last remaining pages of that diary that we may turn to discover the apparent concerns and unhappiness that had descended upon him after the intensity of Chicago where he had crammed a very great deal of study into his two years. Indeed, it might be said that he had hardly lifted his head since joining Dr. Halsted at The University of Texas and now, suddenly, it all seemed to be catching up with him. At the time, he was apparently working hard on lists of axioms for the positive integers and their arithmetic, on which he seemed to have become obsessed at the expense of getting his dissertation into publishable form.

That task had been urged upon him as a matter of urgency by both E.H. Moore and Veblen, but he had yet to complete it. A bout of malaria

---

[7] The letters were returned to him after his father's death and form part of the R.L. Moore Papers in the AAM.
[8] The Ivory Soap pad diary, 19 May 1904, pp. 104–106.

Oswald Veblen

picked up after being bitten by a mosquito while out camping with his brother during the summer holidays of 1905 merely added to his malaise and he had even let drift his correspondence with Halsted about which he had been meticulous in the past. On February 6, 1906, Halsted wrote: 'I have not heard from you in a very long time. Please send me your present address.' Moore responded immediately and recounted his illness, to which Halsted replied on February 13, 1906: 'I was horrified to learn that you had been bitten by a Texas mosquito. It gives me great pleasure to send you ... a complimentary copy of my Poincaré, *Science and Hypothesis*. A new edition of my *Rational Geometry* is called for. I think I will make it rather a simplification attractive to schools, assuming more theorems (the harder ones). I wish you would now look over the book and suggest some better and simpler substitutes all the way through.'

Moore, still affected by the malaria, was not in the mood for such additional labor, however, as his diary entries for February and March 1906 quite clearly indicate:

'Shucks! Is there much use in such diary keeping? This is Sunday. Am I losing both the desire and capacity for continued close concentrated work? Or if not both desire and capacity, at least capacity. Is my mental laziness a result of that malaria and this jaundice or not? In

Chicago, part of the time I don't know but that I had a tendency to loaf, but if so hasn't that tendency *increased*? If so, its intensity approaching a maximum and will it begin to decrease after a while and will I work as hard and as long at a time as ever? At one time, before I went to The University of Texas and for a while after I went there, didn't I have a ... conscious scramble against quitting studying for the day till sundown? Didn't I have that sort of scuffle for at least part of the time during every year there? If so, is that laziness, this disinclination to get right down to hard, long, continued concentration... — a sort of reaction against that sundown business of my earlier days? If so, will things finally adjust themselves and will I before long ... work pretty hard and regularly? I don't know but that it might be hard or impossible to describe exactly my condition but I'm not sure but that I might not suggest [that] procrastination has become one of my traits. For example, sometime last month, I got back my thesis [manuscript] and a letter from Professor Moore who seemed desirous that I should prepare part of my work as a separate paper to be published in *Transactions*, possibly in April and other parts ... to appear possibly in October.

'Now I don't know but that with a certain amount of [strenuous effort] and concentration that the first paper might have been prepared in a week but here it is, March 25 [1906], and I haven't got it ready for him yet. It's true I have, part of the time, been undecided as to how to treat a certain part of the subject,... and I have been vacillating between different methods of treatment. *But* I decide that a certain method seems good and almost no sooner do I feel that it does indeed seem good than I am *liable* to feel ... a strong inclination to postpone for a while the actual working out and writing out of the detail of it.

'Then another day I begin to consider the method and find an objection to it. Think of another method, which has some advantage, but in turn postpone detailed consideration and writing of it.... Procrastination! If I set a certain time limit, then don't know but that I may get pretty busy for a while and work pretty late ... and get something done by about that time. For example Lennes seemed to want me to...'

There, the diary ended and not another word was written in the Ivory Soap pad that had been his companion since he entered The University of Texas in 1898. His mention of Lennes, incidentally, demonstrated a further element of the close cooperation that E.H. Moore has inspired among his students. Although eight years older than R.L. Moore, Lennes received his M.S. in 1903 and then spent a couple of years teaching high school in Chicago before returning to complete his doctoral studies, although he had kept in touch with his former colleagues, Veblen and R.L. Moore in particular, throughout his period of absence. He did not complete his PhD until 1907. It was Lennes who with the other two, arrived at the topological definitions of *connected, arc* and *simply closed* (Jordan) *curve*. It was not available when R.L. Moore completed his dissertation but it became an important aspect of R.L. Moore's later work. He kept up regular correspondence with Lennes for several years, and referred to his work in a number of his own future papers.

Lennes was also one of those Moore consulted during the difficult time after leaving Chicago when he battled with his depression over his desire and capacity for work and sent him copies of his lists of axioms. This struggle within himself was not improved when this first work he had tackled after leaving Chicago failed to reach his intended audience. In it, as R.L. Wilder tells us, one of the axioms was very long and complicated and was the subject of correspondence between R.L. Moore and Veblen to whom he had apparently sent the paper (as he did to Lennes) and which was seemingly then rejected by the *Annals of Mathematics*. In a letter dated 9 April 1906,[9] two weeks after R.L. Moore's despairing diary entry, Veblen wrote:[10]

'Your "lists" of axioms came back from Huntington[11] the other day and I doubt if he understands [the long and complicated axiom]. In consequence he was more impressed by the difficulty than by the value of your work...You ought to write a preamble about your logical aims, condense the proofs as much as possible...The avoidance of ordinal counting and order in any form, the replacing of "class" by "statement", and some account of logic of propositions which you presuppose ought to go into the preface. Have you tried to write postulates of logic?'

In the same letter Veblen urged his friend to press on with other matters, such as a publishable dissertation: 'No doubt you are getting your geometry work into final form as quickly as possible? I am anxious to see that work come out as soon as possible.' Nor was there any encouragement regarding the 'lists' of axioms when he sent them to E.H. Moore. He wrote back to say that he had been too heavily engaged else-

---

[9] As quoted in Wilder, The Mathematical Work of R.L. Moore.
[10] Veblen to Moore, 9 April 1906; R.L. Moore Papers in the AAM.
[11] E.V. Huntington was at that time one of the editors of the *Annals of Mathematics*.

where to make a considered opinion of the work and he, like Veblen, suggested that Moore might do well to consider logic, specifically advising him to read the articles that had been appearing in the *Mathematische Annalen*. The articles were principally concerned with the foundations of the theory of sets, especially with the Axiom of Choice and the Continuum Hypothesis. 'Can it be,' suggested R.L. Wilder in his summary of Moore's mathematical work, 'that we have here and in Veblen's urgings, the origin of R.L.'s dislike for such investigations? As any of his doctoral students can testify, he was a Platonist in regard to the Axiom of Choice, regarding it as an absolute principle and not a matter for research regarding its consistency or admissibility.'[12] Wilder points out that in his later book[13] Moore did not indicate which theorems were dependent upon the Axiom, but stated it as a general principle in his Preface, to be used wherever needed in the text. This was contrary to his custom of giving clear indication in the book regarding which of his set theory axioms each theorem depended upon.

It was also an area in which he and Veblen were at odds, as the latter made clear in Veblen's presidential address before the American Mathematical Society in 1924: 'The conclusion seems inescapable that formal logic has to be taken over by mathematicians. The fact is that there does not exist an adequate logic at the present time, and unless the mathematicians create one, no one else is likely to do so.'

Such matters were of little concern to the reality of life outside the Chicago environment that R.L. Moore experienced upon his arrival to take up his first appointment after graduation. Whether he deliberately sought a relatively low-key job or whether it was the only one on offer remains unclear but it is possible that Moore was seeking a department that was not so well advanced and would allow him to experiment with his teaching ideas. At Tennessee, there were only two members of faculty assigned to the mathematics department, Professor Schmitt and Moore himself. The graduate offerings for the 1906–1907 program were:

Higher Algebra. In this course knowledge of all the preceding courses is presupposed and the principles of algebra are viewed in their

---

[12] In a later reference, R.L. Wilder also wrote that R.L. Moore often expressed a violent dislike of questions concerning logical foundations and recalled one occasion when Wilder asked a student if he had used the Axiom of Choice in obtaining a certain result. Moore turned angrily and exclaimed: 'I thought you'd ask that.'

[13] *Foundations of Point Set Theory,* Amer. Math. Soc. Coll. Pub. Vol. 13, 1932, revised editions, 1962, 1970.

widest applications. Chrystal's *Algebra* and lectures. Senior and Graduate. Fall. Two hours.

Trigonometry of Imaginaries. A course embracing De Moivre's Theorem, hyperbolic trigonometry, and many other subjects not included in courses four and five. Lock's *Higher Trigonometry.* Senior and Graduate. Winter. Two hours.

Modern Geometry. The principles of continuity, duality, inversion, harmonic ranges, poles and polars, and other subjects are considered. Taylor's *Euclid*, V–VI. Senior and Graduate. Spring. Two hours. Further courses, linked to Moore's most recent work, were added to the listings with the proviso 'should there be demand': *Foundations of Arithmetic and Geometry*, including a consideration of various sets of postulates and a study of relations existing between fundamental propositions. (In particular a discussion of the genesis of the ordinary number system from postulates concerning positive integers.) Prerequisite: A certain aptitude for close critical reasoning.

Noneuclidean Geometrics including the Bolyai-Lobachevskian and Riemannian geometries. *Theory of Functions of a Real Variable*, including a study of limits and continuity. *Theory of Functions of a Complex Variable.* It is desirable that there should have been a previous careful study of limits and continuity.

The extent of the response to the offering of Moore's specialties is lost in the mist of time, but from Moore's own recollections years later it would seem the interest was disappointingly slender and none of the additional courses was in fact taken up. Moore was, not unsurprisingly, disillusioned from the outset at what he might well have considered a rejection of his own particular skills and he was especially upset that he would not be given the chance to lead his students into the areas that most interested himself. Not many weeks had passed before he was already reconsidering his position.

His problems were further exacerbated by differences in both teaching style and opinions with Professor Schmitt. These developed into something of an untenable gulf when, at the end of the semester, Moore was slow in handing in his grades. Dr. William Ayres[14] has been quoted[15] as saying that Schmitt approached Moore on the subject. The chairman inquired: 'Professor Moore, I notice that you have not yet turned in your

---

[14] W.L. Ayres studied with Moore at The University of Texas but wrote his PhD thesis under J.R. Kline, Moore's own first doctoral student.
[15] Traylor, p. 65.

grades. Have you finished grading the papers?' Moore said that he had not but did not elaborate further, nor mention that he had not even begun the task. But then that wasn't the question asked of him. A little later, Schmitt became aware of that fact. He called Moore to his office and accused him of lying. The admonishment left Moore depressed and jaded and before the year was out he had left Tennessee to join his colleague Veblen at Princeton. It was, of course, also the alma mater of his mentor G.B. Halsted, although there appears to have been no correspondence between the two on that topic. Halsted may have written but there is little doubt that what was once a lively, frequent flow of letters between the two appears to have all but stopped when Moore was under pressure, attempting to complete his 'lists', make a good impression in his first job and get his dissertation into print. His thoughts were clearly taken up by the desire to leave Tennessee and get to Princeton just as soon as he could. Confirmation finally arrived when he was offered the role of preceptor, under a scheme aimed at giving experience to potential teachers and at the same time providing students with easy and informal access to brilliant young minds. The idea was introduced during Woodrow Wilson's term as university president. Wilson[16], a dyslexic who could not read until he was ten, had graduated from Princeton but gained his PhD in government and history at The Johns Hopkins University. He abandoned law to join Princeton in 1890 and became the most popular and highest-paid faculty member.

In 1902 he was the unanimous choice to become president of Princeton. He launched an ambitious program to upgrade the university both financially and intellectually, and he attempted far-reaching reforms of both undergraduate and graduate education, although several of his proposals ran foul of faculty conservatives and wealthy alumni. The preceptor system, however, was successfully launched into the university administration and although Princeton had fewer mathematical students than Harvard or Yale, it could still boast a larger faculty.

Dean Henry Burchard Fine brought in Veblen as 'one of the new preceptor guys'[17] as they were known. Preceptors were selected for their specialist knowledge in particular disciplines and were taken on in large

---

[16] He won his first term as president of the United States in 1912 and during his second term went into history as one of the founders of the League of Nations.

[17] Deane Montgomery, Oswald Veblen, a memorial in *Bull. Amer. Math. Soc.* 69 (1963), 26–36.

numbers at Princeton. They might teach in classrooms or to small groups of students in informal surroundings, perhaps in rooms on the campus, during unrestricted hours, their work given as an adjunct to mainstream courses. They could discuss assignments and any problems arising, and simply have a closer one-to-one contact with the likes of Moore and other young thinkers of the future who were in the throes of launching their careers. Wallace Carver, from Caldwell, New Jersey, a student whose four years at Princeton were during the Woodrow Wilson era, is on record with this recollection:

'My class was the first to receive and have four years of [Woodrow Wilson's] preceptoral system.... It had something of the Socratic method in it. The president brought 50 new men to faculty recruited from all over the US as preceptors. The Professor in each course had us in the classroom, but one hour a week was subtracted from hours of courses and given to a Preceptor who had no more than five men in his study or room.... In this informal atmosphere we discussed the course [and] we were assigned some outside reading on which we were asked questions later. So we had two men for every course and it was helpful and stimulating. Many of the original 50 rose from the rank of Assistant Professor to Heads of Departments and some to becoming Deans.... It required a larger faculty and that is why Princeton with a smaller number of students than Yale and Harvard had a larger faculty than either of them.'[18]

It was into this preceptor role that Robert Lee Moore arrived at Princeton in 1906 to be confronted by an education system in which Veblen appears to have reveled, in that it became the starting point for his forceful participation in carrying Princeton forward from a slender start to a major mathematics center of excellence.[19] Veblen was to be joined by another evacuee from Chicago, G.D. Birkhoff, who arrived in 1909, became a professor there in 1911 and then moved to Harvard the following year to begin his own illustrious career at that establishment. R.L. Moore, however, was less taken with Princeton than his two colleagues, and the feeling may well have been reciprocal among the university hierarchy. In later years, he would speak of his dislike of some aspects of the Princeton system, in that there existed something akin to

---

[18] Traylor, p. 67.

[19] Deane Montgomery, Oswald Veblen, a memorial in *Bull. Amer. Math. Soc.* 69 (1963), 26–36.

uniform examinations for all calculus sections.[20]

Many courses were taught in more than one section, and were required to cover the same material. Instructors would assemble to grade the examination papers en masse, with each grader responsible for specific problems. Moore apparently believed this would merely encourage examination-driven teaching at the expense of developing a student's overall ability. He was to be preceptor in Plane Trigonometry, Spherical Trigonometry and Applications of Trigonometry as well as Selected Portions of Algebra and Elementary Theory of Equations and Conic Sections, Treated from the Cartesian Standpoint. A further course, Foundations of Geometry, appeared in the Princeton catalog for 1907–1908 and was described as: 'A consideration of various sets of postulates for arithmetic and geometry, and a study of relations existing between fundamental propositions. The development of arithmetic and geometry from a set of postulates. Graduate course, second term, 3 hours a week. Dr. Moore.'[21] But judging from available insight from that period, it does not seem that R.L. Moore exactly lit any fires mathematically. His output was low and so, it seems, was his morale. The Tennessee experience, recurring illness from the malarial infection, and a teaching system at Princeton that he did not favor surely combined, at the very least, to shorten his concentration span, and his temper. He already possessed a trait of speaking his mind quite forcefully. At Princeton, there were disagreements with other more senior members of the faculty and, as at Tennessee, he could not settle. Never one to suffer fools gladly R.L. Moore became notably more impatient with people who irritated him, and some complained that he could become positively aggressive in his general demeanor, pretty well at the drop of a hat.

A number of incidents occurred that quickly passed along the campus grapevine. Among them was a story involving a faculty member with whom Moore shared an office who maintained: 'I was sitting in my office talking with Moore and began talking in a way I knew I shouldn't. Suddenly, Moore got up and I found myself on the floor, and Moore was walking out the door.' Moore had not said a word to his colleague but simply got up, picked the man up from his chair and dropped him on the floor.[22] Moore was not a large man in stature. When he arrived at

---

[20] Mentioned in R.L. Moore, Memorial Resolution, Documents and Minutes of the General Faculty, The University of Texas at Austin, 1975, pp. 11653–11669, available for online viewing at www.utexas.edu/faculty.

[21] As quoted in Traylor, p. 74.

Princeton, his recorded weight was 143 pounds but his frame soon took on muscular additions as he began to participate in a very active manner in boxing, which he took up almost from the day he arrived in Princeton.

Earlier mentions of his interest in the sport are scant. The University of Texas Memorial Resolution to Moore does record that although boxing was not an intercollegiate sport at the university at the turn of the century, there is some evidence that Moore may have been an intramural champion. Princeton, on the other hand, ran a strong and competitive boxing program, linked to the US Military and Naval Academies. Moore got involved, along with Veblen, almost as soon as he arrived on the campus, possibly because he had been advised that some stiff physical exercise, other than the long walks he was so keen on, might help dispel the effects of his recent illnesses. In any event, it was a curious and rigorous routine that he began to impose upon himself, given that up to that point there was virtually no mention of any kind of sporting activity either in his diary, or in his general papers. He embarked on an almost daily schedule of arduous boxing sessions to such an extent that his colleagues might be forgiven for supposing that he was in search of an avenue of aggression through which to release some of the frustrations of his work. He did so with a vigor that verged upon the obsessive and struck up a warm relationship with Princeton's boxing instructor, Spider Kelley, whom he always looked up during his visits to the university in later years. In a letter Moore wrote to his father on 19 March 1907[23] he said that he and Veblen had been spending 'as much as three quarters of an hour on any one day', in the ring. His weight went up (he was 160 lbs when he left Princeton in 1908) but his descriptions of this new-found pastime eventually brought a warning letter from his father that he should not spend so much time in the boxing ring.

He apparently feared that Moore's face, naturally contoured by high cheekbones, was taking on a pugilistic appearance. Moore's enthusiasm for the sport, however, led him to ignore his father's warning and to far greater participation than Veblen. When his friend was not available, Moore would run to the gymnasium almost as soon as he got out of class and take on anyone who was looking for a sparring partner. They would box until the opponent decided he'd had enough, and then Moore

---

[22] Recounted in Traylor, p. 71.
[23] R.L. Moore to Charles Moore, 19 March 1907; R.L. Moore Papers in the AAM.

would sit down to wait for someone else to come along. On occasions, Moore might wait for a third opponent to challenge, sometimes staying in the gym for two or three hours. The Princeton boxers adopted a rather gentlemanly procedure. They would agree to spar with each other on the understanding that they decided in advance whether to 'go easy' or 'go hard'. The latter meant that while they would not attempt to knock each other out, they would not pull their punches. The rules could be changed midway through the bouts only if both men agreed. There was no time limit set to each sparring round, and it was left to one or the other to call a halt. One account records:

'Moore never felt that he should call "stop" but rather, that he should let the other person decide when to stop. One time though, he very nearly called "stop" and was quite relieved when the other man did. He had no idea how long they had sparred, but finally he could not even hold his arms up to defend himself. He let his arms drop to his sides, and still felt that he might be able to deliver some kind of blow by swinging up. Right then, the other man said, "stop" and Moore found that a considerably pleasant sound.' Many of his opponents were larger than himself. One, in particular, was 6'3" tall and had a substantial advantage in reach. This man would invariably want to go easy, but in the midst of sparring, either man might say "go hard" and immediately change the level of aggression. However, this particular opponent would decide, without announcing it, to go hard. Moore learned to go hard with this man not long after he had begun boxing there.[24]

At one point he was practicing a right cross. He wanted to perfect that blow and he was boxing with a man who was less experienced than himself. Moore landed his right and his opponent went down, leaving Moore apologizing profusely. His boxing was temporarily curtailed midway through the year when Moore suffered a further debilitating bout of illness, a particularly painful outbreak of mumps which necessitated his transfer from the university infirmary to a hospital in Trenton, New Jersey. Veblen, on holiday in Stillwater, Minnesota, telephoned to discover his friend's condition but by then Moore had already left for recuperation at his parents' home in Texas. Veblen wrote immediately:[25] 'How are you occupying your time? Did you get a chance to finish your thesis? Have you made a start on that other business? In short, what is the news?'

---

[24] Traylor, pp. 69–70.
[25] Veblen to Moore, 11 July 1907; R.L. Moore Papers in the AAM.

R.L. Moore, relaxing at home

What the 'other business' was is unknown[26], but work on the thesis had slowly come to fruition and Moore, following Veblen's urgings, finally published the first of two papers based on his dissertation, entitled *Geometry in which the Sum of the Angles of Every Triangle is Two Right Angles*.[27] Neither could be considered new work, since they originated back in 1905 when they were presented to the American Mathematical Society in combined form under the title *Sets of metrical hypotheses for geometry*. Possibly prodded by an earlier letter from Halsted and indeed Moore's evident collaboration on Halsted's book *Rational Geometry*, Moore referred to M. Dehn's work[28] which had shown that Hilbert's original axiom sets I, II, and IV, augmented by the assumption S, that the sum of the angles of every triangle is two right angles, were not sufficient to yield III (parallels). In the first paper, he showed that any space satisfying axioms I, II, IV and S must nevertheless be a subspace (via the addition of ideal points) of a space, which III holds. He also stated: 'This result has an interesting connection with our

---

[26] Possibly a reference to an earlier letter in which Veblen had referred to some problems on which Moore was working, in relation to curves. Veblen wrote: 'Why not send me your curve business in its final form? If I can, I will try to look up the literature better than you can in your town.' Cited in Wilder, The Mathematical Work of R.L. Moore.

[27] *Trans. Amer. Math. Soc.* 8 (1907), 369–378.

[28] Die Legendre'schen Sätze über die Winkelsumme im Dreieck, *Math. Ann.* 53 (1900), 404–439.

spatial experience. Statements[29] have been made to the effect that, since no human instruments, however delicate, can measure exactly enough to decide in every conceivable case whether the sum of the angles of a triangle is equal to two right angles (unless the difference between this sum and two right angles should exceed a certain minimum amount), it is therefore impossible to settle the question whether our space is Bolyai-Lobachevskian or Euclidean even though it be granted that it is one or the other.'

Although contact between them had been slight of late, Halsted could still be relied upon to draw attention to his former student's work whenever the occasion arose. He promptly dashed off an article, headed "Even Perfect Measuring Impotent" in which he typically drew upon his colorful and lavish prose for his description of the work, while at the same time taking a side-swipe at E.H. Moore:

'The attention of geometers should be directed to a remarkable article by Dr. R.L. Moore, of Princeton, whose extraordinarily elegant proof of the redundancy of Hilbert's axioms first appeared in the *American Mathematical Monthly*. The new article ... is also a perfecting of the work of the Hilbert school but reaches new results so unexpected, so profound as to be nothing less than epoch-making. We knew that the so-called laboratory method for mathematics, the "measuring" method, was rotten at the core since mathematics is not an experimental science, since no theorem of arithmetic, algebra or geometry can be proved by measurement. Our argument was sufficiently cogent: that the theorems of mathematics are absolutely exact, while no human measurement ever can be exact. But Dr. Moore shows that even granting the impossible, granting the super-human power of precise measurement, we could not thereby ever prove our space Euclidean, ever prove it the space taught in all our text-books.'[30]

The second of his two thesis papers, 'Sets of metrical hypotheses for geometry', was published the following year,[31] dealing with his axioms for euclidean geometry using as primitive notions *point*, *order* and *congruence*. As already mentioned, that too dated in its origination to his 1905 presentation to the AMS. In fact, from the point of completing his

---

[29] Statements by G.B. Halsted, for example. He makes the same point often in his publications as early as 1887.
[30] *Science*, 26, No. 669 (1907) 551.
[31] *Trans. Amer. Math. Soc.* 9 (1908), 487–512, doctoral dissertation.

thesis, Moore passed through the most sterile period of his life in that he published only two more papers before 1916, one of which was 'A note concerning Veblen's axioms for geometry' which is referred to in more detail in the next chapter. R.L. Wilder, a long-time friend and associate of Moore, first as a student and on through his own long mathematical career, surmised that this dearth in output in these years was partly induced by his 'apparent lack of success in finding ... an environment that he could consider satisfactory and permanent.... It can also be surmised during that period that R.L. was coming to the conclusion that the axiomatic foundations of geometry, to which he had devoted so much time, was not as fruitful a field of investigation as he would like'.[32]

He had matured quickly. Finding himself teaching in a setting in which he found little in common with the teaching philosophies of others on the faculty, he produced little mathematics but developed strong physical and mental independence. Professionally, however, he had not settled.

---

[32] Comment and footnote here combined from Wilder's The Mathematical Work of R.L. Moore: Its Background, Nature and Influence, part IIa: Geometry, *Arch. Hist. Exact Sci.* Vol. 26, Springer-Verlag (1982), 73–79.

6

A Settling Experience (1908–1916)

Moore's first major appointment in the tranquility of Northwestern University sees him plotting a routine for his courses, combined with his own research. He begins publishing regularly, and his work in both areas is consolidated by a move to the University of Pennsylvania where he has the freedom to experiment with his teaching method.

R.L. and his uncle, James Willard Moore

*P*rinceton did not detain 26-year-old Moore after his year as a preceptor and it is not difficult to understand why he now moved to a place where he could gather his thoughts. He was still clearly not in a frame of mind to publish, and neglected his correspondence. Letters to Veblen seemed strained, and became less frequent. Even poor Halsted, whose own career was in freefall, could not fire his former protégé into action and his hopes of Moore co-authoring a textbook entitled *Rational Arithmetic* came to naught. All of these factors, when put together, may indicate that Moore was passing through a phase of inactivity, perhaps resulting from a crisis in his confidence. If that was indeed the case, as R.L. Wilder suggests, there was probably no better place to heal himself than his next port of call, Northwestern University, whose setting alone represented something of an idyllic haven.

It will be recalled that it was from here in 1892 that William Rainey Harper plucked a virtually unknown associate professor named E.H. Moore, then 30 years old, to become the first mathematics professor at the University of Chicago. Set in 379 acres of wooded farmland on the shore of Lake Michigan, the university was sited well away from the frontier atmosphere of Chicago. Northwestern dated from 1850 when nine devout Methodists, moved by a combination of religious conviction and secular optimism met above a hardware store in Chicago, and resolved to build an institution of the highest order of excellence. It would serve the people of the Northwest Territory as established by the 1787 Northwest Ordinance, incorporating Indiana, Illinois, Michigan, Wisconsin, Ohio, and part of Minnesota. Although the Act of Incorporation reflected the strong religious beliefs of its founders, faith was not a requirement of students or members of the faculty. Only one of the founders, physician John Evans, was a college graduate. He built railroads, founded the Illinois Medical Society and the Illinois Republican Party, and was a confidant of Abraham Lincoln, who appointed him governor of the Colorado Territory. Evans guided Northwestern until his death in 1897 and the town that grew up around the university was named Evanston in his honor. Northwestern's first building opened in 1855 as a College of Liberal Arts with just two faculty members and 10 male students. By 1900, it had grown to include seven undergraduate and graduate schools with 2,700 students and an annual budget of more than $200,000.[1]

---

[1] A century later, Northwestern had expanded into one of America's leading private research universities, with an annual budget of more than $1 billion, 5,700 employees and 15,800 students enrolled in its 11 colleges and schools on campuses in Evanston and Chicago.

A Settling Experience (1908–1916)

At the time of R.L. Moore's arrival in 1908, Northwestern was still small compared with the major universities of the United States, but it was well on the road to establishing itself as one of the nation's largest private universities, with the Department of Mathematics located in Weinberg College of Arts and Sciences, the largest of eight graduate and undergraduate schools located on the Evanston campus. Even today, the university boasts a quiet setting for the campus, bounded on the east by Lake Michigan and on the west by the stately homes and tree-lined streets of Evanston. Then, as now, it was an excellent environment for academic pursuits with a spacious serenity set in oak groves and gardens linking the new with the old, the latter now highlighted by the graceful limestone spire of University Hall, built in 1869. Moore was hired at a time of change for Northwestern. The Graduate School was about to be established and Northwestern adopted the German model of providing graduate as well as undergraduate instruction, with an emphasis on research as well as teaching. Moore's role was at the more basic level of instructor, teaching plane trigonometry and analytical geometry, algebra, solid geometry and plane trigonometry, differential and integral calculus. An offering of ordinary differential equations was added the following year and a course in noneuclidean geometry appeared in the 1910–1911 catalog, to be taught by Moore on a basis to be arranged. Little other reference material survives from this era to provide any real insight into Moore's stay at Northwestern. There remain, however, some clues that those who came to know him and work with him in later years will recognize as being added to his inventory of experience. For example, the university's corporate seal consisted of an open book surrounded by rays of light and containing two quotations. One is a Greek quotation from the Gospel of Saint John, chapter 1, and verse 14, translating to "The Word…full of grace and truth." Circling the book are the first three words, in Latin, of the University Motto: *Quaecumque sunt vera* (Whatsoever things are true). Both of those quotations might well have been adopted, or at least remembered, in the development of his career and 'truth' is a word that crops up often.

Moore's own mathematics did not flourish at Northwestern but there is little doubt that his experience there from 1908 to 1911 had a settling effect. It was to show itself with one other surprising development, and perhaps to offer a further reason for the slowdown of his mathematical output during those three years. The arrival of Margaret MacLellan Key

*Margaret MacLellan
Key Moore*

into his life was a secret he kept from his diary. Nor is there a mention of her in letters home or in correspondence with friends, not even to Veblen who, in 1908, had married an English girl, Elizabeth Richardson, whom he met when she was visiting her brother Owen Richardson,[2] then teaching physics at Princeton. At the same time, Moore had been conducting a somewhat long-distance courtship of Miss Key whom he first met in the fall of 1901, when he was a fellow at Texas.

Margaret, like Moore, was not one to openly discuss their private life, but she did assert[3] that he first noticed her outside her campus residence. He then selected a room nearby at 2004 Wichita Street, Austin, so that he could see her more often. They were married in Margaret's hometown of Brenham, Texas, in August 1910 and set up their first

---

[2] Later, a professor at King's College in the University of London and Nobel Prize winner.
[3] According to the R.L. Moore Memorial Resolution, Documents and Minutes of the General Faculty.

marital home in Evanston, where they remained for a year. Indeed, they moved only twice more during the next 64 years: first to Philadelphia where they lived for nine years and then to Austin where they set up home on West 23rd Street where they remained for the rest of their lives. Moore had also resumed his own mathematical investigations, if somewhat tentatively. In 1911, he had two papers under way, the first again linked to Veblen's work, although it was not exactly new. "A note concerning Veblen's axioms for geometry"[4] had first been presented to the American Mathematical Society, albeit in rather different form, on 26 October 1907. Moore gave some improvement of Veblen's axioms, reducing the axioms by one. He refined his approach in the published version. A second paper that he sought to publish in 1911 was a departure from his early work in that it was not concerned with axiomatic procedure. It was entitled 'On Duhamel's theorem',[5] but publication was delayed because of the time it took for a referee to pass judgement, as Veblen, by then performing editing duties for *Annals*, mentioned in a letter[6] to Moore that year: 'I am sorry Swift has been so slow with your Duhamel paper. I have tried to shake him up, but his slumbers are very deep.' The work was published the following year and Moore proposed a form of Duhamel's theorem suitable for application to a wider range of problems than that given by Osgood and 'at the same time seems to be even easier to apply'.[7] He also prepared two other papers in 1911, which he subsequently presented to meetings of the American Mathematical Society. They were in abstract form and never published but they demonstrated that Moore was edging his way out of that period which Wilder has described as his most infertile — although the journey was not quite complete. After 1912, he did not publish again for another three years, which is perhaps understandable, given that he had decided to move on.

There seems little doubt that life at Northwestern, along with his marriage, had the desired effect of settling both his temper and his

---

[4] *Trans. Amer. Math. Soc.* 13 (1912), 74–76.

[5] *Ann. of Math.* 13 (1912), 161–166.

[6] Veblen to Moore, 21 November 1911; R.L. Moore Papers in the AAM.

[7] The paper was later used by H.J. Ettlinger, Moore's future long time colleague and friend, at The University of Texas: R.L. Moore's principle and its converse, *Comptes Rendus des Séances de la Soc. des Sc. et de Lettres de Varsovie*, 19 (1927), Classe III, pp. 455–460. Ettlinger gave it generalizations and considered its use in the study of summable functions.

anguish over the loss of his capacity and desire owing to the intense study which he had put himself through. The latter seems to have been all but resolved, although the temper — albeit often mistaken for Texan or Southern ideals — and what became his renowned argumentative trait would barely improve. In the offing was the final stepping-stone in his career. In 1911, at the age of 29, he accepted a position at the University of Pennsylvania in Philadelphia, one of the Ivy League schools, and one of the most progressive. It was founded as a charity school in 1740 by Benjamin Franklin, who became president of the first board of trustees. Nine of the signatories to the Declaration of Independence and eleven of those to the American Constitution were associated with the university. Penn was also one of the first in the country to accept women students and was admitting African-American students by 1879.

The Department of Mathematics dated from 1899. Roxana Hayward Vivian became the department's first woman to earn the PhD there, in 1901, and later becoming Professor of Mathematics and Astronomy at Wellesley College. Two of the first three African-Americans[8] in the entire United States of America to earn their PhD in mathematics did so at Penn, and by an ironic twist of fate became mathematical descendants of R.L Moore, whose own Southern values did not extend to teaching

---

[8] Dudley Weldon Woodard (1881–1965), enrolled in the Graduate School at Penn in 1927, had already accumulated a remarkable set of achievements. He had published his University of Chicago master's thesis in mathematics, *Loci Connected with the Problem of Two Bodies* and had been teaching mathematics at the collegiate level for two decades. At Howard, he had held the post of Dean of the College of Arts and Sciences. A gifted mathematician, he took a scholarly leave in 1927 and spent a year at Penn, working under the direction of John R. Kline and in June 1928, received his PhD. He returned to Howard to establish a graduate mathematics program and in the inaugural year, William Waldron Schieffelin Claytor (1908–1967) emerged as his most promising student. Professor Woodard recommended Claytor for admission to Penn's Graduate School of Arts and Sciences. J.R. Kline agreed to advise Claytor, who was awarded his PhD in 1933. Claytor's dissertation delighted the Penn faculty, for it provided a significant advance in the theory of Peano continua — a branch of point-set topology in which Kline, following his own studies with R.L. Moore, was a recognized authority. On Kline's recommendation, Claytor went to the University of Michigan on a post-doctoral fellowship, where he worked with another of R.L. Moore's doctoral students, R.L. Wilder. Claytor had every reason to expect competing offers from America's leading research universities. But, as Penn's own website now records, 'in that era of pervasive racial discrimination only a predominantly African American institution, West Virginia State College, welcomed him to its faculty and the work of these two men was relatively unrecognized until virtually the end of the twentieth century.'

John R. Kline

African Americans. Both were supervised by Professor John R. Kline, Moore's own first doctoral student, and who remained at Penn throughout his distinguished career. Clearly in Kline's case, the opinions of the teacher on matters outside of mathematical study were not all heeded, nor indeed were they by the many others who followed him under Moore's tutelage. In all other areas, however, Penn represented something of a fresh start for Moore. It was time to build and go forward and, to use a motoring analogy of which he was fond, to put his foot hard on the accelerator when he was appointed an instructor of a strong mathematical department led by Professor George Egbert Fisher, a Penn stalwart of some reputation resulting from a calculus text that he had written. The mathematics faculty included Professors Edwin Schofield Crawley, the more senior, who specialized in analytic and differential geometry and had his eyes on the chairmanship. There were two other professors, Isaac J. Schwatt and George Hervey Hallett, all long servants at Penn. There were five other members of faculty holding rank of assistant professor along with three instructors, R.L. Moore among them. The catalog covered a fairly broad offering, including analysis, algebra, differential equations, geometry, and applied mathematics.

Schwatt and Hallett were more or less opposites in personality. Schwatt, who specialized in real variables and infinite series, was overt and outspoken, even downright rude, in class. He would say, 'Don't you understand? Mr. Smith, there are three kinds of fools: ordinary fools, damn fools, and then there's you.'[9] He chewed tobacco in class and might occasionally spit, much to the chagrin of some students, and an unholy row would ensue. Hallett, an algebraist, was a much calmer man. Moore settled in among them well and in 1912 gave a course entitled "Foundations of Mathematics," and described as: 'The theory of positive integers as a basis for analysis. Rigid motion and correspondence with a number manifold as factors in determining the properties of space. Metrical and non-metrical spaces. A critical study of interrelations between different systems of axioms.' The following year Moore added a course entitled 'Theory of Point Sets: Theory of sets of points in metrical and in non-metrical spaces. Contributions of Fréchet and others to the foundations of point set theory. Content and measure; Jordan curve theory and other applications'.[10]

The two courses became a mainstay in Moore's teaching career and signified not only the advancement of his journey to becoming what has been described as the world-famous 'sage of point set topology' and, crucially, the emergence of his method of teaching. Dealing with the latter first, what became known as the Moore Method, and defined by some as the Texas Method, will be discussed more fully as these pages progress, with interjections on the way from well-known mathematicians who experienced it first hand during the ensuing half century.

As time passed, the Moore Method became generally linked to his Texas era, because that was where he had his greatest successes both in teaching and his own work. There is clear evidence, however, that R.L. Moore began to experiment with these issues almost as soon as he arrived at Penn. In fact, he seemed more concerned with devoting his time and energy to the classroom and his students than in his own research, evident from his list of publications, which showed no new work from the time he joined Penn until 1915. In the classroom, how-

---

[9] Recollections of former student Joseph Thomas, of Durham, North Carolina, July 1971, quoted in Traylor p. 76.

[10] This was entirely topical. In 1906 Fréchet called a space compact if any infinite bounded subset contains a point of accumulation. He extended the concept of convergence from euclidean space by defining metric spaces. He also showed that Cantor's theories of open and closed subsets extended naturally to metric spaces. Further, in a paper to the 1908 International Congress of Mathematicians in Rome, Riesz disposed of the metric approach completely and proposed a new axiomatic approach to topology, based on a set definition of limit points, with no concept of distance. This work, expanded by Hausdorff, allowed the definition of abstract topological spaces.

ever, it was a different story. Three potential PhD students were being guided toward their goal and Moore's method was already the subject of comment and conjecture. An example of this can be seen from the recollections of the second of Moore's own doctoral students, G. H. Hallett, the son of Professor Hallett of the mathematics faculty. He arrived from Harvard to begin studying towards a PhD.[11] He was thus one of the doctoral students who were to experience the Moore style virtually at its inception, though it was doubtless refined over the years. He has described the experience as follows:

'He taught in a very remarkable way. He didn't give us any books. We didn't consult books at all in that course. It was a course in point set theory and he gave us certain axioms to start with and then we were asked at the beginning to prove certain theorems that we were told were true, given those axioms. We would work on the proofs and come back into class and he would ask how many people had the proofs and those who said yes were given a chance, at least one or two of them to give their proofs. And the other members of the class listened carefully to see if they made any mistakes and if a member of the class thought so, he would speak up and say why. And quite often, there were mistakes in the proofs that were caught by the class and sometimes, nobody had the proof the first time and he would let us have another week at that one.... As the course went on though, some of the things given us were more difficult. He would give us a problem and ask what the solution to the problem was without telling us the theorem that was supposed to result. Then we would work on that. I remember that one of the theorems I proved, he said had never been proved until the year before, and I had given him a different proof.... He gave us a problem once which we worked on and which none of us got. He gave us a couple of weeks to work on it and none of us got it. And he said, "Well, I guess you needn't spend any more time on that. This is a problem mathematicians have been working on for centuries and nobody has ever solved it. I just thought you might, just by accident, be able to do something."'

Another aspect of Moore's style that would become familiar to his students down the years was his tactic of inspiring discussion of matters unrelated to mathematics to elicit their opinions on logic. He would open the proceedings quite unexpectedly by speaking for several min-

---

[11] Subsequently gained in 1918 with a thesis entitled: *Linear order in three dimensional Euclidean and double elliptical spaces.*

utes of some topical event, such as the unfolding horrors of World War I. He would then pick up on statements by politicians and comment on the logic, or lack of it, in what they had said. As Hallett recalled: 'He and I were always getting a good deal of satisfaction in picking out the lack of logic in a good many ... statements by important public figures. That was sort of a forerunner to applying the same methods of thinking to public affairs which I went on with after graduation and for the rest of my career. I think this added to the interest of the class and probably made it more attractive.... Dr. Moore's method of teaching brought out what appeared to me to be the two most important faculties in mathematical research: one,... I'd regard as imagination and the second, the ability [of] critical analysis in applying logic to what you think of to try out. And this same criteria, of course, applies to almost everything else. It is a method of thinking. And I think such success as I've had in the work I've done in the field of government probably has had a good deal to do with that.'[12]

A number of key points which were to become a permanent feature of Moore's pedagogical style, and in turn the Moore Method, emerged in his point set theory course at Pennsylvania:

1) There would be no textbooks linked to the course and none was to be consulted. Moore quickly developed his personal 'radar' for spotting those who had accidentally or otherwise become exposed to the work at hand.
2) Students were asked to prove theorems from given axioms and present their proofs in class without seeking help externally or discussing the problem with each other.
3) The rest of the class would then be encouraged to criticize weaknesses or inaccuracies in the presented proofs.
4) He encouraged a strong spirit of competitiveness among his students and devised various means to promote it.
5) From the outset he placed a good deal of emphasis on logic.

---

[12] Hallett had a long career in public service, beginning with a post as Secretary of the Proportional Representation League. In 1937 he wrote *Proportional Representation: The Key to Democracy*. He taught courses in government at several colleges in New York City: Brooklyn, Hunter, NYU, and CCNY. He was given special awards from LaGuardia Memorial Association (1963), New York City Club (1964), and Citizens Union (1947, 1969). During the Second World War he was active in the Committee for Civilian Defense.

Moore's concern to show results in the classroom may be judged from his guidance of, and subsequent long association with, his first doctoral student, John R. Kline. He had arrived at Penn in 1913 from Muhlenburg, PA. He collected his Master's in 1914 and, closely supervised by Moore, won his PhD in 1916 with a thesis entitled *Double elliptic geometry in terms of point and order alone*.[13] Although Moore, like many successful teachers, established a life-long bond with many of his students, always willing to give a reference or advice, Kline was surely a special pupil to Moore.

After Moore left Philadelphia for Texas, Kline himself became a professor of mathematics at Penn and thereafter they maintained a close association through various professional organizations and contacts. It was also with Kline in 1918 that Moore chose to share the honors in the only co-authored paper he ever wrote.[14] They also kept up a regular, if occasionally disputatious, correspondence that lasted until Kline's death in 1955. Letters between them (Moore had by then begun to make copies of much of his outgoing mail[15]) are especially rewarding in their content. Not least among the observations to be made from reading these exchanges is the change that took place in their relationship over the years. Unlike letters between Moore and Halsted, which remained formal to the last, the gradual progress of Kline's confidence in dealing with his doctoral professor is best summed up by the change in the opening address which moved through three distinct phases over a considerable time span: *Dear Professor Moore, Dear Moore, Dear R.L.* Moore's own form of address, on the other hand, never altered.

It was simply: *Dear Kline,* and was always signed somewhat formally R.L. Moore, as indeed it was with Veblen and many others with whom he maintained a lively, if erratic correspondence. One other fact that never altered was Kline's last sentence, in virtually every letter: 'Please give our best regards to Mrs. Moore', a line that appears in virtually all exchanges between Moore and his students. Though apparently not noticeably present in university life, clearly she took great interest in his students, and they all knew it.

---

[13] Published in *Ann. of Math.* 18 (1916–17), 31–44.
[14] R.L. Moore and J.R. Kline: On the most general plane closed point-set through which it is possible to pass a simple continuous arc, *Ann. of Math.* 20 (1919), 218–223.
[15] J.R. Kline correspondence file, R.L. Moore Papers in the AAM.

The Kline-Moore[16] axis in later years provoked a great deal of fruitful contact and exchanges of ideas, mathematical discussion at a personal level and then of students and potential staff between Penn and, The University of Texas at Austin. Several of Kline's early students had moved between the two universities, setting a pattern for the future. W.L. Ayres studied with Moore, then transferred to Penn to become Kline's third doctoral student and, early in an illustrious career as an academician, Ayres returned to Texas for a postdoctoral assignment with Moore. Kline's first doctoral student, H.M. Gehman, his fifth student Leo Zippin and his sixth student, N.E. Rutt, all took postdoctoral work with Moore. The informality and camaraderie that existed in this contact was demonstrated in a letter W.L. Ayres wrote to Moore while in Europe in March 1929 (when Moore was at the height of his productivity): 'I have not had an answer to my letter to you last fall but knowing you as I do it would hardly be correct to say that I was expecting one. I don't suppose you are even writing to Kline this year either, since Rutt is there and you can get news from Kline through him. Maybe that is the reason Kline keeps sending so many fellows to Austin. Otherwise he might never hear anything from you.'[17]

The contact with Kline in this manner was, really, just the beginning of another phenomenon associated with Moore: the future natural growth of a very considerable network of Moore students spanning academic life across the United States. Kline contributed to the start of it,

---

[16] John R. Kline became a well-respected figure in the American mathematical world, not least for his 15-year tenure as Chair of the Department at Penn from 1940 to 1954. He was a Guggenheim Fellow (Göttingen) in 1926–1927 and served at various times as associate editor of the *Transactions of the American Mathematical Society*, the *Bulletin of the American Mathematical Society*, and the *American Journal of Mathematics* and was a member of the editorial board of the AMS Colloquium Publications. He was associate secretary of the AMS from 1933 to 1936 and then secretary from 1936 to 1950. He directed the PhD theses of thirteen students, including National Research Fellows W.L. Ayres, H.M. Gehman, N.E. Rutt, and Leo Zippin. He was also the thesis director of record for Lida K. Barrett, who began her work with R.L. Moore at Texas; the thesis problems were suggested by Kline after her move to the University of Pennsylvania and essentially solved under his direction. After Kline suffered a heart attack, the work was completed under the direction of R.D. Anderson, Moore's former student in Texas, who had joined the faculty at Pennsylvania in 1948. Perhaps Kline's best-known student was Leo Zippin, who wrote with Deane Montgomery the classic monograph *Topological Transformation Groups*. The work described in this treatise, together with work done by Andrew Gleason, provides a solution to the first, and famous, part of Hilbert's Fifth Problem.

[17] Ayres to Moore, 1 March 1929, W.L. Ayres correspondence file in the R.L. Moore Papers in the AAM.

sending students to Moore at Austin in their Fellowship years. The effect was like a ripple from a stone thrown into a pond: students taught directly by Moore remained in contact with him and each other, and this spread into other universities as Moore students took up positions across the land. Eventually second, third and even fourth generation students became part of the network, all connected by a common bond that exists today[18] almost thirty years after Moore's death.

The Moore networking effect went far beyond the personal level of interaction between the circle of his students, and his students' students. It had the effect of spreading the message on topology as a coherent collection of mathematical topics mainly useful to analysis but of sufficient interest in themselves to stand alone. As Ben Fitzpatrick put it:[19] 'A book on the history of general topology which omitted a discussion of R.L. Moore would be like a history of American popular music which failed to mention Irving Berlin.' This contribution to American mathematics at large only began to emerge, however, midway through his time at Penn when at last he showed signs of a revival of interest in getting himself published again, stimulated as he encountered some exceptional students of his own. There is no doubt, either, that he was spurred on by his own competitive trait as he viewed the emergence of some increasingly dramatic work of other members of the group of intense young men who studied under E.H. Moore, Bolza and Maschke in the first decade of the twentieth century, especially from Birkhoff, Veblen, and E.W. Chittenden. They had all contributed to a higher level of research in topology that produced long-lasting results and established it as both respectable and rewarding. Equally important was that they were now producing mathematicians whose work added substantially to the scope and knowledge of topology. As another Moore student, F. Burton Jones noted,[20] it was this sort of exponential growth that estab-

---

[18] An anchor for this ongoing contact and exploration of Moore's educational ideals and method is The Legacy of R.L. Moore Project funded and maintained by the Educational Advancement Foundation, founded in Houston, Texas by Mr. Harry Lucas, Jr. in 1969.

[19] Ben Fitzpatrick, Some Aspects of the Work and Influence of R.L. Moore, in *Handbook of the History of General Topology* Vol. 1, C.E. Aull and R. Lowen (eds.), Kluwer Academic, 1997, pp. 41–61, quote on p. 54. Dr. Fitzpatrick, a doctoral student of H.J. Ettlinger and a former Moore student, was a driving force behind The Legacy of R.L. Moore Project, and the Oral Histories through which much detail about Moore had been gathered prior to Ben's untimely death in 2000. (See author's Acknowledgements.)

[20] F. Burton Jones, The Beginning of Topology in the United States and The Moore School, in *Handbook of the History of General Topology*, C.E. Aull & R. Lowen, (eds.), Volume 1, Kluwer Academic Publishers, 1997, pp. 97–103, quote on p. 98.

lished topology as an essential part of the university curriculum so quickly: 'The quality and usefulness of these results played a part in this development. But there was something entirely different, a sort of historical accident that was the most compelling, at least in set theoretic topology.'

F. Burton Jones identified the gestation period of topology in the United States as between 1905 and 1915, when it began to flourish. R.L. Wilder also noted the importance of the last years of that time frame and that by 1915 Moore had turned towards problems that would lead him into areas that were destined to form his life's work.[21] He was also an active participant in the affairs of the American Mathematical Society. He was often called upon to referee papers prior to publication and from 1913 had been an associate editor of the *Transactions of the American Mathematical Society* and was already taking steps to end his own period of mathematical dormancy. His return to the fray came in 1915 with the first of a deluge of presentations and papers which began to emerge like bullets from a machine gun, exceeding more than thirty over the next decade alone, this in addition to guiding five students towards their PhD in the same period.

He had also resumed his correspondence with Halsted, which took off with a new flourish, bouncing ideas and possibilities on his former mentor who responded enthusiastically, and as ever, fulsome in his praise after Moore had delivered to him a couple of forthcoming articles on analysis situs: 'I believe your work, in quality, is the finest ever done by an American. In Hilbert's assumptions, it was not that they were the most fundamental, but that they were so familiar, that won such recognition. Like Sylvester's, some of your logic is subtle and even elusive. I judge that only a few of your peers can justly estimate what you are doing. But such splendid work, even if recondite, must win recognition and should be an abiding satisfaction to yourself. You must follow your own bent. No one can properly advise a genius.'[22]

Moore had launched this new impetus with three papers in 1915 and two in 1916. In doing so, he published a result that once again drew the admiration of the mathematical community. He established that any plane satisfying Veblen's Axioms I–VIII, XI is a number plane; that any

---

[21] R.L. Wilder, The Mathematical Work of R.L. Moore: Its Background, Nature and Influence, *Arch. Hist. Exact Sci.* 26 (1982), 73–97.

[22] Halsted to Moore, 22 May 1916; R.L. Moore Papers in the AAM.

plane, satisfying those axioms, contains a system of continuous curves such that, considering those curves as straight lines, the plane is an ordinary euclidean plane. Moore stated that any discussion of analysis situs based on Veblen's axioms (as for example, Veblen's proof of the theorem that a Jordan curve divides its plane into just two parts) was no more general than one based on analytic hypotheses. As Traylor points out,[23] Moore's results were remarkable and unexpected by many in that he established a one-to-one reciprocal continuous correspondence between the continuous curves of Veblen's planes and straight lines of an ordinary euclidean plane. From here, Moore moved on to what R.L. Wilder described as the first paper that could be truly placed in the category of point set theory, 'On the foundations of plane analysis situs'.[24] The paper proved a key element in the development of point set topology and, as Wilder states, formed the basis for the subject matter of his advanced courses.

The majority of this work that was to have such a profound effect on his future direction was completed at Penn and most of it while he languished, in terms of his title and salary, at the level of 'instructor'. The fact that there was a potential headline-grabber in their midst had not escaped the attention of some members of the hierarchy at Penn and in 1916 Moore was promoted to the rank of assistant professor. Halsted wrote immediately: 'I am highly gratified to read in the *Bulletin* of your appointment.... I have always held that your work is the most profound, unexpected and epoch-making ever contributed to mathematics by an American. I am proud that you were able to wait for the tardy recognition which at last has come.'[25] His office space remained, however, rather lowly, literally so. It was housed in the original College Hall building, and was a large room in the basement, which he shared with eight members of faculty at instructor and assistant professor level. He was not especially popular amongst his colleagues. According to Joe Frantz,[26] he was already developing a reputation 'as a curmudgeon that would last until his death. No one ... looked on Moore at 95 and talked about what a gentle old man he was'. Moore complained regularly about the office space. It was right next door to a stable where they kept

---

[23] Traylor, *Creative Teaching*, p. 79.
[24] *Trans. Amer. Math. Soc.* 17 (1916), 131–164.
[25] Halsted to Moore, 12 December 1916, R.L. Moore Papers in the AAM.
[26] Frantz, *The Forty-Acre Follies*, pp. 115–116.

the horse that pulled the lawn mower over the campus. It was from this odorous place that he went about the day-to-day business of university life and from which he launched his continual search for likely students to accept the Moore indoctrination for a future in mathematics.

Among those who accepted his approaches — and quite a few, it must be said, did not — was Anna Mullikin who was to become his third doctoral student. She had arrived at Penn from Goucher College. She became aware that some of her fellow students were not enamored of Moore's classroom techniques, especially those at the lower end of the ability scale, or who could not easily adapt to the 'no books' rule or the competitive nature of classroom activity. They apparently felt that they were 'left out' of classroom activity; some even complained that they were omitted from discussion and presentations at the board because of their inability to stay abreast of the work.

Anna Mullikin was one of those who accepted the challenge and benefited from it. She also had first-hand experience of the dismay of those who could not cope or who were given their marching orders because they sought to beat the system: Dr. Moore 'had had his work all published ... and people would go and look it up in the library. He didn't want them to do that. He wanted them to work it out themselves. And he put them out of class if he found out that they were cheating. [In one class] there were three of us, but he put everybody out but me. One was a Catholic nun who tried to get help from me, and he put her out. He said that if she needed help, she didn't belong in his class.'[27] One disgruntled student treated in this manner returned with a copy of the university catalog and, confronting Moore with the complaint that he had been effectively excluded from class, pointed to a paragraph in the catalog which read: 'Students are especially advised not to undertake an unduly large number of lecture courses at one time. The place for graduate study is the library rather than the lecture room. The most that can be done in the lecture is to guide the student into a general acquaintance with the principles of a subject, and it remains for him to broaden his knowledge and develop the details by extensive reading and private study.'[28]

Moore was unrepentant and continued to develop his approach, apparently with the full blessing of the heads of both the department and the Dean of Arts and Sciences. He admitted in later years that it was a

---

[27] Traylor, *Creative Teaching,* p. 87. Recollections of Anna Mullikin, 18 July 1971.

[28] Traylor, *Creative Teaching,* p. 87.

freedom he might not have been given elsewhere. The Moore Method was still in its embryonic stage, although the fundamentals remained unaltered throughout his career in that he would set problems for the students and expect them to tackle them utterly and completely through their own initiative without resorting to learning aids, books or other people.

Those who violated his rules of study were hereinafter considered 'cheats' and were given the cold shoulder treatment described by Anna Mullikin. Those who adhered but struggled in their efforts would be rewarded with his patience in seeing them through to a successful conclusion. As we will see as the application of the Moore Method unfolds in the teaching of 50 doctoral students and countless others, he barely deviated from this premise throughout the rest of his career. And, incidentally, it was as he was reveling somewhat in new acclaim from his recently published works that he received a number of approaches from other universities, offering him associate or full professorships. Penn, which had a full complement of professors on its faculty at the time, was soon to become part of Moore's history.

7

Back to Texas
(1916–1920)

Moore's 1916 paper 'On the foundations of plane analysis situs' was a crucial document, cited by Veblen as a groundbreaking treatise in the development of point set theory. The paper embodied a set of axioms characterizing the euclidean plane and continued the theme in follow up papers. Also, at the end of the decade, Moore returns to Texas.

Moore at his desk

The foundations were laid during his years at Penn for the style and content that would dominate R.L. Moore's research and teaching for the rest of his life and it is worth returning briefly to the impact of his 1916 paper, cited in the preceding chapter, 'On the foundations of plane analysis situs'. It was a crucial document for Moore and the mathematical community. Veblen said as much in a letter to Dickson, then editor of the AMS *Transactions*, recommending that the paper should be published as soon as possible: '[It] seems to me [to be] a valuable one (1) as a contribution to the foundations of point-set theory along a direction which is bound to receive attention in the future and (2) as containing new and interesting methods of proving theorems in this domain.'[1] It was also the point at which Veblen and Moore diverged in their research and development. Veblen followed Poincaré while Moore developed his own school based on analysis situs, later to be called topology. That first major document in 1916 embodied a set of axioms characterizing the euclidean plane. In a follow-up paper, he showed the axiom system to be categorical, and further down the line applied it in a way that was to become the forerunner to the creative use of the axiomatic method, soon to become more popular. As Veblen noted, Moore's ideas and methods were new to America and gave rise to greater development by Moore himself and later by his students who were inspired by the ongoing themes that arose from the work. It also moved him on, virtually at a stroke, from the ever-present tag of the Chicago school (and thus the linkage to E.H. Moore) into the realms of becoming an established mathematician of some stature in his own right. Like Veblen, Birkhoff, Bliss and others, he was pushing his own boundaries outward and, better still, had set his sights on new areas of research that were to significantly alter his own aspirations, and those of many others. He began producing substantial mathematics and it would soon be recognized that the most important of these related to his move away from pure geometry toward the concepts of point set topology. The publication of the first of these inspirational new papers in 1916 was followed by his becoming a member of the Council of the American Mathematical Society in 1917. There was also, even at that stage, a wider interest being shown in his 'method' in the classroom, word of which had reached the mathematical grapevine.

---

[1] Letter from Oswald Veblen to L.E. Dickson in Chicago, 15 November 1915; R.L. Moore Papers in the AAM.

The teaching aspect will be the subject of considerable input from those at the sharp end of it as these pages progress. The subtlety of his research, however, is a more complex issue. R.L. Wilder makes a clear distinction between the two aspects of these developments although he insists that the paper on plane analysis situs formed the basis for both. Wilder was well placed to make an early[2] on-the-spot assessment, and have the benefit of reflection in later years when he and other Moore students published their own work, inspired by Moore's ideas and methods. It is as well, therefore, to utilize Wilder's own summary of events:[3] 'In most cases, a significant result in mathematics has a complex and intricate background, no particular element of which can be assigned as *the* influence that motivated the result. The best one can do is to indicate significant areas of mathematics in which the influence of Moore's research is clearly indicated, even though not necessarily the sole reservoir from which the basic ideas involved were derived. The situation is further complicated by the fact that Moore was, after all, only one of a growing number of mathematicians in the United States, Poland, Russia, Germany etc., who took part in the building of the foundations of topology.'

Wilder adds that the general format of that paper was by now classic: Primitive terms (a class S of elements called *points* and a class of sub-classes of points called *regions*), axioms, development of the theorems therefrom, and finally independency examples for the axioms. The axiom system used for the proofs was denoted by Σ_1; in a final section he discussed modification of Σ_1 denoted by Σ_2 and Σ_3. Moore stated: 'The notions point, line, plane, order, and congruence are fundamental in euclidean geometry. Point, line and order (on a line) are fundamental in descriptive geometry. Point, limit-point and regions (of certain types) are fundamental in analysis situs. It seems desirable that each of these doctrines should be founded on (developed from) a set of postulates (axioms) concerning notions that are fundamental for that particular doctrine. Euclidean geometry and descriptive geometry have been so developed. The present paper contains two systems of axioms, Σ_2 and Σ_3, each of which is sufficient for a considerable body of theorems in the domain of plane analysis situs. The axioms of each system are stat-

---

[2] R.L. Wilder became a student of R.L. Moore at The University of Texas in 1921.
[3] In his paper The Mathematical Work of R.L. Moore: Its Background, Nature and Influence, *Arch. Hist. Exact Sci.* 26 (1982), 73–97, quote on p. 87.

ed in terms of a class, S, of elements called points and a class of subclasses of S called regions.'

The definitions and axioms he produced in the paper were to be the standard-bearers in point set topology for decades hence. Moore did admit to a 'certain amount of resemblance' between Axiom I and Veblen's Postulate of Uniformity included in his 1905 paper, 'Definition in terms of order alone in the linear continuum and in well-ordered sets'. However, Wilder draws attention to the fact that while Veblen's postulate may have influenced Moore's formulation of his Axiom I, the implications were far stronger and, in particular, Moore's Axiom I implies the separability of the space, whereas Veblen's does not. Veblen postulated separability in a separate axiom. E.W. Chittenden was to point out[4] that the importance of the regular and perfectly separable, therefore metric, spaces in the analysis of continua is indicated by the fact that nine years before the publication of the discoveries of Urysohn, R.L. Moore assumed these properties in the first of a system of axioms for the foundation of plane analysis situs.[5] In fact, the stature of Moore's work soon became well recognized and he had begun to use the 1916 paper as the basis of his pedagogical pursuits even before its publication. As already noted, the first of his 50 doctoral students, J.R. Kline graduated that year. He followed up his thesis with some prolific writing, and his second paper, in 1917, established a converse to the Jordan theorem: Suppose K is a closed set of points and that $S - K = S_1 + S_2$, where S_1 and S_2 are non-compact point sets such that (1) every two points of S_i ($i = 1, 2$) can be joined by an arc lying entirely in S_i; (2) every arc joining a point of S_1 to a point of S_2 contains a point of K; (3) if O is a point of K and P is a point not belonging to K, then P can be joined to O by an arc having no points except O common with K. Every point set K that satisfies these conditions is an open curve.[6]

---

[4] In his paper, On the metrization problem and related problems in the theory of abstract sets, *Bull. Amer. Math. Soc.* 33 (1927), 13–34.

[5] Pavel Urysohn, a young assistant professor at the University of Moscow, lectured on the topology of continua. He published a series of short notes on this topic during 1922. The complete theory was described in a paper published in the *Comptes rendus* of the Academy of Sciences in Paris. This attracted international interest and Urysohn published a full version of his dimension theory in *Fundamenta mathematica*. He wrote a major paper in two parts in 1923 but they did not appear in print until 1925 and 1926 by which time he had sadly died.

[6] J.R. Kline, The converse of the theorem concerning the division of a plane by an open curve. *Trans. Amer. Math. Soc.* 18 (1917), 177–184, quote on p. 178.

Milton Brockett Porter, chair of the Pure Mathematics Department at UT

Moore was impressed and decided to hand Kline the honor, and such it was at the time, of joining him as co-author of a paper entitled, 'On the most general plane closed point-set through which it is possible to pass a simple continuous arc'.[7] Published in the fall of 1919, it followed Moore's own current train of thought in point set topology and carried an opening introduction as follows:

'A set of points is said to be totally disconnected if it contains no connected subset consisting of more than one point. In 1905 L. Zoretti showed that every closed, bounded and totally disconnected set of points is a subset of a Cantorean line. In 1906 F. Riesz attempted to show that every such set of points is a subset of a simple continuous arc. Shortly thereafter, Zoretti pointed out that Riesz's argument was fallacious. He, however, left unsettled the question whether Riesz's theorem was true or false. In 1910, in an article that contains no reference either to Riesz or to Zoretti, Denjoy indicated that this theorem could be proved with the use of certain ideas contained in a former paper of his own. We have not, however, succeeded in determining from his meager indications just what sort of argument he had in mind. At any rate, in order that a closed and bounded point-set should be a subset of a simple continuous arc, it is of course not necessary that it should be totally disconnected. In the present

---

[7] *Ann. of Math.* 20 (1919), 218–223.

paper we will establish the following result: Theorem 1: *In order that a closed and bounded point-set M should be a subset of a simple continuous arc, it is necessary and sufficient that every closed, connected subset of M should be either a single point or a simple continuous arc t such that no point of t, with the exception of its endpoints, is a limit point of M − t.*'

The direction of Moore's work was clearly established, and over the four years following his paper on the foundations of plane analysis situs, from 1916 to early 1920, Moore produced eight further papers while still at Penn. This was in addition to supervising the studies of two more doctoral students, Hallett and Mullikin, although the latter did not complete her studies before Moore left for Texas, and subsequently followed him there for the final furlong. Both, as with Kline, were seemingly influenced by Moore's own current research direction.[8] Mullikin in particular attracted considerable attention and all three went on to significant achievement in their respective careers. Kline, as indicated in the previous chapter, became a national figure as a university academician, was a long-time secretary of the American Mathematical Society until his death in 1955 as well as, variously, an editor of the *Transactions*, the AMS *Bulletin, the American Journal of Mathematics* and a member of the editorial board of AMS Colloquium Publications. Hallett, as we have seen, was to choose a career in public service while Mullikin became a high school teacher of mathematics in Pennsylvania.

Only Kline was in a position to adopt in modified form, elements of the Moore Method for which the three PhD students, and others in their courses during Moore's ten years at Pennsylvania, had become the guinea pigs as he established his teaching ideals. More than 40 years later, Moore was to recall those days at Penn when he first began to put them to the test:

'I had stated some axioms and theorems and proposed that [the class] prove the theorems from the axioms. Sometime in December, I stated Theorem 15. J.R. Kline presented a proof the following April. Another member of the class did not want to give up and listen to Kline's proof after he had tried for so long to get a proof of his own and he walked out of the classroom. Towards the end of the course, he said he thought

---

[8] J.R. Kline's thesis was entitled *Double elliptic geometry in terms of point and order*; G.H. Hallett followed with *Linear order in three dimensional euclidean and double elliptical spaces*, while Mullikin came in with *Certain theorems relating to plane connected point sets.*

he could prove this theorem if he could only "get off the boundary" as he expressed it. More than 12 years later when I was on a visit to a university where he had become a professor, I said, "Have you got Theorem 15 yet?" And he answered No.'[9]

This scenario would be repeated time and time again over the coming decades as the full force of his unique pedagogy was unleashed at The University of Texas which, against competition from other universities, won his acceptance to go back to his alma mater as an associate professor of mathematics in 1919 and he returned to Texas the following year. Before going on to recount his arrival there, another remnant of Moore's thoughts at the time may provide some enlightenment on the question of how his experiments with banning the use of books at Penn developed into an unswerving, unrepentant teaching style at Austin. The following passage in his own hand is contained in one of 121 brown school exercise books in the collection of his papers at the Archives of American Mathematics. The books contain a huge quantity of mathematics, built up over the years. At the back of one of them is a curiously written and undated treatise that reads almost like a defense of his style, as if he were standing in a courtroom to answer charges of failing to teach his students in the recognized manner, a Spencer Tracy-like figure in *Inherit the Wind*. The passage was chosen by the present author from a mountain of material scanned during research as an appropriate text with which to demonstrate Moore's intentions as he headed back to Texas:

'Some months ago, I raised with my class ... the question whether or not there exists on the x-axis a closed and bounded point set M, containing no interval, such that every point of M is a limit point of M. It looked for a while as though the course was going to come to an end without the question having been settled by anyone enrolled in it. I hoped it would not turn out that way but I feared that it would. Are you saying, "Of *course* you would not want it to turn out that way? Certainly you would not want your class to *remain* ignorant concerning such a thing as this." Remain *ignorant*? I certainly believe that *some* and I *hope* that *all* of its members would rather remain ignorant than to be *told* the answer. Then why did I want this question settled by one of them? If no one of them had settled it, that would have seemed to indicate that there

---

[9] From a draft presentation prepared by R.L. Moore for a film about his Method sponsored by the Mathematical Association of America in 1966, entitled *Challenge in the Classroom*, in the R.L. Moore Papers at the AAM.

was no one in the class whose ability to work things out for himself had been sufficiently developed for this purpose and whether this would have been a reflection on them or on me, I would certainly not have liked it. Are you saying that if it were a reflection on me and not on any one of them, then it would have been my fault for not having furnished them with sufficient *tools* for this purpose? If so, I do not agree with you at all, and I feel *tempted* to say you miss the whole point of all this. If it is just a question of supplying *tools*, who couldn't do that? And who couldn't use tools if supplied with suitable ones for this purpose? I would feel that if it had been my fault it would have been my fault for not ____ and here I leave a blank for the lack of suitable words.

'*However*, to emphasize this: How many, if any of you who hear me, ever *tried* without any *assistance* or previous knowledge concerning this matter to settle the question whether such a point set exists? I imagine that at least 95 per cent of all who know the answer either read or were told the answer and were therefore deprived of the opportunity to try to settle the question for themselves. Is a person who was told the answer to such a question without ever having had the opportunity to work out the answer a good judge of the difficulty of the task? In this connection, I am reminded of the statement: "It is only the clearest of minds that are the first to think of something which once stated is clear to everyone".'[10]

He would encounter many detractors, doubters and opponents over the years, and most would mount similar challenges posed by Moore himself in that rationale. Among those who could be classed a doubter was none other than L.E. Dickson. One summer in the early 1920s, the issue came up during a visit by R.L. Moore to the University of Chicago. He had joined Dickson and E.H. Moore for lunch, during which R.L. Moore explained his method. E.H. Moore was intrigued and urged his former student to go into greater detail. R.L. Moore told them how he would not allow the use of books, and described how he challenged his students with questions and theorems, insisting that they settle them entirely through their own devices, without reference to any published work or assistance of any kind. E.H. Moore was quietly fascinated.[11] Dickson, who had

---

10 R.L. Moore, *Challenge in the Classroom* notes; R.L. Moore Papers in the AAM.
11 It will be recalled that in their joint biographical memoir of E.H. Moore (cited in Chapter Three) Dickson and Bliss wrote that he influenced radically the methods of undergraduate instruction in mathematics at the University of Chicago. He gave courses in beginning calculus himself, casting aside textbooks, and concentrating instead on the fundamentals of the topic and their graphical interpretation.

observed at close hand E.H. Moore's own pedagogical experiments, was not impressed by R.L. Moore's more overt approach to dispensing with conventional teaching aids and said as much. E.H. Moore told R.L. Moore that he felt the approach had a good deal of merit,[12] which was not an unexpected reaction, given that, as Dickson and Bliss later wrote, E.H. Moore had often defied many established rules of pedagogy. Joe Frantz took this further, presumably from conversations he had with R.L. Moore at Texas and maintained that E.H. Moore changed his teaching style to conform, to some extent, to that described by his former student.[13]

By then, R.L. Moore had already established himself, and his method, at The University of Texas. He moved back to Texas at the age of 37, taking up the role of associate professor in the fall of 1920. He was able to bring to his new appointment a creditable track record and a reputation for some outstanding mathematics. Such qualities were needed in the Department of Mathematics at The University of Texas as it launched into the third decade of the twentieth century. Expansion was finally under way for the campus of the main university, which originally consisted of the 40-acre tract on College Hill. In 1910 George W. Brackenridge gave the university a tract of 500 acres on the banks of the Colorado River, but a proposal to move the university to that site was bitterly contested and the site was eventually used for life-sciences research. In 1921, however, the legislature appropriated $1,350,000 for the purchase of additional land adjacent to the original 40 acres.[14]

The university was expanding rapidly when R.L. Moore arrived but, for some years, budgetary restrictions and indecision over new build-

---

[12] Recounted by Traylor, from conversations with R.L. Moore, in *Creative Teaching*, p. 92.

[13] Frantz, in *The Forty-Acre Follies*, p. 116.

[14] Because of the constitutional prohibition of the use of general revenue for buildings, temporary frame structures had to be erected to house the growing student body, especially after World War I, and the university became famous for its "Shakeresque" architecture. Oil was discovered on university land in 1923 — the gushers of wealth presciently anticipated by Ashbel Smith, though perhaps not in the form he envisioned. The resulting savings in the available fund made passage of a constitutional amendment possible in 1931, so that a bond issue for construction of fireproof buildings could be passed. The university also acquired the land and buildings of the former Blind Institute in 1925, the Cavanaugh homestead on Waller Creek in 1930, and the grounds of Texas Wesleyan College and property on Whitis Avenue in 1931. Other lots have also been acquired from time to time to total about 350 acres. The J.J. Pickle Research Campus, a 476-acre site eight miles north of the main campus, houses research organizations in engineering, science, and the social sciences. The Montopolis Research Center is located on 94 acres in southeast Austin.

Harry Yandell Benedict

ings meant that the university wasn't exactly setting the world alight in mathematics. The departure of Halsted, for all his quirks and egocentricity, had left a gaping hole at Austin. But now, the department was in the hands of Milton Brockett Porter, a gifted mathematician who had been Halsted's first protégé, returning to The University of Texas as head of Pure Mathematics in 1903. One other difference Moore found on his return was that there were two departments of mathematics, and that suited him fine. The situation had arisen out of budget controls that determined there should be only one professor in each department, so mathematics was split into two so that Porter could be hired in addition to H.Y. Benedict, who had also gained his doctorate at Harvard. Joe Frantz reckoned this had left students unnecessarily confused as to which they should join:

'Faculty advisers didn't understand the difference either. The situation was rooted in personalities, which happens in armies, corporations, political machines and churches. When Halsted left under duress, Milton B. Porter was brought back as the new professor. Also in the department was a future president of UT, H.Y. Benedict ... obviously a young man on the rise. He was about the same age as Porter. Should he be forced to spend the next 40 years on the faculty as a subordinate, although he was undeniably brilliant and deserving? Some practical mind came up with a solution: give Benedict his own department. So it was done. Porter was professor of Pure Mathematics, Benedict of Applied Mathematics.'

This situation remained during the first decade of the twentieth century until Benedict was promoted to dean of the College of Arts and Sciences and from 1913 until 1920 he was also dean of men. In 1927 he was named president of the University. He was the first ex-student to become president and he occupied that office longer than anyone else in the University's history.[15] The two mathematics departments remained separated for the next forty years, reinforced by the incoming associate professor in Porter's department, Dr. R.L. Moore, who subsequently fought tooth and nail at various stages during that time span to protect the purity of his domain while at the same time, says Frantz, he was 'staffing both departments with his own graduates'. To some extent, the system offered a degree of flexibility that benefited both teacher and student, and this was especially so as Porter took advantage of the impending expansion plans for the university, which took a decade or more to come to fruition. He began to build up the mathematics faculty with men of proven, if unconventional, ability. Porter, himself a man of versatile interests, a high level of intellectual achievement and a whimsical humor, sought men of character to join him. He was well connected, and undoubtedly drew upon those connections in formulating his plans for seeking new blood for the faculty as and when he was permitted to take them on. His professors at Harvard, Maxime Bôcher and William Fogg Osgood, became personal friends and he was fond of saying, 'They made a pet of me as soon as I got there.'[16] He had a close friendship and contact with many other luminaries of the era, including G.D. Birkhoff, L.E. Dickson, James Pierpont and the world-famous Josiah Willard Gibbs at Yale. During a sojourn abroad in 1908–1909 he added to his list of contacts and friends such figures as Felix Klein and Henri Poincaré. It was against this background that Porter joined with Benedict in the dedicated mission to build up the mathematics department in its contribution to the mathematical community at large and, in particular, to national and international research and publishing.

---

[15] During his administration, President Benedict oversaw a major building construction program that added 15 new buildings to the campus. He did not live to see the completion of the McDonald Observatory, a project the University started with an $850,000 gift from the estate of banker William Johnson McDonald.

[16] M.B. Porter, Memorial Resolution, Documents and Minutes of the General Faculty, The University of Texas at Austin, 1961, pp. 7670–7673.

The doctorate was conferred for the first time at The University of Texas in 1915, and the output in the next five years was sparse. Porter's interests in further graduate study were well known, and in 1920 he joined a committee of three appointed by President Sidney Mezes to formulate plans to encourage graduate work so that the university might compete more favorably with other institutions. Progress was slow and it was not until 1925 that under Porter's chairmanship, the committee's recommendations were implemented for the foundation of the Graduate School. Porter's influence on these developments was substantial and he brought to the task his personal knowledge of those who were to lead the march forward. Over time, he and Benedict put together a team of significant quality who were to form the backbone of mathematics at The University of Texas for the next fifty years or more, aided and abetted, of course, by newcomers[17] as the years passed but led principally by three outstanding mathematicians who were all 'characters' to boot: Hyman Joseph Ettlinger, Robert Lee Moore and Harry Schultz Vandiver, the latter arriving in 1924. To that list was later added the name of H. S. Wall, although his arrival did not occur for another 20 years. Ettlinger had connections with Porter's own Harvard background. He had majored in mathematics there, studying under Osgood, Birkhoff, Bôcher and Julian Lowell Coolidge, at the time exceedingly impressive names to appear on any CV. He joined The University of Texas for the 1913–1914 year to begin what turned out to be a 73-year association. A great athlete and former All-American football hero at Washington University, St. Louis, (where he gained his B.A.) Ettlinger doubled as an assistant football coach to the varsity football team at Austin for several years.

R. L. Moore arrived in 1920 into what was still a relatively small university, but was destined for fast growth. He was confronted by an interesting and somewhat difficult situation at Austin. Mary E. Decherd who, it will be recalled, 17 years earlier had been hired as a teacher instead of R. L. Moore, was still in place. Moore exacted a modicum of revenge upon his return. He discovered that Decherd was listed to teach a course entitled "Foundations of Geometry" and Moore took immediate action. One of her intended students, Blanche Bennet Grover, gave this recollection of events:

'It was scheduled for Miss Decherd … and she didn't show up, but Dr. Moore did. That's what happened. And he changed the course all

---

[17] These included another important stalwart figure in UT mathematics, Hubert Stanley Wall, who joined the faculty in 1946.

Back to Texas (1916–1920)

Paul Mason Batchelder

around. It was supposed to be a geometry course and it just opened up a whole new world for me. I had never been taught like that. He came in and he wrote some definitions on the board. He talked to us awhile, and then gave us some axioms. Then he gave us some theorems to prove. He kept giving theorems and there were only about five in the class ... when you said you had a proof, he would ask you to go prove it. If there was someone in his class who thought he was pretty smart and tried to pretend he knew more than he did, Dr. Moore really would pick on him.'[18]

In 1920–1921, Moore's first full year, there was a total enrollment of 6,888 students, a figure that increased at a rate of almost 1,000 students a year, rising to 10,461 by 1925. As the population of the university grew, the old ruling of only one professor to a department was finally dropped and in 1922 Moore learned that he was to be promoted to a full professorship in Pure Mathematics. The faculty would now consist of professors Porter, Dodd, Moore, with associate professor Ettlinger, adjunct professors, Dercherd, Paul Mason Batchelder and Harry Vandiver, assisted by four instructors. It was a competitive line-up of

---

[18] Quoted in Traylor, p. 94.

talented teachers and it is worthwhile now to introduce Batchelder and Vandiver into the scenario, two figures who, like Moore and Ettlinger, would remain at Austin until their retirement.

Batchelder, four years younger than Moore, had gained his M.A. in physics at Princeton, and studied mathematics under Birkhoff. He was so inspired he switched from physics to mathematics and followed Birkhoff to Harvard in 1912, took two years out to become an instructor at Northwestern and then returned to Harvard for his doctorate, with a dissertation entitled *The Hypergeometric Difference Equation*. He joined The University of Texas in 1916 as an instructor, and there began the Legend of Cosine Red. It was a nickname he acquired in his early trigonometry classes for his red hair and it stuck for the rest of his working life, all of which was, like R.L. Moore, spent at The University of Texas. He retired in 1954 with the rank of associate professor. Batchelder was a great character whose appeal to mathematics students is confirmed by the statistic that between 1926 and 1953, he supervised the masters' theses of 47 graduate students.

He was a precise and diligent man about whom there was a favorite anecdote. Three instructors one Sunday afternoon were in the dome of the Texas State Capitol when through binoculars one of them spotted Batchelder walking across the campus in front of the west wing of the Old Main Building. Some time later, one of them challenged him with the question: 'Where were you last Sunday afternoon at 3:20 PM?' The professor looked at his watch and replied: 'Let's see ... I must have been in front of the west wing of the Old Main Building.'[19]

The next arrival under Milton Porter's expansion scheme was Harry Schulz Vandiver, in 1924. He was the same age as R.L. Moore and a curious man, described by many as a brilliant teacher, but about whom there were three facts that are, in recollections of him, often placed over and above his substantial mathematical achievements:

1) He shunned public school education, left at an early age to join the family business as a freight agent, and received no degree until the University of Pennsylvania made him an honorary Doctor of Science in 1946, when he was sixty-three.
2) He never owned, and rarely lived in, a house. He and family spent their lives in hotel and apartment rooms, to which he trailed his substantial collections of classical records and books.

---

[19] Professor P.M. Batchelder, Memorial Resolution, Documents and Minutes of the General Faculty, The University of Texas at Austin, 1971, pp. 11466–11470.

3) Although quite friendly for a while, a bitter feud later developed between himself and R.L. Moore and they spent years not talking to each other.

The origins of the dispute were unclear. Vandiver, annoyingly from Moore's point of view, acquired considerable recognition without completing any undergraduate program apart from irregular attendance at some graduate course at Penn. He taught himself mathematics by studying problems presented in the *American Mathematical Monthly*. He went on to produce a number of papers, one of them in 1904 was written in collaboration with G.D. Birkhoff, entitled "On the Integral Divisors of $a^n - b^n$," which was published in the *Annals of Mathematics* and was still being quoted two-thirds of a century later. He also edited Chapter 26 in Volume II of L.E. Dickson's *History of the Theory of Numbers*. But Vandiver, essentially a research mathematician, was best known for his life-long work on Fermat's Last Theorem, for which he won prizes for papers on the subject and in 1952 achieved further fame when he used a computer to prove it for all primes less than 2000.

Vandiver had been brought in directly by Porter who, according to the faculty Memorial Resolution for him, 'knew that the mere possession of a doctoral degree (or any other degree) was small indication of ability'.[20]

---

[20] H.S. Vandiver, Memorial Resolution, Documents and Minutes of the General Faculty, The University of Texas at Austin, 1974, pp. 10926–10940.

8
A Rewarding Decade (1920–1930)

An exceptional period, with Moore prolific in his research and publications, which launched the work of many of his students with outstanding results. His teaching style was now also in full flow and early beneficiaries included R.L. Wilder and G.T. Whyburn.

R.L. Moore walking down
Guadalupe Street in Austin

*I*n the decade of the 1920s Moore's personal work in research progressed at a fast pace in both quantity and quality and a clear pattern began to emerge vis-à-vis Moore's *teaching* and his *research*: the number of students he graduated were, not unexpectedly, fewer at times of his own high output. At the same time, however, his own research results formed the basis of his teaching and if they had no other justification their use would have been sufficient. In fact, a large number of his students used them as the basis for their own extended research, and with excellent and enduring results.

It was also an important era at a personal level when any lingering doubts about his ability to sustain himself in the increasingly competitive field of American mathematics were completely removed. Now, the pendulum swung entirely in the opposite direction, as demonstrated by a note from the office of the Editors of *Proceedings of the National Academy of Sciences* in 1925 in relation to a paper R.L. Moore had submitted for consideration. Normally, submissions were published only after a strict system of refereeing by a member of the Academy. The measure of respect for Moore's work could be gleaned from the letter which read: 'Your article for *Proceedings* can, as I understand it, be submitted under a more or less blanket authorization of E.H. Moore, our editor of mathematics, to the effect that you would only offer material worthy of our acceptance.'[1] This recognition was reflected across the American mathematical community, and no wonder. Almost half his lifetime total of 64 papers was published between 1919 and 1926 — 32 papers in all — and within the next six years, he published a further six. In addition, the first edition of his book, *Foundations of Point Set Theory,* appeared in 1932. Undoubtedly, the flexibility of the two mathematics departments at Austin, in spite of his personal animosity towards those in Applied Mathematics, aided Moore's research work, as it did in his avowed and unremitting application of the Moore Method of teaching his students. Many observers have commented that the 'no-books' style of teaching would not only have been frowned upon in many establishments at that time, it would have been banned. That it

---

[1] E.B. Wilson, National Academy of Sciences, to R.L. Moore, 3 July 1925, R.L. Moore Papers in the AAM. Albert Lewis also points out that non-members of the NAS had to be sponsored by a member to make a submission. At this time, it does seem that papers may not have been further vetted or refereed, a practice which later led to some reforms. Vandiver was known to casually sponsor some papers later shown to be at least bordering on the fraudulent.

went on and ultimately became an established and famous aspect of mathematics teaching in general at The University of Texas, drawing students whose specific aim was to study with Moore, was in no small measure due to the aura of freedom of expression and experimentation created by Milton Brockett Porter at Austin to meet the post-World War I expansion of the university.

The expansion, along with Moore's own developing interests, is reflected in the courses taught at Austin as the decade of the 1920s progressed. By 1924–1925, the Department of Applied Mathematics, on the one hand, included in its catalog courses that were mainly designed for engineering students, but were 'open to academic students and count toward academic degrees'.[2] By the end of the decade, the Pure Mathematics Department faculty had extended to 14 and during that period, four of the five instructors hired for the Department of Pure Mathematics had earned or would earn their doctoral degrees under Moore and another of his doctoral students had been appointed an adjunct professor in the Department of Applied Mathematics.[3] The Pure Mathematics Department in its re-organized catalog offered 30 courses for the year 1924–1925, the year Vandiver joined the faculty.

Moore was to teach Introduction to Foundations of Geometry, Foundations of Mathematics, Point Sets and Continuous Transformations, and Theory of Functions of a Real Variable. A year later, Moore increased his assignments to include calculus and Research in Point-Set Theory. Calculus was to become an important development in the progression of the Moore Method, and his teaching practices in general. It was at that point that he really began to develop his courses sequentially to allow students to progress under his direction throughout their college and university career (see Chapter Sixteen for further discussion). As outlined in the first chapter, it was here that he clearly began to link the attraction of his course offerings to his search for mathematical talent that he could exploit to the full. He did this against the background of the newness of his own research, so that the work he

---

[2] The University of Texas Bulletin, College of Arts and Sciences Catalog, The University of Texas, Austin,1924–1925; as quoted in Traylor, p. 112.

[3] Those who joined the faculty as instructors during or after their doctoral studies in the 1920s were Anna Mullikin, R.L. Wilder, R.G. Lubben and G.T. Whyburn while C.M. Cleveland joined the Department of Applied Mathematics where he remained throughout his academic career. Lubben also remained at The University of Texas until his retirement (see Appendix Two).

was producing became the basis for further study and exploitation by his students. Students would be presented with theorems only recently proved, and confronted with his own ground-breaking ideas and methods, they found themselves on the frontier of research as his production of papers was beginning to catch a wide and international audience.

At the start of the decade, for example, Moore was working on a series of papers concerning continuous curves. Although he would return to it later, this phase of the work more or less culminated with his 1923 expository paper, 'Report on Continuous Curves from the Viewpoint of Analysis Situs',[4] whereupon he paid less attention to this area, leaving the field open to his students and especially, as Wilder points out,[5] to Gordon T. Whyburn (see below) who propagated the notion of cyclic elements of a continuous curve, which proved to be a useful device for analyzing the structure of curves.[6]

At the time, there was a mounting interest in work on topology of general spaces, and especially in the theory of continua. In Poland, Sierpiński and his colleagues had founded a school of set-theoretic topology and their work attracted much attention. By 1924, Moore was pursing a similar vein of research and that year published two papers that extended one of Sierpiński's theorems. Apart from being published in America in *Proceedings of the National Academy of Sciences*, they were also taken up by a new mathematical journal, *Fundamenta Mathematicae* launched in Poland in 1920. The journal began to publish a series of papers by Moore and his recognition internationally was such that some of his work was better known outside America than within it.

Meanwhile, Moore began to roll out the welcome mat for students he reckoned would be ideally suited to his teaching method, and to some degree the subject matter of his own research. Results over almost half a century were such that his successes demonstrated that he was a teacher *par excellence* but also careful about whom he took on. For the development of both these aspects we can turn to the reflection of his first doctoral student at Texas, Raymond L. Wilder, for his views on this issue. Certainly, Moore was always the recruiter and would regularly sit

---

[4] *Bull. Amer. Math. Soc.* 29 (1923), 289–302.

[5] In R.L. Wilder's The Mathematical Work of R.L. Moore: Its Background, Nature and Influence.

[6] See B.L. McAllister, Cyclic elements in topology; a history, *Amer. Math. Monthly* 73 (1966), 337–350.

Fall 1928
Back row, left to right: W.T. Reid, J.H. Roberts, C.M. Cleveland, Norman Rutt, J.L. Dorroh; Front row, left to right: Lucille Whyburn, G.T. Whyburn, R.L. Moore, R.G. Lubben.

at the mathematics desk during registration period when students were shopping for courses. It was said he could spot likely candidates even before they spoke, young people he could turn into mathematical stars. He would literally entice them into his world from there and wherever else he could find them. Wilder was a classic example. He had earned both his Bachelor's and Master's degrees at Brown University and had come to Texas specifically because he wanted to study for his PhD under a particular teacher. And it was not R.L. Moore.

Wilder, a multi-talented young man who in his youth played piano to silent movies in the local movie house, liked mathematics, history and anthropology but had set his cap at becoming an actuary. He had worked out that this would be the most profitable route for a career and he sought entry to The University of Texas, which had an actuarial specialist in Professor E.L. Dodd. He had an excellent reputation for producing capable people for that profession. He was, in fact, one of the few draws that Texas possessed at the time for people coming from any distance out of state. Wilder decided that whilst he was at Texas, he would also like to become better acquainted with pure mathematics, and found a course to his liking under R.L. Moore's offerings — only to find that Moore did not at first want him. So had the great mathematical talent spotter failed in his technique for identifying a future talent, as

Wilder duly became? 'Not at all,' Wilder would confess later. 'It was simply because I was a Yankee.'[7]

It was a tongue-in-cheek reference on the part of Wilder, and although Moore's Southern principles had been re-invigorated by his removal back to Texas, it was not the reason for Wilder's rejection. He could, of course, take instantly against people for reasons that may never have been made known to the unfortunate soul in front of him. In Wilder's case, the truth was that Moore was not enamored of his choice of career. He never considered 'statistics to be part of mathematics'[8] and he was certainly not going to accept Wilder in his course on the basis that topology was a secondary subject. In fact, Moore remarked that he very much doubted that Wilder would have the stamina for proving theorems and perhaps would not be much good at it even if he did.[9] Wilder persisted that he really was interested in pure mathematics and answered some of Moore's questions, including 'What is an axiom?' Moore relented and allowed Wilder to enroll but then totally ignored him for the first few weeks. He changed his approach, however, when Wilder proved some difficult theorems and, impressed by his talent, he then began to take a greater interest.

This grew to outright excitement when Moore discovered that Wilder had solved a problem that had baffled even J.R. Kline. Consequently, Moore literally challenged Wilder to change course and take his PhD in topology, although the deadline for taking language and qualifying exams had passed. Wilder agreed and as R H Bing describes, 'Moore cut through the red tape and arranged for Wilder to take the exams after the deadline.' He graduated with his doctorate in 1924 with a thesis *Concerning Continuous Curves* and never did become an actuary.[10] Like J.R. Kline, Wilder remained in life-long contact and subse-

---

[7] Remarks at the presentation breakfast of The University of Texas at Austin Mathematics Award honoring the memory of Professors R.L. Moore and H.S. Wall, San Antonio, 24 January 1976.

[8] R.E. Greenwood, History of the Various Departments of Mathematics at the University of Texas at Austin 1883–1983, p. 45. Unpublished manuscript, R.E. Greenwood Papers in the Archives of American Mathematics.

[9] From R H Bing's article in *Amer. Math. Monthly* 80 (1973), 117–119, on the Award for Distinguished Service to Professor R.L. Wilder.

[10] R.L. Wilder went on to become one of America's outstanding mathematicians. As well as his election as a member of the National Academy of Sciences, he served as President of the American Mathematical Society in 1955–1956 and as President of the Mathematical Association of America in 1965–1966. The bulk of his academic career was

quently in a paper on Moore, he examined the teacher-student relationship, which, in his experience, was quite unique: 'An essential part of the method was Moore's ability to search out and recognize creative ability among the multitude of students who presented themselves at The University of Texas. It was Moore's custom to teach five courses... consisting of calculus, an intermediate course such as advanced calculus, and three courses which began with point set topology ("Foundations of Mathematics") and culminated in a research course. Frequently he would find a promising student in his calculus class, and from then on that student would become a major project. Moore would carry him on through one course after another of the above sequence to the PhD. If any proof were needed that the *capability* of doing creative work in mathematics is not the rare genetic accident that it is commonly considered, Moore certainly gave it during his career as a teacher.'[11]

Not all his students were found this way. Some were "caught" by happening to take one of his advanced courses as an elective to round out their main objectives (which were as various as chemistry, medicine, public school teaching, applied mathematics, to name but a few). Wilder maintained that once a student with innate mathematical talent came under Moore's influence, his destiny was virtually decided. As Moore's reputation spread, advanced students from other institutions came to Texas to seek entrance to his courses. This involved careful selection on Moore's part. Students who had already taken courses in function theory and who therefore presumably knew the basic topological properties of the linear continuum and the complex plane might be discouraged from taking Moore's course in Foundations. Students already in Foundations were virtually forbidden to take courses con-

---

at the University of Michigan, where he directed twenty-five PhD students. In his honor the University of Michigan established the Raymond L. Wilder Professorship of Mathematics. Before going to Michigan he taught at Ohio State for two years, and after retiring from Michigan he taught at UC Santa Barbara. He was the AMS Colloquium Lecturer in 1943 (*The Topology of Manifolds* was published in the Colloquium series), and the Gibbs Lecturer in 1969. He wrote *Introduction to the Foundations of Mathematics, The Evolution of Mathematical Concepts; an Elementary Study,* and *Mathematics as a Cultural System.* He was a recipient of the Distinguished Service Award of the MAA and the Lester R. Ford Award. He was also especially noted for his campaigning efforts to allow the immigration of leading foreign mathematicians into America just prior to the outbreak of World War II.

[11] R.L. Wilder, Robert Lee Moore, 1882–1974, *Bull. Amer. Math. Soc.* 82 (1976), 412–427.

taining related material. Reading books or papers relating to the material of the course was also ruled out. This naturally implied a relationship with other members of his department that might be very difficult to achieve in other institutions. The aim of the Method, in Wilder's view, was to develop research ability, not knowledge alone: 'To some, knowledge may mean power; but to quote a statement Moore made to a prospective student from another university, "What does *information* amount to compared to *power*?"[12] Putting the matter in capsule form, it was a unique method employed by a unique man in a unique situation.'

Wilder, and many other future students, often utilized his method as a basis for modified usage. Moore, however, never permitted any detraction from the firm rules he set for his students, which he fully expected them to abide by without question. Wilder, or at least his publishers, should have known better therefore when in September 1951 Moore received the following letter from Edward Moody, of John Wiley and Sons, New York publishers, announcing that Professor R.L. Wilder had submitted the manuscript of a textbook for his course Foundations of Mathematics: 'I am writing to inquire whether you would be willing to let me know if such a textbook would be useful in your course on this subject and, if so, whether you would be apt to adopt it ... and how many students a year would be inclined to buy it.'[13] Moore dashed off an immediate reply:

'In the last twenty years I have never required *any* [underlined by Moore] student to buy any book for my Foundation of Mathematics course. Indeed, time after time, I have indicated that I would prefer that while taking this course no students read *anything* in print too closely *related* to it and I have even gone so far as to remove, or have removed, from our university library a copy of my own book *Foundations of Point Set Theory*. On one occasion this university copy was lost or misplaced and I was inclined to be sorry that it was afterwards found and again shelved in our library. I suppose you will be inclined to say, "If you don't want anyone to read your book why did you write it." I didn't say I do not want *anyone* to read it. I don't want it to be read by any one who is at the same time taking this particular course of *mine* [double

---

[12] A quotation Moore used in a letter to future student, Miss M.-E. Hamstrom, 7 May 1948. From the R.L. Moore Legacy Collection in the AAM. See Chapter 15.

[13] Moody to Moore, 26 Sept. 1951, R.L. Moore Papers in the AAM.

Gordon T. Whyburn

underlined] or who is going to take up the course of *mine* [double underlined] afterwards.'[14]

Another example of Moore's determination to guide his discoveries in the manner of his own choosing came with the arrival at Texas of a man who, although he didn't know it then, was to be his next doctoral student, Gordon T. Whyburn. He turned out to be another outstanding mathematician who had no intention of pursuing such a career. He had traveled to The University of Texas from a small community in Denton County, north of Dallas, specifically to study chemistry. Whyburn also decided at a late stage to take calculus with R.L. Moore, who immediately recognized Whyburn's potential and urged him to abort his aspirations in chemistry and take mathematics instead.

Whyburn, a shy and gentle man (although a tough administrator in later years) respectfully declined Moore's suggestion and pressed on with his Master's degree in chemistry, awarded in 1926. Moore did not give up. He kept up the pressure, insisting at every opportunity that he had ability in mathematics that ought not to be wasted. Whyburn was finally persuaded that it would be foolish to continue with chemistry

---

[14] Moore to Moody, 7 Oct. 1951, emphasis in original; R.L. Moore Papers in the AAM.

when he had already completed high quality research in mathematics. He agreed to change direction and went on to earn his doctorate with a thesis entitled *Concerning Continua in a Plane*.

Whyburn also followed Moore's deepening devotion to topology and it was to become the topic of research throughout his life.[15] He presented his first paper on cyclic elements for locally connected plane continua, at the Western Christmas Meeting of the American Mathematical Society in Chicago in December 1926 and after receiving his doctorate became an adjunct professor of mathematics at The University of Texas, a post he held until 1929. In that year, R.L. Moore received a communication from the administrators of the Guggenheim Fellowship, stating they were considering an award to Gordon Whyburn and requested a reference. Moore wrote:

'He has marked originality and power as an investigator. I do not think it likely there exists anywhere in this country another man of his age[16] (or within five or six years of his age) whose contribution to mathematics as represented by papers either published or accepted for publication equal to those of Whyburn in both quality and quantity. The notion *cyclic element* which he had introduced into mathematics ... is in my opinion a notion of fundamental importance and I think he is particularly well qualified to extend and apply it.'[17]

Whyburn was subsequently awarded the Guggenheim Fellowship and took off with Lucille for Europe. He spent the academic year 1929–1930 mostly working with Hans Hahn in Vienna but also established links with Kuratowski and Sierpiński in Warsaw. On his return, he was appointed associate professor of mathematics at The Johns Hopkins University but in 1934 moved to the University of Virginia where he remained until his retirement.[18] His research work was to attract great

---

[15] G.T. Whyburn's elder brother, William, a doctoral student of H.J. Ettlinger, was also at The University of Texas at the same time but unlike Gordon he was studying mathematics in the first instance. He went on to become Chairman of the Mathematics Department at the University of California, Los Angeles and then at the University of North Carolina.

[16] He was then 25 years old.

[17] Moore to the Guggenheim Foundation, 21 January 1929; R.L. Moore Papers in the AAM.

[18] Whyburn was elected to membership of the National Academy of Sciences in 1957, and became president of the American Mathematical Society in 1953–1954. He was chair of the mathematics department at the University of Virginia from 1934 until 1966. He was the AMS Colloquium Lecturer in 1940 and his Colloquium book *Analytic Topology* received international attention. He was also awarded the Chauvenet Prize of the Mathematical Association of America.

attention, commencing with his first contributions on cyclic elements and the structure of continua. This work aimed at examining a locally connected plane continuum and the regions of the plane created by it.

The theory was based on cyclic elements, that is a region C such that any two points of C are contained in a simple closed curve of C. He began working on homology theory and studied different notions of convergence in the space of all subsets of a compact metric space. The publication of his book *Analytic Topology* in 1942 was the result of continued work to that which he had presented as the American Mathematical Society's Colloquium Lecturer in 1940. Although well recognized as a man of brilliance, Whyburn perhaps became best known among his contemporaries for the incredible speed he possessed in research, yet he never disposed of the rudimentary elements learned with R.L. Moore, those of continuity and patience. The result was a lifetime of achievement.

While they could not undermine the brilliance of his work, some considered Whyburn to be one of the Moore 'products' — students he hijacked, some from courses that would have taken them in another direction, and was then taught and coached by Moore in all his major mathematics. It was this aspect that, for a number of years, became the subject of an ongoing debate at UT about any student being allowed to graduate after having studied under only one professor, although the signers of dissertations might be several. R.G. Lubben, on the other hand, came to Moore through the conventional route. He took all of his undergraduate work at Texas and studied with Moore in a course entitled 'Point Sets and Continuous Transformations'. He was captivated and remained under Moore's guidance for his doctorate, awarded in 1925 with a thesis that indicated the prevailing influence of Moore, entitled *The Double-Elliptic Case of the Lie-Riemann-Helmholtz-Hilbert Problem of the Foundations of Geometry.*[19] Lubben gave the solution to the then last remaining problem in the foundations of geometry. He was an independent discoverer of maximal compactifications of completely regular spaces, although priority in publication is assigned to Stone and Čech. He was a National Research Fellow (Göttingen) in 1926–1927, and subsequently joined the mathematics faculty at UT, where he too remained until his retirement in 1968.

---

[19] Moore himself had authored a paper entitled On the Lie-Riemann-Helmholtz-Hilbert problem of the foundations of geometry, *Amer. J. Math.* 41 (1919), 299–319.

R.L. Moore, October 1930

There was one further element that arose with Moore's involvement of his students in his own research, one that would continue down the decades. It was in the future relationship between Moore and his students. As already touched upon in Chapter Six, the relationships, contact and networking that developed among his students resulted in the interchange of ideas and discoveries and the movement of promising students from one university to another. News, views and mathematical discussion passed between them, along with sightings of potential talent, thus enabling students to shift from one place to another, either during their graduate study or following the formal completion of it. There were benefits to both students and teachers and Moore never held back in praising the contribution of his students in his own research, especially in the period leading up to the publication of his book, *Foundations of Point Set Theory*.

The emergence of highly talented men and women from his 'stable' did not go unnoticed in the great and growing community of American mathematicians. It was undoubtedly a contributory factor in the growth of his own reputation, aided by his prolific output of thought-provoking papers throughout the 1920s. Although by nature he was not a joiner of associations and mathematical groups, he took an active part in the affairs of the American Mathematical Society and toward the end of the decade honors and accolades began arriving in abundance. In 1929, he was appointed Research Lecturer at The University of Texas.

The Graduate Council of the University established the position in 1924 when the possibility of attracting visiting lecturers to Texas was difficult because of travel and budget restrictions. The council decided to use the skills of its own staff to encourage research among the members of the faculty of the University and, in particular, to bring home to students the importance of research. The position was rotated annually and the lectureship required the holder to deliver up to five lectures in a chosen field of investigation. These lectures were then published by the University and would be given publicity and distribution. The Department of Pure Mathematics managed to place four of its members in this elite squad, the first being Milton Brockett Porter. Moore was the second, followed in later years by Dodd and Vandiver. Moore chose a series of lectures aligned to the current thrust of his work, point set theory.

In that same year of 1929 he was selected to deliver the prestigious Colloquium Lectures for the American Mathematical Society at Boulder, Colorado, in August. Among those in the audience was one of Moore's most recent doctoral students, Gordon Whyburn (whom we discussed above), who wrote a preliminary report for the AMS *Bulletin*, prior to full publication of the lectures, and in this brief form it provides a glimpse into the direction and newness of Moore's thoughts and ideas at that time:

'The lectures delivered by Professor R.L. Moore were remarkably well attended ... the subject [being] "Point set theory" ... with which a considerable number of American mathematicians were not familiar; advantage was taken by many of these of this opportunity to gain insight into some of the underlying concepts of this most fascinating and fundamental branch of mathematics. The treatment of the subject, based on a set of axioms which was given, is to a large extent original with the lecturer, a most noteworthy contribution of this type having been published by him in the *Transactions of the American Mathematical Society* in 1916. In addition to his own work, however, the material covered embraced that of a number of other mathematicians both in this country and in Europe.

'On the basis of a system of seven axioms stated in terms of undefined concepts *point* and *region*, it is first shown that any space satisfying these axioms is homeomorphic with the surface of a sphere or with the euclidean plane according as it is or is not compact. Some of the main features of the treatment up to this point are the facts 1) that at no point, up until the very last when coordinates are introduced, is it necessary or apparently even of any advantage to introduce or make use of the notion of distance, and 2) that a very large body of propositions follow from very small sub-systems of the whole system of axioms, such as from Axioms 0 and 1 or Axioms 1 and 2, for example.'[20]

Publication of Moore's 1929 lectures made a considerable impact, both in America and Europe. Whyburn, on his visit to Europe under the Guggenheim Fellowship, met numerous eminent mathematicians and everywhere he went he was questioned about Moore's work. He relayed the fact to Moore, along with some specific queries which Moore, for once, answered promptly to clarify points raised by Bronislaw Knaster.[21] He also informed Whyburn: 'Work on my book is progressing slowly but I hope to send off the manuscript before Jan. 1, 1931. I have been delayed considerably since about March 15 by various matters including (1) the preparation and delivery of five University Research Professor Lectures and (2) Cleveland's[22] and Dorroh's[23] theses. Dorroh passed his oral examination Monday of this week and Cleveland passed his the next day. They are scheduled to receive their degrees in June. Roberts[24] is to [be] back here next session. Vickery[25] and Klipple[26] are doing quite well. I have considerable hopes for both

---

[20] *Bull. Amer. Math. Soc.* 35 (1929), 775–77.

[21] Bronislaw Knaster was to be cited in Moore's Colloquium book and includes the reference to 'Remark on a theorem of R.L. Moore' 1926. There is also some discussion of his work in Ciesielski, Krzystof, and Pogoda, Zdzislaw: The Beginning of Polish topology, *Math. Intelligencer,* 18(3) (1996), 32–39.

[22] Clark M. Cleveland, Moore's eighth PhD student, awarded in 1930 with a thesis entitled *On the existence of acyclic curves satisfying certain conditions with respect to a given continuous curve.* He joined the faculty of Applied Mathematics and Astronomy at UT and remained there throughout his career.

[23] Joe L. Dorroh, Moore's ninth doctoral student (1930), with a thesis title, *Some Metric properties of descriptive planes.* A National Research Fellow at Cal Tech in 1931, he later joined Texas A&I where he remained until his retirement.

[24] John H. Roberts, Moore's seventh PhD student (1929), thesis: *Concerning non-dense plane continua.* He joined Kline at Penn in 1931 and then went on to Duke University where he remained until his retirement, directing 24 PhD students whom he saw as Moore 'descendants'.

[25] Charles W. Vickery, Moore's tenth PhD success (1932), thesis: *Spaces in which there exist uncountable convergent sequences of points.* Moore was somewhat disappointed when he chose a career as a statistician in government service.

26 Edmund Klipple, PhD supervised by Moore (1932), thesis: *Spaces in which there exist contiguous points.* Joined Texas A&M in 1935 and remained there until his retirement.

A Rewarding Decade (1920–1930)

Taken at the Mathematical Association of America/American Mathematical Society/American Association for the Advancement of Science meetings in Cleveland, Ohio, December 1930. Left to right, disregarding row: Wilfrid Wilson, J.W. Alexander, W.L. Ayres, G.T. Whyburn, R.L. Wilder, P.M. Swingle, C.N. Reynolds, W.W. Flexner, R.L. Moore, T.C. Benton, K. Menger, S. Lefschetz

of them. Kindly remember me to Drs. Knaster, Kuratowski, Mazurkiewicz, Sierpiński and Zarankiewicz.'[27]

The response to Moore's 1929 lectures was such that the following year, the Council of the American Mathematical Society made it known that its members had nominated Moore for the unique honor of Visiting Lecturer for the Society for the year 1931–1932. He was the first native-born American to be offered the position, previously filled by eminent mathematicians from Europe. There were apparently some difficulties in regard to paying his salary during his time away from Austin in the dark and dire Depression gripping America. The chairman of the lectureship committee, G.D. Birkhoff, makes this clear in a letter urging Moore to accept the appointment:

'It is extremely important for the general mathematical body in America that men (like yourself) who have not only done important research but who have a point of view which is influencing current mathematics, should be able from time to time to visit our American mathematical centers and explain that point of view and its possible

---

[27] Moore to Whyburn, 22 May 1930, R.L. Moore Papers in the AAM.

development in an intimate way.... The importance of such interchange is recognized in Europe where men frequently go from one country to another to give a series of lectures at different institutions. Unfortunately, on account of various factors, it has not been possible to do the same thing here in America. The Visiting Lectureship of the American Mathematical Society is designed to remedy this deficiency.... I believe our American universities will wish to make arrangements so that a Visiting Lecturer taken from their Faculties may accept the appointment without loss of salary. I know this would be the case at Harvard University ... [and] I should hope the university would undertake the responsibility of making suitable arrangements for someone to carry on the duties of the Lecturer in his absence.'[28]

Moore accepted and among those who wrote to applaud his decision was Solomon Lefschetz of Princeton, a great rival with whom he carried on verbal jousting over several decades. Lefschetz had the good grace to write: 'I have just heard that you have accepted the visiting professorship of the society. All hail. This means that you will stop here at Princeton during the course of 1931. In any case, good luck.'[29] The lecture tour was an unqualified success with substantial plaudits as he traveled to some of the nation's finest universities, including the four where he had taught: Chicago, Princeton, Pennsylvania, Northwestern, Harvard, Swarthmore, the University of California at Berkeley, the University of Southern California, Stanford University, the universities of Washington, Minnesota, Michigan, Cincinnati, Buffalo, Iowa, Duke and the Rice Institute (now Rice University).

Simultaneously, Moore was informed that he had been elected to membership in the National Academy of Sciences and thus, as the lecture tour ended, he entered 1932, the year of his own half century, as one of America's most respected and influential mathematicians. It was also a year marked with sadness over the death of one of his great mentors,

---

[28] Birkhoff to Moore, 15 September 1930; R.L. Moore Papers in the AAM.

[29] Solomon Lefschetz, 30 November 1930, R.L. Moore Papers in the AAM. Lefschetz was a Russian born, Jewish mathematician who was the main source of the algebraic aspects of topology, an area into which R.L. Moore did not venture at all. He had two artificial hands over which he always wore shiny black gloves. First thing every morning a graduate student had to push a piece of chalk into his hand. His students at Princeton made up a rhyme about him, that might also have applied to Moore: *Here's to Lefschetz, Solomon L., Irrepressible as hell, When he's at last beneath the sod, He'll then begin to heckle God.*

E.H. Moore, at the age of 70. However, the same year saw another milestone with the publication of his groundbreaking book, *Foundations of Point Set Theory* by the American Mathematical Society as Volume 13 in the Colloquium Publications Series. It was to become a major point of reference for many subsequent investigations, continuing for several decades afterwards and revived with the publication of a revised edition in 1962.

It apparently pleased Moore no end to receive the following letter from Caroline Seely in the publishing department of the AMS after publication of the first edition, providing him with a note of charges for corrections: 'I think the bill is very reasonable; certainly the proof corrections are an "all-time record low". The previous record was held by Jackson, with $10.50 on a 200-page book but yours is $12 on 500 pp. And Lefschetz has $450 on about 400 pp.'[30]

---

[30] Seely to Moore, 25 June 1932; R.L. Moore Papers in the AAM.

9
A Change of Direction
(1931–1932)

International recognition, publication of his book followed by other accolades, and the completion of his half century of published papers while at the same time embarking on a controversial hunt for mathematical 'talent' that was to produce some of America's finest mathematicians.

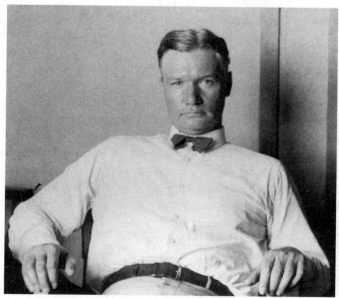

R.L. Moore in his office

The publication of R.L. Moore's book, *Foundations of Point Set Theory*, marked a distinct turning point in his mathematical career and indeed another apparent change of direction. As R.L. Wilder noted, the book represented a culmination of his work on topology that had emerged from his 1916 paper, 'On the foundations of plane analysis situs' which, in turn, had inspired a flow of additional papers through the decade of the 1920s and ancillary research by his students. His future output in terms of published research, however, now went into quite a dramatic decline.

By 1932, the year his book arrived in print, he had seen the publication of 50 papers and had made countless presentations and lectures to the American Mathematical Society and other bodies and universities, whereas in the rest of his life (which lasted a further 42 years) he published only 16 papers, the last in 1953. Whether it was a conscious decision, or one that simply evolved over time is not clear, but from his fiftieth birthday onwards he evidently decided to direct his energies into teaching and the production of doctoral students, of whom 43 were awarded their PhDs in 1930 and beyond.

Furthermore, a great deal of his teaching and much of the work completed by his students after 1930 was based on the ideas in his book, which in itself already reflected the results of research by the eleven successful doctoral students he had supervised so far. The point was made in a fine review of his book, published in the *American Mathematical Monthly*, which began: 'The book is based on a set of axioms for a space S, which make that space topologically equivalent to an ordinary two-dimensional sphere, or to a plane according as S is or is not compact. As the axioms are introduced, one or two at a time, the consequences of the new axiom or axioms together with some of the old ones are developed. As a result, theorems in the early chapters apply to very wide classes of spaces, including for example, Hilbert-space in Chapter II. The thoroughgoing investigation of the properties considered is due largely to Moore and to his students. This thoroughness is evident in the present volume, which contains many results hitherto unpublished.'[1]

The acknowledgement of the contribution and progress of Moore's students in this review was no less than Moore himself had done,

---

[1] *Foundations of Point Set Theory*, review author A.B. Brown, *Amer. Math. Monthly* 40 (1933), 100–101.

although there is little doubt that he was disappointed when his doctoral students, on leaving his cloisters, did not pursue a career in pure mathematics. It later became a matter of some conjecture as to whether his success as a teacher was in part due to allowing students to take part in a reenactment of his own ideas. At the time, however, there was no doubt that Moore was on the road to establishing that his method of teaching, already highly successful, would be ongoing and eventually put him high among America's leading echelon of doctoral producers. The road towards this accomplishment, using the axiomatic approach for both research and teaching, began in earnest in the 1930s, although not without detractors in respected quarters.

Poincaré had already expressed a view in an article on 'The Future of Mathematics' that, while conceding the excellence of Hilbert's use of the axiomatic method, the problem of providing axiomatic foundations for various fields of mathematics would be very restricted. In 1931 Hermann Weyl pursued the same line when he stated:

'I should not pass over in silence the fact that today the feeling among mathematicians is beginning to spread that the fertility of these abstracting methods [as embodied in axiomatics] is approaching exhaustion. The case is this: that all these nice general notions do not fall into our laps by themselves. But definite concrete problems were conquered in their undivided complexity, single-handed by brute force, so to speak. Only afterwards the axiomaticians came along and stated: Instead of breaking in the door with all your might and bruising your hands, you should have constructed such and such a key of skill, and by it you would have been able to open the door quite smoothly. But they can construct the key only because they are able, after the breaking in was successful, to study the lock from within and without. Before you can generalize, formalize and axiomatize, there must be a mathematical substance. I think that the mathematical substance in the formalizing of which we have trained ourselves during the last decades, becomes gradually exhausted. And so I foresee that the generation now rising will have a hard time in mathematics.'[2]

In any event, the warnings of both these eminent mathematicians proved to be overstated. R.L. Moore, in particular, had by then unlocked the door to considerable research with the use of his axiom

---

[2] H. Weyl, Emmy Noether, *Scripta Math.* 3 (1935) 1–20, as quoted in R.L. Wilder, Axiomatics and the Development of Creative Talent, p. 475.

Raymond L. Wilder

system for plane analysis situs. More importantly, he was now demonstrating, with very positive results, that it could be used for the discovery and development of creative talent. But this was not due to the axiomatic approach alone. What Moore did, and others did not, was to combine it with his own teaching regime which he had honed over time, going back all the way to his days at Northwestern and Penn. As he entered the decade of the 1930s, it was running to perfection. R.L. Wilder, in his paper 'Axiomatics and the development of creative talent',[3] provides this summary of the stage Moore had reached:

'The use by Moore of the axioms for plane analysis situs in his teaching had many elements in common with the Socratic method as revealed in the *Dialogues*, especially in the general type of interplay between master and pupil.... He set up a course, which he called "Foundations of Mathematics", and admitted ... only students he considered mature enough and sufficiently sympathetic with the aims of the course to profit thereby.... He based his selection of students, from those applying for admission, on either previous contacts (usually in prior courses) or, in the case of students newly arrived on the campus, on analysis via personal interview.... The amazing success of the course was no doubt in some measure due to this selection process.... There

---

[3] Published in *The Axiomatic Method with Special Reference to Geometry and Physics*, North Holland, 1959, pp. 474–488, quote from pp. 477–479.

was no attempt to cater to the capacities of the "average" student, rather was the pace set by the *most talented* in the class. Now I grant that there seems to be nothing sensational about this.... The noteworthy fact about Moore's work is that he began finding the capacity for mathematical creativeness where no one suspected it existed! In short, he *found* and *developed creative talent.* I think there is no question but that this was in large measure due to the fact that the student felt that he was being "let in" on the management and handling of the material. He was afforded a chance to experience the thrill of creating mathematical concepts and to glimpse the inherent beauty of mathematics, without having any of the rigor omitted in order to ease the process. And in their turn, when they went forth to become teachers, these students later used a similar scheme. True, they met with varying success — after all, a pedagogical system, no matter how well conceived, must be operated by a good teacher.'

Wilder concluded his address to the symposium with a precise summary of the basic essentials of the application of the Moore Method in advanced classes as it had evolved by the mid-1930s:

1. Selection of students capable (as much as one could tell from personal contacts or history) of coping with the type of material to be studied.
2. Control of the size of the group participating, from four to eight students probably the best number.
3. Injection of the proper amount of intuitive material, as an aid in the construction of proofs.
4. Insistence on rigorous proof, by the students themselves, in accordance with the ideal type of axiomatic development.
5. Encouragement of a good-natured competition.

The Moore Method, then, had progressed considerably since the descriptions of it by his students, first at Northwestern and then at Penn. What had transpired was the way in which he combined the axiomatic method with his practical teaching, and the involvement of his students in the vast work program he had undertaken. This was no better demonstrated than in those first ten years at Texas when the direction of his own research became more focused. As will be seen as this chapter proceeds, this subsequently brought some quite outstanding follow-on research from students who were exposed to it.

The search for the right caliber of students, as required by Moore, did in fact suffer from time to time by circumstances outside of his control, such as the outbreak of World War II. In the early 1930s, the effects of the economic depression also hit university enrollment figures and the arrival of 'suitable' young men and women for his courses went into decline. In the year of 1933–1934, for example, student enrollment actually declined for the first time in the university's history. This meant there was something of a scramble to 'capture' those students who showed the kind of ability that Moore and his colleague professors were seeking. At The University of Texas, as no doubt in other establishments, a definite system of talent-spotting came into play along with some intense rivalry between the Pure and Applied Mathematics Departments of The University of Texas. Competition for the cream of students became more intense as the years passed, especially when, by the early 1930s, Pure Mathematics was offering 40 courses, twice as many as Applied Mathematics. Moore especially went a-hunting and according to Professor Robert Greenwood, assistant professors and instructors who taught freshman classes were on the lookout for any student with mathematical aptitude. The senior professors relied on their tip-offs and so developed something of an unseemly race to attract the best into their courses. Greenwood saw it first-hand in varying degrees of competitiveness over a 50-year time frame, first as a student and then as a member of the mathematics faculty: '(I) while a freshman noticed the talent-seeking proclivities of [my] two freshman teachers. [As someone once described it,] these instructors were equipped with "seining" nets and they seined their classes for "talent".'[4]

Not content with relying on their spy network, some professors, and R.L. Moore was a leader among them, would venture into what was then the old auditorium that became a meeting place where the students gathered. They would sit around, usually in groups related to the nature of their studies, and talk. Some of the professors noticed this and began to go out to join the students. Although Joe Frantz described him as an 'ivory tower sort' as far as the rest of the faculty was concerned, Moore began to wander nonchalantly into the auditorium where he would ease himself into discussions and then talk about anything but mathematics. Eventually, it became a regular occurrence to see him holding forth with students in the afternoons and most days made himself available in his office to anyone who wished to see him, and perhaps join his class. As Lucille Whyburn recalled from her time there:

---

[4] R.E. Greenwood, *History of the Various Departments of Mathematics at The University of Texas at Austin: 1883–1983*, p. 54. Unpublished manuscript, R.E. Greenwood Papers in the AAM.

A Change of Direction (1931–1932)

Robert E. Greenwood

'I don't remember ever taking a class from him in the afternoon, but he was there in his office or out in the auditorium. And if you wanted to go and see him, he was always available. I never knew him to refuse to see a student. Now, he may have done so, but he certainly never refused to see me. His help was always given openly and freely and, I think, had a great deal to do with your feeling of comfort about mathematics, and instilling in you a desire to really understand it. So many people now talk in what I call a supercilious fashion about mathematics and as long as they can keep the conversation on a monologue, they just go along fine. They just sort of startle you with their rendition but then, if you just horn in on them a little about their depth of understanding of a notion, very often you will find it's shallow. They don't have any depth of understanding and this was part of Dr. Moore's real success as a teacher.'[5]

The situation described above was not merely of benefit to the students, but to Moore himself as he scouted for suitable candidates for what Robert Greenwood described as his 'indoctrination' and to some, that is what it was. Although full of praise for the outstanding results achieved by Moore at Texas, as will unfold in abundance in ensuing

---

[5] Lucille Whyburn, Oral History interview with Douglas Forbes, ca. 1970; R.L. Moore Legacy Collection in the AAM.

chapters, Greenwood was forthright enough to discard the rose-tinted spectacles and admit that the Moore Method did not then, and would not in the future, work for everyone. Greenwood says he was certainly influenced by Moore 'but two exposures to probability, statistics and combinational analysis in World War II and in the Korean incident made more lasting patterns of thought in [me]'.[6] The 'seining' situation described by Greenwood went on for years, and for a long period descended into quite bitter acrimony between members of the faculty.

Robert Greenwood recalls that in the early 1940s, Moore (then turned 60) and Associate Professor Edwin F. Beckenbach were vying for graduate students: 'Reliable sources say that one heated confrontation led to fisticuffs. Professor Beckenbach subsequently accepted an offer [at UCLA]. He wrote a letter to the President of UT, presumably offering advice on how to improve the mathematical situation there but the contents of the letter were never made public.'

These character traits were part of the make-up of R.L. Moore, aligned to the uniqueness of the Method he used for teaching. His students came to know of them in due course and some of them were not easy to accommodate or live with. Many felt unable to do so, and Moore too was not one to allow an uneasy situation between himself and a student to affect the rest of his class. He would merely suggest that the student might care to consider removing himself to another place, or indeed arrange for him to receive an offer he simply could not refuse. When he spotted a 'talent' Moore, in his customary and well-established predatory fashion, did his level best to secure him and so often this effort paid handsome dividends, as past, present and future entrants into his courses proved with great distinction. None did more so than F. Burton Jones, another chemistry student who, like Wilder and Whyburn before him, was hijacked into mathematics by Moore. Jones was a native of Cisco, Texas, where his father was a pharmacist and local politician. He qualified for a Regents Scholarship to The University of Texas where he started the pre-law program but switched to chemistry. The mathematics courses included calculus, which brought him under the influence of Moore. Jones's fate was more or less sealed. Although he graduated in chemistry in 1932, Moore immediately offered him a part-time job as a math instructor. At a time when the

---

[6] R.E. Greenwood, *History of the Various Departments of Mathematics at The University of Texas at Austin: 1883–1983*, p. 54. Unpublished manuscript, R.E. Greenwood Papers in the AAM.

employment situation was dire, Jones accepted with alacrity. Moore convinced him that his future lay in mathematics, and assured him that he would guide him toward his doctorate, which he achieved in 1935 with a thesis entitled *Concerning R. L. Moore's Axioms 1–5*. The recollections of Jones[7] provide a further and intriguing insight into Moore's *modus operandi,* specifically in the years immediately after the publication of his book. He would begin his graduate course in topology as usual, by carefully selecting the members of the class. His idea was to have a class as homogeneously ignorant (topologically) as possible.

He would usually caution the group not to read topology and especially not to read his own book, but simply to use their own ability, seeking to establish a level playing field for all his students.[8] Recalling that in the 1920s and 1930s very few students had had any exposure to topology or topological ideas, Jones tells us that competition was one of the driving forces used by Moore to spur on his students, staged in the form of a challenge that they felt obliged to accept and fulfill. Jones arrived when the Moore Method was already well established:

'He would ... state the axioms that the class was to start with (Axioms 0 and 1 of his book,... omitting part (4) of Axiom 1). An example or two of situations where the axioms could be said to apply (e.g., the plane or Hilbert space) would be given. He would sometimes give a different definition of region for a familiar space (e.g., euclidean 3-space) to give some intuitive feeling for the meaning of an "undefined term" in the axiomatic system. Of course, this was part of his own personal philosophy and he considered it part of the motivation of the subject. After stating the axioms and giving motivating examples to illustrate their meaning he would then state some definitions and theorems. He simply read them from his book as the students copied them down.

'He would then instruct the class to find proofs of their own and to construct examples to show that the hypotheses of the theorems could not be weakened, omitted, or partially omitted. When the class returned for the next meeting he would call on some student to prove Theorem 1.

---

[7] F. Burton Jones (1910–1999), The Moore Method, *Amer. Math. Monthly* 84 (1977), 273–278.

[8] The exception, by then, was his Foundations of Geometry course for which he made no attempt to select or exclude students because he assumed that all of them were uniformly ignorant of the Hilbert-Veblen-Moore axiomatic approach to the subject. High school teachers and post-graduate scholars were encouraged to attend the course but less experienced students and even non-mathematical majors were also accepted.

'After he became familiar with the abilities of the class members, he would call on them in reverse order, and in this way give the more unsuccessful students first chance when they *did* get a proof. He wasn't inflexible in this procedure but it was clear that he preferred it. When a student stated that he could prove Theorem x, he was asked to go to the blackboard and present his proof. Then the other students, especially those who hadn't been able to discover a proof, would make sure that the proof presented was correct and convincing.'

Jones states that Moore took a stand against heckling, although it was seldom necessary because the whole atmosphere of the classroom was one of serious community effort to understand the argument.[9] When a flaw appeared in a 'proof' everyone would patiently wait for the student at the board to 'patch it up'. If he could not, he would sit down. Moore would then ask the next student to try or if he thought the difficulty encountered was sufficiently interesting, he would save that theorem until next time and go on to the next unproved theorem (starting again at the bottom of the class). Occasionally theorems got left over indefinitely but nearly all of these would be proved in some subsequent year:

'Quite frequently when a flaw would appear in a proof everyone would spend some time (possibly in class) trying to get an example to show that it couldn't be "patched up", i.e., a counterexample to the argument (even though the theorem might be correct). This kind of experience is seldom encountered in courses or in any place outside of one's own research work. Yet this kind of activity is vitally necessary for the research worker.'

Moore was generous in his praise of students, and especially so when, as occasionally happened, one discovered an improvement of one of his own theorems. He thereafter referred to that theorem with the student's name and in the 1962 revision of his book a number of names were included in the text, sometimes with comments concerning the origin of some proofs and concepts in the appendix. Although the improvement to his theorems might not be particularly significant, the encouragement given by his recognition of it was considerable.

---

[9] R.L. Wilder, on the other hand, stated in his address at a symposium on the axiomatic method, 'I could give you many interesting, and amusing, accounts of the by-play between teacher and students, as well as between the students themselves; good-natured "heckling" was encouraged. However, the point to be emphasized is that Moore *put the students entirely on their own resources* so far as supplying proofs was concerned.' From his Axiomatics and the Development of Creative Talent, pp. 478–479.

A Change of Direction (1931–1932)

Spring 1931
Karl Menger, Milton B. Porter, J.H. Roberts, and R. G. Lubben

Jones was adamant that his experience with Moore and exposure to the Moore Method had given his own career a kick-start in a direction that he had never previously considered, or even thought possible. He became a brilliant teacher and a researcher of considerable repute in his own right. He was noted for his contribution to important wartime developments at the Harvard Underwater Sound Laboratory to which he was attached for almost three years.[10] Wilder also credits him with introducing the term 'Moore Space' for a topological space that satisfies Axiom 0 (which states that every region is a point set) and parts (1), (2) and (3) of Axiom 1 in Moore's book. It seems to have been first used by Jones[11] in the mid-1930s when he began producing important results that were to have a considerable impact on the development of point set topology and continuum theory in the last half of the twentieth century.

---

[10] He was part of a six-man team that developed scanning sonar for the Navy.
[11] See his paper Concerning normal and completely normal spaces, *Bull. Amer. Math. Soc.* 43 (1937), 671–679.

Jones told Wilder[12] that he became interested in the problem of metrization of Moore spaces after reading a paper of fellow Moore student J.H. Roberts[13] which raised the question of whether Moore spaces were metrizable. Jones subsequently published his result which proved that every separable, normal Moore space is metrizable. He also went further on the issue of whether every normal Moore space was metrizable. During the ensuing 50 years, around 350 papers specifically relating to Moore spaces were published, to the point where, in 1959, a separate classification number was assigned to Moore spaces by the abstracting journal, *Mathematical Reviews*. These ongoing references, still applicable at the beginning of the twenty-first century, were themselves stimulated by Moore students, and then their students and their students' students as well as other authors around the world. Another of Jones's best-known papers[14], which was published as early as 1938, demonstrated that if the phrase 'simple closed curve' in Axiom 5 of Moore's Foundations was replaced by the phrase 'compact continuum', then the first four axioms plus the weakened Axiom 5 implied the original Axiom 5.

In the revised edition of his book published in 1962, Moore replaced the original Axiom 5 with the modified version of Jones and recorded the fact that the change was a major one. In this, and in several other similar, if lesser cases, Moore's students were achieving results that edged towards, and occasionally bettered, Moore's genius. Burton Jones was one of those people and it is an interesting diversion to discover how he, one of America's most respected teachers of his era, also took on board the Moore Method. He utilized a modified version of it in his own classrooms first at The University of Texas where he remained until 1950, and then at the University of North Carolina where he directed five PhD dissertations and chaired the Department of Mathematics. In 1962 he moved to the University of California at Riverside, where he helped launch the doctoral program in mathematics and directed ten more PhD dissertations before his retirement in 1978.

The most important modification he made in applying Moore's Method was to curtail the elitist tendency and to make his classes less

---

[12] Mentioned in Wilder's The Mathematical Work of R.L. Moore: Its Background, Nature and Influences, p. 88.

[13] J.H. Roberts, A property related to completeness, *Bull. Amer. Math. Soc.* 38 (1932), 835–838.

[14] F. Burton Jones, Concerning R.L. Moore's Axiom 5, *Bull. Amer. Math. Soc.* 44 (1938), 689–692.

forbidding to those of a slower disposition. 'People generally tend to be embarrassed,' he wrote, 'by mistakes, especially public mistakes and care must be taken not to make them feel that criticism is ridicule.'[15] He also incorporated a textbook into his courses, albeit well after his courses had commenced. Around Christmas time, he would allow his students to take Kelley[16] as bedtime reading, although it was never discussed in class. The course would then continue as before, with students working out their own proofs. Like Moore, Jones would begin with a sequence of definitions and statements of theorems in his graduate course in topology and move on from there accompanied, Jones always stressed, by a large input of patience on the part of the instructor. Professor Louis McAuley, his first PhD student, in a 1969 tribute to Jones, spoke of his 'magical powers in the classroom, a master who breathes the very life of mathematics into his students'.[17]

Others remember Jones as a source of encouragement and of knowledge about problems. His role was that of the kindly uncle, who moderated differences between those trained in the Moore school and those who had taken their training elsewhere. He also discovered that Moore's old rival, Lefschetz, who had beaten him to the presidency of the American Mathematical Society (his term of office was 1935–1936) had tried out the Moore Method. One of Wilder's students at Michigan was Norman Steenrod who had since moved to Princeton. He wrote to Wilder on 28 February 1937, in the following terms:

'I've been running a seminar in topology and doing it in the spirit of the R.L. Moore school. So far I've been following the notes to the course you gave at Michigan. The class contains about six willing workers. They have proved already 19 theorems and 16 lemmas. There are 10 lemmas left before it is necessary to introduce Axiom VI. I hesitate to do this. I've been stalling them off by giving them some propositions

---

[15] F. Burton Jones, The Moore Method, *Amer. Math. Monthly* 84 (1997), 273–278, quote from p. 276. 'Ridicule' is an interesting choice of word here in that on occasions Moore did ridicule some students, but appears to have done so quite selectively, as a goad to students who he judged could take it. On the other hand, a later doctoral student, W.S. Mahavier, has stated in his recollections for the R.L. Moore Oral History Project, that Moore seemed to sense that in the first few undergraduate courses criticism of Mahavier's work while at the board would have been fatal to his continuing the course.

[16] John L. Kelley's *General Topology*. He was a doctoral student of G.T. Whyburn.

[17] L. McAuley, Dedication, Proceedings of the Auburn Topology Conference (Auburn University, 1969; dedicated to F. Burton Jones on the occasion of his 60th birthday), pp. iii–viii.

on the theory of sets to prove. The thing I've noticed about Princeton is that the students who come here without having done research work have considerable difficulty in learning how to do it. They have to pick it up for themselves. So I talked up the virtues of the Moore system to Lefschetz. He finally agreed that it was worth trying.... So far he has been satisfied ... and was very pleased with the way things were going. He was so pleased in fact he is starting a seminar in algebraic geometry to be conducted in nearly the same fashion as possible. Perhaps now you see my difficulty with Axiom VI. It is just the kind of thing Lefschetz would find objectionable. In a way I sympathize. The axiom is hard to remember. Could it be replaced by merely requiring that the boundary of a region is a simple closed curve? Perhaps you know of some other set up which would be more satisfactory.'[18]

Wilder's response to Steenrod's dilemma is not recorded, but what it does demonstrate, apart from the difficulties some people experienced in aligning Moore's Method to his axiomatic system, was that the message was traveling across America. Moore was in great demand. Lefschetz wrote insisting that he call in at Princeton whenever he was nearby and many major universities invited him to give lectures on his work in the years following the publication of his book. For example, the University of Chicago wrote to request that he deliver four lectures on specific aspects, (1) Foundations of Point Set Theory (2) Continuous Curves (3) Upper Semi-Continuous Collections and (4) The Structure of Continua. The Rice Institute published a pamphlet on a lecture he delivered to them, and Gordon Whyburn at the University of Virginia invited him to address a conference of point set topology to be held in conjunction with Duke University.

The invitations continued to pour in, even from Europe, although there is no indication that Moore ever left the United States of America in his entire life. The demands of his time were already too great, not least of which was his commitment to the American Mathematical Society, serving on committees involved in administration, membership and offices. In February 1937, it was announced he had been elected as president of the Society. He thus, somewhat belatedly, joined a throng of mentors and friends to hold the office, including E.H. Moore (1901–1902), L.E. Dickson (1917–1918), G.A. Bliss (1921–1922),

---

[18] Steenrod to Wilder, 28 February 1937, from the R.L. Wilder Papers, Archives of American Mathematics, Center for American History, The University of Texas at Austin.

Oswald Veblen (1923–1924) and G.D. Birkhoff (1925–1926).[19] Birkhoff was among the first to write with congratulations saying that he was delighted: 'you thoroughly deserve the honor and I hope you will be able to work effectively for the cause of research and the general advancement of the society.'[20]

As the incoming president of the American Association for the Advancement of Science, Birkhoff also took the opportunity to enlist Moore's help in forging greater links between the two organizations. His letter contained a warning over the apparent reluctance of the AMS to hold joint meetings with the AAAS, citing reasons that he considered vital for the future. Moore was known to be among the doubters:

'American mathematics can only become national and powerful in the country at large if it maintains its place with dignity with other sciences. There are very definite efforts to be seen in certain quarters to create the opinion that mathematicians in general are so aloof from their fellow scientists and so naïve in their essential character that they are not really very important after all. If at this juncture … the American Mathematical Society were to serve notice that it is not interested in co-operation, I believe this opinion would gain considerable headway. This I would regard as most unfortunate.'[21]

The issue went to the core of debate arising in schools, colleges and universities across America at a time of controversy over teaching methods, and some general misgivings were highlighted in an article in *Science* magazine, written by Professor Otis F. Curtis, of Cornell University. Although not a mathematician, in fact at the time he was president of the American Society of Plant Physiologists, his article, condensed from an address he gave to his society, was clipped by R.L. Moore and survives in his papers. Moore had marked a large number of salient points in the professor's address with which he was quite clearly in agreement and which, to some extent, would not be out of place in an educational seminar at the start of the twenty-first century, either. Inasmuch as R.L. Moore had taken the trouble to mark up the article for future reference, it is worth making a fuller examination of its content, which served to reaffirm his own views on the subject of education, and

---

[19] Three of Moore's own doctoral students later followed him: G.T. Whyburn (1953–54), R.L. Wilder (1955–56) and R H Bing (1977–78).
[20] Birkhoff to Moore, 4 February 1937; R.L. Moore Papers in the AAM.
[21] Birkhoff to Moore, 4 February 1937; R.L. Moore Papers in the AAM.

R.E. Basye and E.C. Klipple, students of Moore in the 1930s

which would certainly be applied to the teaching of mathematics then (and now!):

'It seems that much of the training in our grade schools and high schools and even in universities is of [the] authoritative or dogmatic type. The pupil is not trained to think for himself, for the major emphasis is placed on learning so-called facts.... more and more information is drilled into the minds of our students. They are led to accept some text or individual as authority and are not encouraged to form their own opinions or to use their own judgement. However, to accept blindly the opinions and statements of others, to accept authority, does not lead to understanding by the pupil. In fact, the teacher who teaches authoritatively teaches answers, and is not helping the student to learn for himself or to arrive at a real understanding.... Vitality leaves a subject when it is carried on by authority, no matter what the subject....'

A section in the article specially marked up by Moore went on:

'The student should learn to evaluate evidence and draw his own conclusions, and not merely accept answers given by a book or a teacher. I am afraid we teachers too often discourage independent thinking. We expect correct answers or answers in the same words as we ourselves have formulated them, and pay too little attention to the type of reasoning which leads to the answer. Often a minor slip leads to an unsatisfactory answer when the main line of reasoning is the student's own and is perfectly sound. In attempting to get the answer satisfactory to the teacher, the student is often forced to learn by rote. He may then present a suitable answer but have no clear idea as to the underlying principles. I knew of a high-school teacher of geometry who forced her students to use the same lettering as the text in proving a theorem. If the letters were changed she could not prove it herself. She had learned by authority, had memorized the formula and was passing it on verbatim. She did not herself understand the proof.'[22]

Another section of the professor's address also caught Moore's eye, and he was known to have quoted it on several occasions, years hence. It was a reference to Andreas Vesalius, the Brussels physician who revolutionized the study of biology, and whose original studies of the human body were considered sacrilegious. One of his students said that his teacher not only demonstrated new truths to him, but 'so taught that the student could discover new truths for himself and could develop beyond what he was taught'. The whole tenor of the article clearly captivated Moore's interest. As has already been demonstrated in this chapter, the majority of his students felt themselves liberated by his method in that they were taught to seek the truth for themselves. It also summed up the direction Moore was taking in his own life, one in which he was to dedicate himself to teaching above all else.

---

[22] Professor Otis F. Curtis, Education by authority or for authority? Are science teachers teaching science? Condensed version from the address of the retiring president of the American Society of Plant Physiologists at Richmond, VA, 28 December 1938, published in *Science*, 90, August 1939, pp. 93–101, quotes from p. 96 and 97.

10
Politics and Persuasion (1933–1938)

A period of consolidation: in an era dominated by Roosevelt and New Dealers at home (against which he railed vociferously), and the arrival on US shores of many Jewish mathematicians and scientists from Europe, about which he was also outspoken; culminating in his election as president of the AMS for the two years of 1937–1938, and the foundation of the Mathematical Reviews.

R.L. Moore, ca. 1935

The years prior to Moore's term as president of the American Mathematical Society were challenging and demanding both in terms of the history of the Society itself and the world at large. As active head of the Society, he would be called upon to lead it into its 50th anniversary celebrations while at the same time overseeing important, and controversial, matters arising from developments in Europe. Since Hitler's rise to power in 1933, many of the world's most respected mathematicians had sought asylum in the United States and this in turn had, by 1937, when Moore took over as president, already had a dramatic impact on the mathematical sciences in the country. From the beginning of the 1930s, the incoming enrichment of talent and scholarship soon became apparent at every level. In due course, they would also contribute to an unparalleled mobilization of effort drawn together from the whole of the American scientific and mathematical community for war-related non-combat tasks in which, incidentally, a number of Moore-trained students would figure prominently. However, long before the war, and America's eventual involvement in it, became a reality, Moore was confronted by a number of issues arising from these developments, some of which did not sit well with his personal beliefs and attitudes. Certainly there were many in his profession who were concerned that the flow of incoming people became more acute at a time when there was still a good deal of unemployment in the United States, resulting from the Depression. It is worth recalling that in the same month of March 1933 when Franklin D. Roosevelt was sworn in as President of the United States, Hitler took power in Germany. As the Nazis began their onslaught against Jews and liberals, Germany's scientific and mathematical communities were among the groups targeted. The pace of exodus quickened as various scientists and mathematicians, along with students and administrators, took up the task of identifying to the Storm Troopers 'undesirable' people in their professions. The result was that expelling the Jews succeeded in virtually eliminating the possibility of further German contributions to a number of branches of research.

Roosevelt meanwhile was confronted by a dire economic situation. Most banks had been shut down, industrial production had fallen to just 56 percent of its 1929 level, at least 13 million wage earners were unemployed, and farmers were in desperate straits. The new President's famous '100 Days' and subsequent actions brought him the honor of becoming simultaneously one of the most loved and most hated men in

American history. On the one hand, he was seen as the savior of his country and on the other the scourge of free-market capitalism, unconstitutionally expanding the powers of the federal government, and transforming the nation into a welfare state.

R.L. Moore's view of things undoubtedly fell into the latter category and the arrival on American shores of émigrés from Europe merely strengthened his opinion, given that so many of them were seeking work and shelter in educational establishments. Burton Jones, then studying with Moore for his PhD, recalled that Moore saw it as something of an 'invasion'.[1] He did not welcome them with open arms, not even the likes of Albert Einstein. After the Nobel prizewinner for Physics renounced his German citizenship and fled the country in fear of his life, he eventually joined Oswald Veblen in a full-time position as a foundation member of the School of Mathematics at the new Institute for Advanced Study in Princeton. Burton Jones states that Moore's reaction to this news was: 'Well, he thought he was a physicist.' This was true, of course, but Moore was well aware that Veblen had turned his attention toward differential geometry soon after the appearance of Einstein's general theory of relativity and had produced some substantial and acclaimed work in that direction. Since then, Veblen had established Princeton as one of the three leading centers in the world for topological research,[2] aided and abetted by Solomon Lefschetz who became a Henry Fine Professor there in 1933, the same year that Einstein left Germany.

At worst, Moore seems to have been selective in his attitude towards Jews in American society and Jewish mathematicians. He did not support an open door policy but, at the same time, according to Burton Jones,[3] Moore held a certain amount of respect for them as a group, in that they had to 'work harder and do more' to establish their place in society. Moore, said Jones, had expressed the view that Jewish mathe-

---

[1] A comment made in his recollections of Moore, during a taped interview with R.D. Anderson, Ben Fitzpatrick and Charles Hagopian recorded for The Legacy of R.L. Moore Project, 25–27 July, 1997; transcript in the R.L. Moore Legacy Collection, Archives of American Mathematics, Center for American History, The University of Texas at Austin (hereafter referred to as the R.L. Moore Legacy Collection in the AAM).

[2] Princeton, the Polish school and Moore and his students.

[3] From the taped interview with R.D. Anderson, Ben Fitzpatrick and Charles Hagopian recorded for The Legacy of R.L. Moore Project, 25–27 July, 1997, R.L. Moore Legacy Collection in the AAM.

Possibly taken in the Moore home, ca. 1935
Left to right, back row: R.E. Basye, E.C. Klipple, F. Burton Jones;
Front row: C.W. Vickery, R.L. Moore, R.G. Lubben

maticians, were by necessity 'crafty and smart'. As has already been noted, he had a life-long friendship with his Jewish colleague at The University of Texas, H.J. Ettlinger, and directed his son Martin in his MA studies. The son went on to an illustrious career, a professor at Rice and then at the University of Copenhagen, always readily acknowledging Moore's contribution. R.D. Anderson remembers that Moore 'used the expression Northern Jews when he wanted to stew Ettlinger, [senior]' in a somewhat light-hearted manner, nor did he show any apparent reluctance to accept Jewish students. If there was a streak of anti-Semitism in him he was certainly not alone in the world of science and mathematics, both in the United States and the United Kingdom. Many well-known figures held stronger views, overtly displayed, on the issue of incoming scholars, Jews in particular, from Europe. G.D. Birkhoff, who was president of the American Association for the Advancement of Science while Moore was president of the AMS, seems to have had tendencies in this direction. One of Birkhoff's former students, Chandler Davis, wrote: 'He systematically kept Jews out of his department, but apparently relented late in life and favored appointing one by the 1940s ... Though his record is mixed and some were more implacably anti-Semitic than he was, his actions in this regard are important because of his very great influence.'⁴ This unpleasant era of mathematical history, marred by the minority, had deeper connotations more to do with politics than mathematics.

The Depression left a scar on American universities, with many well-qualified and established people out of work, or having to take work for a pittance that did not match their ability. Moore's politics were firm and outspoken, and still steeped in the Southern principles by which he was raised. He would have no truck with American leftwingers. He deplored Roosevelt's determination as he ran for a second term in 1936 to push forward with further New Deal reforms. The president's well-quoted speech, 'I see one-third of a nation ill-housed, ill-clad, ill-nourished' did little to melt the hearts of Southern Conservatives. It was surely Moore's personal loathing of Roosevelt that had something to do with an embarrassing incident during the fiftieth anniversary celebrations of the American Mathematical Society, when Moore was at the helm.

Dr. Thornton C. Fry, one of the organizers of a celebratory dinner to be held during the national meeting of the AMS in September 1938, had written to Moore enclosing a letter addressed to President Roosevelt at the White House. Fry asked that Moore should sign the letter and forward it on to the President. It outlined a brief history of the Society and informed the President that 'it would contribute greatly to the dignity of this occasion if you could honor us with a message of greeting, which could be read at that time.... it is generally recognized that we are in the front rank among leading nations of the world in both the quantity and the quality of the mathematical work done here ... I sincerely hope you will find it possible to send us a word of greeting on this occasion which is so important to us and we believe in a modest way to the country at large.'⁵

Moore was having none it. On the same day that he received the letter drafted by Fry and his organizing committee, Moore replied: 'I do not see that the reading of solicited greetings from President Roosevelt on the occasion of the celebration of the fiftieth anniversary of the founding of the Society would add at all to the dignity of the occasion. And I am not sure that it is not beneath the dignity of the American

---

[4] Quoted in a biographical summary of G.D. Birkhoff by J.J. O'Connor and E.F. Robertson in the Mac Tutor History of Mathematics Archive, University of St. Andrews, Scotland, on www-history.mcs.st-andrews.ac.uk/history.
[5] Letter to the President, dated 26 July 1938, from the R.L. Moore Papers in the AAM.

Mathematical Society to solicit such greetings. At least, I do not care to have such a solicitation made over my signature.'[6] Nor did Moore's opinion of Roosevelt mellow during the war years, as is revealed in a letter from the man who had looked after his insurance needs for many years, R.B. Robbins. The letter was sent to Moore at a time when Roosevelt was seeking re-election for an unprecedented fourth term. Although it provides only one side of the story, the long and painful letter from which brief extracts are given below provides a fascinating glimpse of the passions that engulfed both sides. Robbins wrote:

'My good friend,

One of the members of my office has just told me of your visit this afternoon while I was out of the office. He has said to me that he thinks you were in earnest when you said that if I intended to vote for Mr. Roosevelt for re-election this year, you would place your business, which I have valued very highly through the years, with some other Agent.... Certainly, I am not in complete agreement with Mr. Roosevelt in a good many things that he has done but it is my feeling, which I believe my friends will tolerate and respect, that the most important thing I can contemplate at this time in the disturbed affairs of the whole world is that I have a daughter and granddaughter who will live in some sort of world for the next thirty-five or forty years. (If) ... it is still your feeling that you can no longer entrust the handling of your insurance matters to one who, after feeling his way through as cautiously as he has known how to, has decided that he should vote for Mr. Roosevelt ... I shall mark up another very great disappointment in my life. (But) ... a long time ago, I discovered I have no energy or ill will to spare in feeling hard toward anyone...

Sincerely, Your friend, R.B. Robbins

'PS: At the risk of bad judgment and perhaps poor taste in indulging in a lengthy postscript I am influenced to say that for some reason which I do not fuss about, life has been very hard for me.... I envy my friends who can always be sure that they are right.... It does not seem to me that men like you and I could find it in our hearts to be intolerant of the position and opinions of others. (For example) when the Ku-Klux was born, a number of the prominent outstanding businessmen of Austin came to me and threatened me if I did not join the [Klan]. They stated to me that

---

[6] Moore to Dr. T.C. Fry, Bell Telephone Laboratories, New York City, 27 July 1938, from the R.L. Moore Papers in the AAM.

my business would go to the 'bow-wows' if I did not associate myself with the best people of the community and join the Klan. I replied that I could not endorse a move that would foster the idea of the government being administered by a clandestine group. I further stated that if, after living in Austin for thirty years and trying my dead-level best to earn and retain the confidence and high esteem of my fellow-men, it was all to be dumped to the four winds unless I joined a secret organization, then I should move into another community. I lost some good business at that time, which I very much needed, and yet I would have been dishonest to the best that there is in me, if there is anything good in me at all, if I had aligned myself with any such organization.'[7]

There is no evidence in his papers of a response by Moore, nor is there any indication as to whether he continued to trade with his long-time friend Robbins. There were, however, other issues arising in the chronology of events leading to World War II and beyond that provide further insight into Moore's attitudes and beliefs at the time. They also demonstrate the extremes of his vision, and in particular the ability, as president of the American Mathematical Society, for example, to deal with controversy in statesman-like impartiality while in his personal beliefs he could swing violently towards a firm and conservative viewpoint from which nothing would detract him. An example of the former was to arise during his presidency of the Society when he was, through his office as president, required to oversee a delicate mission to ensure the continuation of international exchanges of research and ideas in the face of the rising tensions immediately prior to World War II. With the mathematical community by then virtually split by the inroads of Fascism into Germany and Italy, Moore formed a committee under the auspices of the American Mathematical Society, which was ultimately to lead to the formation of a journal that is today at the heart of one of the most complete databases of mathematical literature in the world.

Their task was to rescue the international exchange of views among mathematicians across the globe, which had been seriously undermined and tainted by the edicts of Hitler and Mussolini. At the center of this operation was the highly respected reviewing journal *Zentralblatt für Mathematik und ihre Grenzgebiete (Zbl)*, overseen by an editorial board made up of distinguished mathematicians from Europe and America,

---

[7] Letter from R.B. Robbins, the Robbins Company, Austin, 4 May 1944, from the R.L. Moore Papers in the AAM.

and published from Germany by Julius Springer. American representatives on the board included Richard Courant, J.D. Tamarkin, and Oswald Veblen. Others outside Germany included the famed Italian differential geometer T. Levi-Civita, a professor at the University of Rome, who was among the first to be ousted from his university when Mussolini began Italy's own purge against Jews. The journal had grown rapidly in reputation and circulation since its foundation in 1931 under the editorship of its creator, Otto Neugebauer, a professor at the University of Göttingen, and was recognized as an indispensable tool for all mathematicians. Within two years, however, the political landscape had altered dramatically. Neugebauer made no secret of his opposition to the Nazis and in 1934, under the threat of arrest, he accepted a position at the University of Copenhagen from where he continued to edit *Zbl*. Soon afterwards, Springer was required to ban all Russian contributions to the abstracting journal.

At the beginning of 1938, further restrictions were imposed by those monitoring such publications under Dr. Joseph Goebbels' new Ministry of Public Enlightenment and Propaganda, which required that no refugee Jews be allowed to act as reviewers and sought a written guarantee from the publishers that this policy would be adhered to without exception. Levi-Civita was immediately ejected from his consultative role. Neugebauer resigned in protest and several members of the editorial board, including all those from the United States, followed his lead. Neugebauer also destroyed all records in his possession, except for the cumulative index.

Oswald Veblen, realizing the importance of the index, found him a chair at Brown University and along with others persuaded the American Mathematical Society to take the initiative and support Neugebauer. On the strength of this, and the prospect of the formation of a new journal in sight, Neugebauer set sail for the United States, bringing with him the complete index of *Zbl*. There was one last-minute hitch, however. The index was confiscated by US Customs as being potentially subversive.[8] Fortunately, it was saved and survives to this day. Meanwhile, senior figures in the AMS were examining the possibility of replicating the journal he had created and an informal meeting

---

8 Biographical summary of Otto Neugebauer, J.J. O'Connor and E.F. Robertson in the Mac Tutor History of Mathematics Archive, University of St. Andrews, Scotland, on www-history.mcs.st-andrews.ac.uk/history.

R.G. Lubben, C.W. Vickery, F.C. Biesele, ca. 1936

of the AMS Council at Williamsburg agreed that the AMS should consider sponsoring a new abstracting journal in mathematics.

President Moore, acting on the formal recommendation of his Council, appointed a committee to investigate, comprising C.R. Adams, chairman, G.D. Birkhoff, A.B. Coble, T.C. Fry, Marston Morse, and G.T. Whyburn. Tamarkin and Veblen were not considered 'for diplomatic reasons',[9] because they had so recently been on the *Zbl* editorial committee. Initially, the possibility of buying *Zbl* was considered and a meeting was arranged with a representative from Springer. He insisted, however, that the German government required that *Zbl* remain in German ownership. Springer made it clear that *Zbl* would continue 'as an impartial abstract journal on the present high level'. By then, however, the availability of competent reviewers was hugely diminished by racial and political restrictions and the refusal of many other mathematicians outside of Germany to contribute. It has to be said, however, that the latter view was by no means universal, a fact which the AMS committee discovered as it began its deliberations. The possibility was raised that the setting up of a new journal would forge a rift with

---

[9] E. Pitcher, *A History of the Second Fifty Years; American Mathematical Society 1939–1988*; AMS, 1988, p. 70.

German and Italian mathematicians, and damage international goodwill in mathematics when plans were well advanced for the International Congress of 1940.

G.D. Birkhoff led the opposing voices, stating that it would be an 'unfriendly act' to launch a similar journal and pointed out that the reviewing process would fall heavily on the shoulders of young American mathematicians. He suggested an alternative of a monthly listing of papers, carried out at a clerical level, accompanied by reviews of selected papers. His proposal was rejected and the upshot was that R.L. Moore's committee recommended the formation of *Mathematical Reviews* sponsored by the AMS and other mathematical societies, with additional funding from the Carnegie Corporation and the Rockefeller Foundation. Otto Neugebauer and J.D. Tamarkin were appointed joint editors, under the direction of an editorial committee chaired by Veblen, and the journal came into being in January 1940.

The first volume of *Mathematical Reviews* ran to 400 pages and 2120 reviews. The pagination remained constant throughout the war years but thereafter rapidly increased, reaching 2548 pages by the time of its twenty-first anniversary. The following year two volumes appeared, and thereafter further substantial increases in pages and reviews occurred year after year, demonstrating both increase in the amount of published mathematics and the indispensability of *MR* to the mathematical community. It was therefore a creditable aspect that the journal was born as a result of actions taken during R.L. Moore's tenure as president of the AMS, although it must be said that there is no indication of his personal views on a task he performed as the formal hand of confirmation in his presidential role. The speed with which this operation was undertaken, something of a miracle in itself, meant that this essential tool was available during the years when scientists and mathematicians collaborated in an unprecedented manner in support of the defensive and offensive capabilities of the West. The need became urgent and vital when the Japanese joined the two Fascist factions to form the Axis alliance that encompassed formidable military might, backed by some of the finest scientific, mathematical and technical brains in the world.

In this respect, there was a further interesting development that does indeed provide insight into R.L. Moore's attitudes. Having, on the one hand, contributed to the foundation of *MR,* he was to find himself at loggerheads with proposals to enforce the mobilization of the scientific, mathematical and technical communities. Scholars of every discipline were being marshaled on both sides of the Atlantic. Mussolini established a National Council of Research under Guglielmo Marconi in 1936. Hitler assumed that German scientific ability was unbeatable, but had made the mistake of chasing dozens of brilliant minds out of Europe, thereby enriching the brain pool of Britain and the United States. Stalin had dispatched thousands of scientists to the Gulag out of the fear that technical experts might turn to political opposition, and there they were forced into their research. Thus, persecution and constraints on intellectual freedom meant that research and development in these nations fell behind the Free World nations, rather than beating them.

Churchill established a Scientific Advisory Committee under L.A. Lindemann while Roosevelt entrusted the American effort to Vannevar Bush's Office of Scientific Research and Development which gave out contracts of $1,000,000 or more to over 50 universities during the war. And so, whether under duress in the Communist or Fascist states, or as part of a program of planned research and development in the Free World, scientific and mathematical input formed a substantial part of the armory of state power, while at the same time demonstrating that an overpreponderance of state control did not necessarily bring better results.

Given Moore's own opposition to state intervention into free markets in the US, whether labor or business related, it was not surprising to discover that he launched something of a personal campaign against what became known as the Science Mobilization Bill, or S.702.[10] It was introduced in February 1943, by Senator H.M. Kilgore, a member of the Committee on Military Affairs. The principal aims of the Bill were more or less a duplication of Bush's OSRD mandate. In its introduction, Kilgore declared that its purpose was to 'mobilize the scientific and technical resources of the Nation, to establish an Office of Scientific and Technical Mobilization, and for other purposes'.

The latter included the introduction of punitive legal powers to effect the purpose of the Bill, which included the mobilization of scientific and technical facilities, requisition of patents and technical materials that might prove useful to the war effort, and the purchase of the capital stock of any corporation deemed by the USA to be in control of such facilities.

---

[10] Seventy-eighth Congress, First Session, in the Senate of the United States, 11 February, 1943, Mr. Kilgore introduced the Bill S.7092 which was read twice and referred to the Committee on Military Affairs.

Moore obtained a copy of the proposed Bill and, in his usual manner, had underscored those areas that he particularly opposed. He particularly marked a section that stated that no person would be excused from complying with the requirements of the Bill because of his privilege against self-incrimination. The Bill so incensed him that he wrote to Kilgore: 'I am utterly and unqualifiedly opposed to the Science Mobilization Bill. I hope that an Office of Scientific and Technical Mobilization as described therein may never be established in this country.'[11]

Kilgore replied politely thanking Moore for his letter but stated that if he had had the opportunity of attending hearings on the Bill, 'you could hardly doubt the necessity of this legislation'. As the Bill came closer to being adopted, Moore unsuccessfully attempted to enlist the weight of the American Mathematical Society in opposition. In August 1943, he wrote to Professor Arnold Dresden, secretary ad interim: 'On page 151 of the August 13 1943 issue of *Science*, Senator H.M. Kilgore has had the unmitigated effrontery to say, "More than a thousand letters have come to me about S.702. From these letters I find that men of science favor the bill. The vested interests, and those who are influenced or controlled by the vested interests, are against it, and they are most unscientific in their attacks upon it."[12] I propose that the Council of the American Mathematical Society go on record as being unqualifiedly opposed to the Kilgore Science Mobilization Bill (S.702) and that the Senate of the United States should be informed of this action.' Moore's objection seemed pointed towards the mechanics of Kilgore's Bill, that is the controls and legal machinery he included to enforce its objectives. And, of course, mobilization of scientific and mathematical personnel from across the United States did proceed in substantial manner with $3,850,000,000 being spent in the United States alone on research and development, almost half of which went on the Manhattan Project. History now records this great contribution to the war effort, and the implications for the future of mathematical sciences, which grew out of the military requirements of World War II, whichever side they were on.

A number of Moore's students were engaged in that effort and predominant among them was F. Burton Jones who, it will be recalled, had joined the faculty at The University of Texas as an instructor in

---

[11] R.L. Moore to H.M. Kilgore, 17 May 1943; R.L. Moore Papers in the AAM.

[12] From assorted correspondence on S.702, between 17 May and 29 August 1943 in the R.L. Moore Papers in the AAM.

*F. Burton Jones
and R. G. Lubben*

September 1935 immediately after being awarded his doctorate. Over time, Jones's research produced results that had a considerable impact on the development of point set topology and continuum theory, many of them proceeding on from Moore's own work, and giving rise to hundreds of papers in topology and set theory as well. The first of Burton Jones's lifetime total of 67 papers appeared in 1935 on separable spaces, while his third in 1937 on normal spaces became recognized as one of his most important. He also wrote many papers on homogeneous continua.

He had been promoted to Assistant Professor in September 1940 and he and Moore were a formidable duo in the Department of Pure Mathematics. Indeed, it was because of Jones's popularity and his qualities as a teacher that Moore kept a close eye on him, and insisted on having the last word as to which courses Burton Jones could teach. In 1942, Jones left the university to participate in the mobilization of scientific and mathematical personnel. He was co-opted into the Harvard Underwater Sound Laboratory where he helped develop scanning sonar used in anti-submarine warfare. The importance of his contribution can be gauged from a letter received by R.L. Moore in June 1945 from the director of the Underwater Sound Laboratory, Harvard University, by which time Jones had returned to his post at The University of Texas and that very summer had been penciled in by Moore to teach in summer school. The HUSL director wanted Jones back at Harvard for July and August and he wrote to R.L. Moore to request his release:

'I think you know we have a very high regard for Burt's abilities and that we are joined by the Naval officers who are familiar with his work at HUSL in holding in high esteem the contribution he made to our development program. Since Burt left us last fall, several of his colleagues have been vigorously occupied in preparing a full report on the development program, which had been carried out under his direction ... we believe it would be very valuable if its scientific integrity could be further assured by having Burt work over the copy with our editorial staff.'[13]

Moore's view on mobilization had already softened to an appreciation of what had taken place. In a way, some areas of research and development, as in the case of Jones, could be seen as a tribute to himself, in this case that one of his protégés had made such a name for himself in the annals of wartime research and development. It was one more element in the indisputable fact that Moore students were moving ahead on the basis of research work which he instigated and which they pursued and, coincidentally, from which he was beginning to take something of a backseat as his own research work went into decline.

He published only half a dozen papers in the entire decade of the 1930s, although the work was of considerable note. He seemed more intent on providing a springboard for future development by his former and current students. Of particular interest, as R.L. Wilder pointed out in his biographical tribute to Moore,[14] were his papers entitled 'Foundations of a point set theory of spaces in which some points are contiguous to others'[15] and 'On the structure of continua'.[16] Wilder noted that they displayed a system of axioms whose primitive terms, in addition to *point* and *region,* contain the term *contiguous to,* denoting a relation between points. In particular one point can be contiguous to another.

---

[13] Letter from Dr. F.V. Hunt, director Underwater Sound Laboratory, Harvard University, 14 June 1945; R.L. Moore Papers in the AAM.

[14] R.L. Wilder, Robert Lee Moore (1882–1974), *Bull. Amer. Math. Soc.* 82 (1976), 417–427.

[15] Rice Institute Pamphlet 23, (1936), pp. 1–41.

[16] Ibid, pp. 58–74.

A key aspect for introducing this notion appeared to have been related to its application to structural properties of a continuum in terms of specialized subsets. Wilder further noted that it was curious that this material did not create more subsequent research than it did, since certainly the notion of contiguous points should prove fruitful, not only as a mathematical concept, but as a physical notion.

Moore continued his researches into the structure of continua in three further papers published between 1937 and 1943[17], making special use of such concepts as continua of condensation and upper semi-continuous collections of continua.

He had introduced upper semi-continuous collections as far back as 1925, in his paper 'Concerning upper semi-continuous collections of continua'[18] in which it was shown that if such a collection, G, of mutually exclusive bounded continua fills up a plane E^2 and none of its elements separates E^2, then it is itself a plane in terms of the elements of G as "points" and with "limit point" suitably defined. A similar theorem holds true for the 2-sphere, S^2. This theme was continued in 'Concerning upper semi-continuous collections'.[19] Moore showed that if the elements of G are allowed to separate S^2, then the resulting configuration, C is a *cactoid* (a continuous curve whose maximal cyclic elements are 2-spheres). Wilder further stated that in view of the prescribed definition of limit for the elements of an upper semi-continuous collection, these elements are the counterimages of points of C under a monotonic continuous mapping of S^2 onto C. This theorem was not only generalized to 2-manifolds and higher dimensional configurations, but the notion of monotone mapping proved very fruitful in later set-theoretic investigations.[20]

---

[17] Concerning essential continua of condensation, *Trans. Amer. Math. Soc.* 42 (1937), 41–52; Concerning the open subsets of a plane continuum, *Proc. Nat. Acad. Sci. USA* 26, (1940), 24–25. Concerning continua which have dendratomic subsets, *Proc. Nat. Acad. Sci. USA* 29 (1943), 384–389.

[18] *Trans. Amer. Math. Soc.* 27 (1925), 416–428.

[19] *Monatsh. Math. Phy.* 36 (1929), 81–88.

[20] Cf J.H. Roberts and N.E. Steenrod, Monotone transformations of two-dimensional manifolds, *Ann. of Math.* 30 (1938), 851–862; R.L. Wilder, Monotone mappings of manifolds, *Pacific J. Math.* 7 (1957), 1519–1523, and Monotone mappings of manifolds, II, *Michigan Math. J.* 9 (1958), 19–23.

11

Moore the Teacher: A New Era (1939–1944)

The beginning of a new phase of his long career, devoting himself almost entirely to teaching, and those he drew in made a significant contribution to the scientific effort of the war years and beyond in the mathematical community as a whole; it was the beginning of Moore's 'golden age' for discovering mathematical talent.

Moore at his desk

The end of Moore's two-year term as president of the American Mathematical Society in 1939 marked the beginning of what may be seen as the third phase of his long career, one in which he now devoted himself almost entirely to teaching. The decision appears to have been a conscious and deliberate act, having now passed through his period of intense research and the time-consuming duties connected with the American Mathematical Society and working on the mathematics committee of the American Association for the Advancement of Science. These additional demands on his time had taken their toll, especially in the periods immediately prior to and following his presidency of the AMS at one of the busiest and potentially most difficult times in its history. Whereas in the first half of the 1930s, Moore had produced six doctoral students, no other names came through with his signature on their thesis until 1941. That situation was about to change dramatically, but at the same time the prolific nature of his research work continued to decline.

He returned to his university duties with a flourish after all remaining matters from the AMS presidency had been accomplished and from all available records and the recollections of his students from that period, he stood on the brink of what might be termed his golden age. It was the beginning of the most productive period of his life in terms of 'capturing' and guiding a remarkable succession of men and women who in turn went on to make a considerable contribution to the international mathematical community in quite diverse ways. In spite of the obvious problems encountered during the coming years of World War II, the period from 1940 to his retirement saw no fewer than 35 students gain their doctorates under his personal supervision, along with a number of others who came specifically to work with him while studying at other institutions.

Among them were six students who, like Wilder and Whyburn, became president or vice president, or both, of either the American Mathematical Society or the Mathematical Association of America. Two of them, Wilder and R H Bing became president of both those organizations, a third, E.E. Moise became vice president of the AMS and president of the MAA while a fourth, R.D. Anderson, also became vice president of the AMS and president of the MAA. They and their colleagues, apart from the acclaim that came with high office and their respective achievements in mathematics, also became well known for their graduation from the Moore School, which had gained notoriety in universities and educational institutions across the land.

Moore had, at the turn of the new decade, embarked on this final phase: the dedicated quest to produce mathematicians to his own exacting standards. It was a quest that eventually surrounded him with plaudits, devotion from many of his students (although by no means all) and controversy. He resumed his tactic of acquiring likely subjects wherever he could find them, and that search was not necessarily restricted to the halls of The University of Texas. This is immediately highlighted as we travel back to that era through the eyes and recollections in this chapter of two of the future presidential candidates mentioned above, Richard D. Anderson[1] and Gail S. Young[2] who also became, incidentally, close friends (as did many of the Moore students). It is through these oral histories drawn from many of those students who worked under Moore from 1940 onwards, that these remaining chapters now benefit with a considerable input of personal recollections. As will be seen in the ensuing pages, they allow us to explore in depth not only Moore's role as a teacher, but his personal attitudes, opinions and disposition that made him such an intriguing and forceful 'character' while at the same time, taking a cursory glance at the students who came through his courses, men and women who in turn add their own characteristics to this story.

Anderson, one of twin brothers, was born in Connecticut. His father was a psychologist on the faculty at Yale. There were five children, the male twins, then one daughter and then another set of twins, a boy and a girl. The family moved to Minneapolis when Anderson was three and there they remained for a stable and comfortable upbringing in those depressed times. Richard Anderson met Moore quite by chance in the late summer of 1941 when as a 19-year-old student who had just graduated from the University of Minnesota, he attended a mathematics meeting in Chicago:

'I had applied for some assistantships but I hadn't gotten any. I think the reason was that I had graduated in two years and when I applied to

---

[1] Extracts from interviews in this and in ensuing chapters concerning Dr. Anderson with the author in January 2001 and with the late Dr. Ben Fitzpatrick, for The R.L. Moore Oral History Project. Tapes and transcripts of the latter are maintained in the R.L. Moore Legacy Collection in the AAM.

[2] Extracts from a taped interview with Dr. G.S. Young (1915–1999), conducted by Douglas Forbes in 1970 for his PhD gained at the University of Wisconsin, 1971, with a dissertation entitled *The Texas System: R.L. Moore's Original Edition*. Tapes/transcripts of this and other interviews conducted for the dissertation are in the R.L. Moore Legacy Collection in the AAM.

Richard D. Anderson

places like Berkeley, I just never heard from them. They obviously didn't think I was going to graduate. Minnesota couldn't offer me one because my father was on the faculty, and there were nepotism rules at that time. In early September, my father had a meeting in Evanston at the same time as an AMS math meeting in Chicago. I had no particular reasons to go, but anyhow I rode down with him and went to the math meeting. While I was there, I met Moore. He knew one of my professors from Minnesota and essentially on his recommendation, he offered me a job on the spot, as a part-time instructor at Texas.

'At the time, I knew nothing about R.L. Moore. He looked to me like he was sort of a country farmer. He didn't talk about himself, never mentioned his work and certainly made no mention of his achievements. He talked almost exclusively about other members of the faculty, like Dodd, Ettlinger and Porter and gave the impression of being proud to be on the faculty with these people. It was only later that I discovered the truth about Moore's background and the great respect people had for him. Two weeks later I was on the train to Austin, and that sort of changed my life, and I was happy about that. If I'd stayed in Minnesota, I'd have probably ended up as a statistician!

'I found Moore to be an inspirational kind of teacher, a man totally dedicated to his students, more so than any other teacher I've known.

Why? I have my own personal theories as to why a mathematician who had spent much of his life immersed in research should redirect his activities — as he had quite clearly done when I joined him in 1941 — towards his students. He was almost sixty, he had done his work, he'd written his book, a major book, he'd been president of the AMS, some of his students were already high-fliers, members of the National Academy of Sciences and the like. His intuition from where I sat was to build on that with positive results through teaching. Whether he thought of it as some kind of legacy, I don't know. But in a way that's what happened.'

Anderson arrived and, like everyone who came into Moore's vicinity, was soon acquainted with the Method. No textbooks, no previous exposure to the subject matter, no furtive study in libraries, or any other place that might give clues to the resolution of some of the tasks that were soon to confront his new young students. What they embarked upon was this now-established journey of discovery in which they were forced to challenge themselves, seriously and honestly, and stare their ability in the face. Moore would guide them through it in a variety of ways. There were four in the class that Anderson first entered, but one 'bombed out completely'. He was a good student, and eventually gained his degree elsewhere, but the way Moore taught, it all went straight over his head. Anderson went on:

'Also at Austin at that time, although more advanced, were three others who were to gain their PhDs with Moore: Gail Young, Ed Moise and Ed Burgess who all became life-long friends and, coincidentally, Burgess served on the same ship as myself in the Navy during the war — although at different times. I suppose there were a number, maybe as many as 15 students, who could be identified as part of the system at that time. We took classes with Moore, Ettlinger, Vandiver and Dodd.

'Ed Moise was a very bright guy. He had been an undergraduate at Tulane, and was essentially brought to work with R.L. Moore by his brother-in-law, Gail Young who married Ed's sister. Gail was an earlier student, a year or so ahead of Ed in school, and Gail had come over to Texas to work with Moore and to have a half-time instructorship as a senior finishing his degree while working toward a doctorate with Moore. Gail graduated in 1942. Moise, Ed Burgess and myself had to wait until we came back from the war.'

Anderson enlisted in the Navy the day after the Japanese attack on Pearl Harbor, although he was not called up until April 1942. Before he left Austin, Moore gave him an autographed copy of his book. 'I consider that he gave the book to me as a positive endorsement of my potential at Texas and potential in topology. I think it was intended that way. I worked my way through the first chapter. It took me about three months, because I read the theorem and then tried to prove it myself, in typical Moore style.' Anderson also took a companion class with Vandiver who was essentially a research mathematician. During a lifetime at Austin, he produced only five doctoral students, three of them in the 1950s. Anderson recalls that Vandiver, by then a distinguished number theorist, was at the time attempting to teach the way Moore taught. He would pose the question and seek to guide his students towards a result:

'But Vandiver didn't realize that Moore had a very carefully organized structure sequence in his questions, with prompts in between so he didn't just send us off and tell us to see what we could do. He was definitely leading students towards more and more sophisticated thinking, towards research with the goal of developing research mathematicians, people who were really creative.

'Vandiver, on the other hand, would just come in sort of casually and ask things and eventually he gave up on that and went to reading books, chapters from Albert's *Algebra* and from Vandiver's own books. It was not a successful course but particularly interesting in terms of the caliber of the four people who were in it — Gail Young, Ed Moise, myself and Joe Diaz (the latter eventually gaining his PhD under Ettlinger before leaving Texas to go to Brown). There have been very few classes anywhere where the minimum level of recognition given to students, and their eventual recognition in the academic community, could rival that of those four students.'

Gail Young, meanwhile, graduated in the spring of 1942 with a thesis entitled *Concerning the outer boundaries of certain connected domains*. As well as Moore, Burton Jones was among the signers of his PhD before he too left for his service with the Underwater Sound Laboratory at Harvard. Young's recollections for the Forbes Dissertation interviews provide a valuable insight into his time at Texas, and some further crucial elements concerning the Moore Method. What especially becomes apparent is that although Moore had devised his style of teaching and a program of course work that remained much the same over many years, he was nonetheless continually honing the process, finding new tricks to inspire, excite and sometimes derogate.

Whether it be Anderson, Moise, Burgess or whoever, the experience was at the very least demanding and at worst demoralizing as Young

Moore the Teacher: A New Era (1939–1944)

Gail S. Young

recalled: 'I just finished my junior year at Tulane, and I had been working fifty hours a week as a proofreader. I had never heard of R. L. Moore and I had never heard of topology but one of my professors, Buchanan,[3] had been talking to Moore who had asked if he had any good students for Texas. We discussed taking the next year with him. Buchanan talked to Moore, and I was pleased because that meant to me that he would do something and I would get my degree. Well, I never did get a chance to talk to Moore. Buchanan, who himself produced 27 doctoral students and four presidents of the AMS, must have told Moore that I had potential, and that was good enough for him. I was at work one night when I got a phone call from my wife, and she told me that a telegram had arrived from Texas telling me to be there Monday morning. Oh, boy! I went over there, arriving Sunday night and met with Moore the next day.

'Moore very soon realized I had trouble with exact mathematical statements. Every day for about a month he came into class and put two statements down on the board, like continuity and uniform continuity, and asked me, nobody else, if those were the same or different. I got to

---

[3] This person has not been further identified.

where I could tell by his ears which the right answer was, but my brain wouldn't give me that answer, so I'd know I was answering wrong. I was the only one in my class with that trouble, so I got the treatment for a month. It was very effective. You notice it's just you being the idiot, and eventually avoidance reaction sets in.

'After class had been going on for about a month, I asked Moore, "What is the plan for this?" It was a big mystery to me then especially as there were two guys in class who were good and they set a terrific pace. Poor me, I didn't know any of what was going on.... I was worried about logic, and I asked him some questions about some very unsatisfactory answers. I didn't get rebuffed, but he just wasn't going to straighten these things out. He just sat there pleasantly, sort of smiling and saying uh huh, uh huh, but I wanted about a 15-minute lecture, the kind of thing that they have in the beginning of so many pre-calculus books. I wanted a check like that very badly, and Moore wasn't going to provide it. After that I think I went to his office about five times in my time there, three and a half years. The classroom was the place for it and that, of course, is time for the competition. We were all fighting for approval.

'I worked terribly hard and nothing at all happened until nearly Thanksgiving, when Moore stated another theorem.... I went home and tried to prove it. The next day Moore asked around and nobody had it proved. He came to me and I said sort of distantly no, but I had an example whereupon Moore stopped everything. I got up and gave the example, and it worked and Moore said, "That's pretty good, Mr. Young, that's pretty good." I was almost hysterical. He then proceeded to give a history of the question, and this person and that person had worked on the question. He explained that this was Miss Anna Mullikin's example, that she was the first to solve and publish it for her dissertation[4]. There was always recognition like that, when someone had done something bright and he wanted to show them just how bright it was. You'd found out all by yourself something they had done.

'The next year the two fast guys just dropped out; went elsewhere. There were others in the class who were never going to make it and once it became clear they weren't really going to prove theorems, they

---

[4] Anna Mullikin, it will be recalled, was Moore's third successful doctoral student back in 1922, at the University of Pennsylvania, with a dissertation entitled *Certain theorems relating to plane connected point sets*. As an aside, Gail Young added that Moore was 'heartbroken' when she went on to become a high school teacher instead of pressing on with a career in mathematics.

could sit there, they could take the final examination and get a grade on it, and that was okay with Moore. There might be four mediocre students; one of them that year just tuned out altogether. That also was all right with Moore. If, however, you were a dud and didn't know it, and kept on trying to do things, he'd get annoyed especially if you kept going to the board and talked nonsense. He would let it run for quite a long while, give that person a fair try.

'He was incredibly patient, but [if] you were a self-confident idiot, Moore couldn't take it. The year before mine, there was a kid who never got axiom zero straight. That first semester Moore gave him an F, and the second semester he gave him a G, which was a grade designed for people who took the first semester course and not only failed, but weren't allowed to go on to the second half. You couldn't give a lower grade than G. I keep on thinking of one of the more important devices that he had which was the order in which he called on the students: it was his Inverse Order of Likelihood. It became obvious that the one he called on last was usually the one that had it. He gave the duds some sort of formal opportunity for them to say something first, as he did with me when I first got there. Well, it took quite a while for me to be the last person called on because there were others still ahead of me. But then it hit, and I was no longer a dud. It was a real battle getting to that point. There was one guy who hadn't really done any mathematics to speak of, after graduation, and it was very clear to us that Moore thought this was pretty much inexcusable ... he came close to a nervous breakdown and was told by a doctor to stay away from mathematics for a time. He got over it, though, and ended up a professor.

'I would really like to have arranged for a psychological study of each of Moore's students. Impossible, of course, but one thing that I've noticed was that a number of them seemed to have had problems with their own father. Now, here is this incredible father substitute that came along and I've wondered how much emotion came about through something like that.[5]

---

[5] There is evidence that it was still present in the psyches of many of his students years later. When Douglas Forbes was conducting interviews with Moore students for his dissertation, he asked those who had agreed to participate to complete a questionnaire. During his interview with Forbes, Gail Young said 'I don't think you got an answer from Ed Moise, did you?' Forbes confirmed that he did not. Young replied: 'He and I talked about this and he said that he had sat down and written a long thing, and then he realized it was terribly self-revealing, and he wasn't about to send this back.'

In response to Young, Forbes noted that 'Moore instinctively seems to have been able to hone in on those characteristics or personality traits that stood out in any given individual and capitalize on them. If they weren't natively this or that, then he just found something that they were, and I would suspect that he would do best on those characteristics, such as a father image where he was most at home. But, he certainly would capitalize on lots of others: humility and stuff like that.'

Then there were the sub-currents in Moore's classes, which in many ways echoed Moore's experiences with Halsted. His first mentor, it will be recalled, often used to talk in class about everything and anything, and often a whole period would pass without any mathematics being accomplished. He used current situations in life to spark a discussion that his students might later discover impacted on their studies. Moore did much the same, especially when no one had any proofs to present. Young recalled that Moore would fill in by first just talking about general subjects, or even topics that may be in the headlines, or he might concoct examples of possible real-life situations just to get the class started, and then turn them into a discussion. The ulterior purpose was to show a perspective on life tied to mathematics, that mathematics was life and that, according to Gail Young, there was no such thing as a middle position or reasonable position, merely a definitive position. In exposing students to his ideas, his own social bias came bubbling to the surface and much of it was openly rejected by those he was addressing. This was especially so on one of his favorite topics, Roosevelt and the New Dealers,[6] and Moore did not always get an encouraging response, as Young describes:

'I had a very a strong wish at the time that he'd talk about something else because we were all New Dealers. I think he knew where we all stood, but those who did tackle him did not seem to have met much resentment. Some did. Moise, for example, was already committed on

---

[6] When Gail Young arrived at The University of Texas in 1939, the ramifications of Roosevelt's Second New Deal were still a national debating point. Especially controversial among conservatives were the Social Security Act, the Works Progress Administration (WPA), and the Wagner Act on labor relations. The Second New Deal established for the first time a 'safety net' benefit for all Americans, especially the unemployed, disabled and elderly. Between 1935 and 1941, the WPA employed a monthly average of 2.1 million workers on a variety of civil projects as well as natural-resource conservation and artistic and cultural programs such as painting public murals and writing local and regional histories. In addition, Congress passed new tax measures, labeled by its opponents as a 'soak-the-rich' tax that raised tax rates for persons with large incomes and for large corporations.

many social issues of his day, and was a very good speaker who could express his thoughts well. After a while there were a number of things they wouldn't talk about. There was another guy, Walter, who was foolish enough to argue, and Moore didn't mind. He always wanted to argue, particularly with him, but he laid traps. One time he came in and he said he was going to say a lot of bad things about Roosevelt. He wanted to read us the political platforms, and after each one he'd say, "And it seems to me that Roosevelt has done this pretty well, hasn't he?" At the end, when this student was thoroughly hooked, Moore [would demolish the arguments], and just said, "That's the socialist party for you."

'He came into class one day upset, he'd obviously been talking to somebody, and he didn't even ask if there were any theorems. He began right away: *Suppose you were out and you came home and you found a strange man in your house lying on your couch, and you told him to get out, and he wouldn't get out. Would you shoot him?* Complete silence, except for Walter who said, "No. You wouldn't." Then Moore said theorem so and so. The next day he came in again, and he doubled up a little more: *The man had muddy boots, and when you told him to get out he swore at you. Would you shoot him?* Walter still wouldn't shoot him. There was a still stronger version. The man was now drunk and breaking things. Finally, Moore came in and said: *Suppose you're walking down the street, and you pass a round building with one window. You look into that window and you see a mad man...over fifty women and children, and the mad man had a machine gun, and you've got a rifle. Would you shoot him?* Walter said, "No, I'd climb into the window and try to take the machine gun away from him." Moore glared, walked out, slammed the door, and didn't come back that day. His students picked up surprisingly few of his ideas and thoughts in that regard. They seem to have almost out of hand rejected that sort of an attitude but My God, it was a surprising bunch of ideas and at that time, in Austin, Texas, they were by no means as bizarre as they seem today.'

There is little doubt, however, that much of the classroom controversy was a deliberate act on Moore's part to keep the general atmosphere on the boil. The extrovert nature of his opinions became a favorite topic of discussion and gossip in the halls, and from there became lodged in the University's history. No distinction was made between views he held true and strong, as indeed he did with many, and those which he deliberately exaggerated for the purpose of inspiring discussion in class whenever the problem arose of what to do when 'no one has anything'.

left to right: Robert Sorgenfrey, Robert Swain, Bernadine Sorgenfrey, Mary Ruth Coleman, Walter Coleman, Harlan Cross Miller

The latter event was certainly one that taxed future Moore Method instructors as they faced the challenge of filling the gaps without unsettling and distracting their students with new problems. Burton Jones in his paper on the Moore Method recounted three main topics which he relied upon as useful side issues: (1) Problems about set theory, well-ordering and cardinality can be taken up and worked out by the students on the spur of the moment, (2) how some of the theory simplifies if one assumes the space to be metric, (3) the history of some of the ideas and the personalities of some of the people involved. Jones added that he made up questions involving the application of theorems already proved (or even those yet to be proved) which can be settled in ten or fifteen minutes each. 'At such times one can introduce the beginnings of notions, usually in connection with examples, that will be useful on theorems to come. That is, one can somewhat randomly (and out of context) put into the subconscious minds of the students pictures, ideas and notions which will resurface weeks (or even months) later in a proof. There should be no hint as to what the ideas are really for and in fact the student when he later uses one of the notions will have the feeling that he discovered it himself. In this way, as well as the actual statement of applicable lemmas, the proofs of quite difficult theorems can be made accessible.'[7]

---

[7] F. Burton Jones, The Moore Method, *Amer. Math. Monthly* 84 (1977), 273–278.

Burton Jones also identified other issues. Not least among them were the 'duds', as Gail Young described those with a lesser appreciation of the subject matter, along with the timid who were loathe to go to the board to present their proof and finally those who had real difficulty in reading mathematics and were thus often unable to compete in the company in which they found themselves. Students affected by these issues could indeed present problems for the instructor and, if not handled with due diligence, seriously undermine classroom activity, not to mention a student's self-belief. As both Jones and Young point out, dealing with these situations required considerable patience from both the instructor and the more able students. This in turn placed the onus on the instructor to be able to recognize a student with latent talent, perhaps not yet blossoming forth, who might otherwise be inadvertently thrown off course by over-zealous comments or lack of understanding in the early stages.

A proof 'shot down' was a common cause of embarrassment and Jones makes the point that there was a fine line between criticism and ridicule. It was not difficult for the instructor to fall into the latter trap, and certainly not unknown with Moore if he was in a black mood. Mary Ellen Rudin, whom we will encounter more fully in Chapter 13, made an incisive assessment of this aspect of the Moore Method, and in particular his insistence on calling up first the students he thought least likely to have a correct solution. She recalled a class in 1945 when those who had left for war service had just returned: 'I started with R.D. Anderson, R H Bing, Ed Moise and Ed Burgess. Actually, in our group there was another, a sixth whom we killed off right away. He was a very smart guy — I think he went into computer sciences eventually — but he wasn't strong enough to compete with the rest of us. Moore always began with him and then let one of us show him how to solve the problem correctly and boy, did it work out badly for him. It builds your ego to be able to do a problem when someone else can't but it destroys that person's ego. I never liked that feature of Moore's classes. Yet I participated in it.'[8]

Moore is on record as answering this criticism when he described his reaction to a former Army officer whom he always called first because he did not do well and 'because I expected him to make some ridiculous statement'.[9] This went on for some time until, finally, the student began

---

[8] From an interview published in *More Mathematical People,* Donald J. Albers, G.L. Alexanderson, and Constance Reid, Harcourt Brace Jovanovich, 1990, pp. 283–303.

[9] Moore responding to a question in the MAA film *Challenge in the Classroom,* 1966.

to get it right and became one of the best students in class. This, Moore indicated, was justification and in a student's search for a resolution to a problem, he would rarely, if ever, step in to help or point out the obvious, although occasionally he was not averse to some diversionary tactic that might enable a student to have second thoughts. 'We all know how difficult even the obvious is before it becomes obvious,' Jones wrote. 'The instructor must simply be willing to wait for the student's mental chemistry to work. It helps if the instructor feels rewarded when the student does finally see how to put together a few ideas correctly.'

Conversely, it was also true that many Moore students simply would not have accepted help, preferring to stick to the golden rule of working it out for themselves even if that meant going out of the classroom when someone who claimed to have the proof went to the board to make his presentation before they arrived at their own. Staying out was also a problem in itself, because too many students taking that option too often would, naturally, affect their own ability to stay abreast of class progress. In handwritten notes among his papers, Moore addressed the problem, noting that he had occasionally resolved it by telling a student, 'If you don't want to listen to someone else proving a theorem that you have not yet proved then how about trying to be the first one to prove it and if you fail to be then paying the penalty for not being [the first] by staying in and listening to someone else's proof?'[10] In regard to talent, he had a calculus class, a junior-senior class and two summer courses to fish in, and in the words of Gail Young, he remained '...a darn good screener of talent.' Generally he was looking to the long term but some of those students who stayed on in the shadow of the master found difficulty in getting out into the sunshine themselves.

Renke Lubben was a perfect example. He gained his PhD under Moore in 1925 with a thesis that solved the then last remaining problem in the foundations of geometry. He was an independent discoverer of maximal compactifications of completely regular spaces, although priority in publication is assigned to Stone and Čech. He was also National Research Fellow (Göttingen) in 1926–1927. In a faculty that included R.L. Moore, his close colleague Ettlinger, and the genius of Vandiver, Lubben and the rest of the department had to take the crumbs. Burton Jones, in later life, admitted that he had taken rather longer than he

---

[10] Handwritten notes on *Challenge in the Classroom*; R.L. Moore Papers in the AAM.

ought to have done to realize that he should get away, far away from the presence of Moore. He admired him deeply, but there was no room for an identical model, which in some respects Jones sought to become. Moore simply would not allow it, at least on his own turf.

Douglas Forbes, in his interview with Gail Young, reckoned there were several instances where students Moore considered would make good teachers were going down the pure research track before he headed them off. In other words, when he knew that they would be offered a job paying very highly in terms of research for companies, say in physics or some related fields, he would make a subtle but concerted effort to get the students to change course, by telling said student he would make a brilliant teacher. There is ample evidence in his papers to show that he would frequently write to professors in other universities or use the growing network of his own former students now established elsewhere to promote the abilities of particular students.

'I think he probably hoped a lot of his students would use some of his methods,' said Forbes. But teacher or researcher, Moore's students became recognized for their mathematical ability even if, according to some, their base was not broad enough. Gail Young made the point that algebraic topology, for example, was to Moore 'the work of the devil'. R.L. Wilder had interestingly attempted to guide a number of students toward filling this gap in their studies and indeed Young and E.E. Moise both subsequently joined him at Michigan.[11] Yet while Moore, on the one hand, seemingly had great respect for one of the finest protagonists in that field, Solomon Lefschetz, at the same time he spoke in derogatory terms about his work in class. 'His relations with Lefschetz were very odd,' said Young. 'He talked about Lefschetz constantly, very critically, and also poking a certain amount of fun. The first time I met Lefschetz was with Moore at the summer meeting of 1941. From the way Moore talked, I simply assumed that he wouldn't want me to meet him. However, he

---

[11] Young gained his PhD from The University of Texas in 1942 with a thesis entitled *Concerning the outer boundaries of certain connected domains*. He later held appointments at Purdue, Michigan, Tulane, Rochester, Case-Western Reserve, Wyoming, and Columbia. As well as becoming president of the Mathematical Association of America, he won the Distinguished Service Award of the MAA in 1987. He worked with the School Mathematics Study Group and with the Committee on the Undergraduate Program in Mathematics. He directed fourteen PhD students, with one of whom (John Hocking) he wrote the successful textbook *Topology*. Another of his students was the African-American Beauregard Stubblefield, who became a Moore Method descendent having studied at Michigan.

dragged me over to Lefschetz and insisted on introducing me. Moore was always anxious to find out negative things about mathematicians of other schools personally and academically really, I guess. In that sense, he had me go and hear [Norman] Steenrod[12] talk at a meeting. I was to listen to Steenrod and come back and tell him what I thought of him. I came back and told Moore that Steenrod was obviously a dope.' Moore enjoyed the comment but Young, of course, soon revised his assessment, as Moore probably knew he would. He later made a point of confessing his derogatory remarks to Steenrod, along with a fulsome apology! They remained good friends for years afterwards.

As to Moore's own PhD stars of the previous two decades, it is worth recording that the output of some of those who had by now made a name for themselves in both American and European circles was substantial in terms of papers and their general contribution to the mathematical community. J.R. Kline, R.L. Wilder, G.T. Whyburn, J.H. Roberts and F. Burton Jones in particular were by the 1940s among the most significant contributors to the mathematical press and strong supporters of both AMS and MAA activities. Through them, and Moore himself, combatants raised in the style of Moore, directly or indirectly, were on the move.

---

[12] Norman Steenrod was, at the time, the star PhD student under Solomon Lefschetz. He had studied at the University of Miami at Oxford, Ohio in 1927 and later moved to the University of Michigan at Ann Arbor taking courses in physics, philosophy, and economics. He also took one mathematics course, which sealed his future. He took a topology course given by R.L. Wilder and graduated from Ann Arbor in 1932 but did not obtain a fellowship to allow him to undertake research. He took a year out to work on topology problems that Wilder had given him and working on his own, wrote his first paper, which led to several offers of fellowships. He chose Princeton, where Wilder was spending a year working with Lefschetz. He was awarded his PhD and continued to work at Princeton as an instructor before accepting an appointment to the University of Chicago in 1939, returning to the University of Michigan in 1942. He went back to Princeton in 1947 and remained on the faculty for the rest of his career. After his work on point-set topology Steenrod, too, was guided by Wilder toward algebraic topology. He is best known for introducing the Steenrod algebra. His important paper on this topic was published in 1942 in which Steenrod squares were introduced for the first time. He continued to work on homology theories (*Foundations of algebraic topology* written with Samuel Eilenberg and published in 1952). Steenrod received many honors for his major contributions to topology. He gave the American Mathematical Society Colloquium lectures in 1957, joining Wilder, Lefschetz and Moore in his election to the National Academy of Sciences.

12

Blacklisted!
(1943)

Matters that grew into national headlines when The University of Texas was under the microscope over actions by its Board of Regents and local politicians who charged that the Texas faculty was riddled with Communists and homosexuals. An interesting tale of its time, with some outspoken input from Moore.

R.L. Moore during registration at UT

R. L. Moore's record as an active participant in committees or as a joiner of organizations was nothing compared with the work of some of his students, and similarly he had never shown any great enthusiasm to involve himself in faculty business, other than in a cursory fashion. For years, his appearance at faculty meetings was seldom as part of any consortium of professorial views. If and when he did attend, he came to support what usually turned out to be a minority viewpoint — his own. Towards the end of the 1930s, however, his attendance became more frequent when politics, this time much closer to home, drew R.L. Moore into fierce rhetoric in a controversy that eventually captured national headlines. In what developed into one of the most unsavory periods in the history of The University of Texas, bitter infighting developed between the Board of Regents and academia.

The university already had a dire reputation for divisions created by the interference of local politics. At stake this time were a number of vital issues as the Regents attempted to alter tenure arrangements, sack New Dealers and take greater control of faculty and courses. When the then University president Homer P. Rainey resisted, they sacked him. What began as a local dispute roared into one of national importance during this long-running saga, resulting ultimately in the University being censored by the Southern Association of Colleges and Secondary Schools and blacklisted by the American Association of University Professors for nine years. The struggle between the warring factions knocked the real war off the front pages of local newspapers and attracted some colorful reportage at the national level, such as this quotation from the much-respected educator, historian, critic and novelist Bernard De Voto, writing in *Harper's Magazine*:

'Education is no longer education in Texas. The University of Texas can no longer seek the truth, discover the truth or teach the truth. It has been taken over by a dictatorship.... (but) what has happened in Texas has happened to us all. When Texas has lost its freedom we have lost ours. Dr. Rainey has been fighting our war. So have the thousands of Texans who have been roused to support him. It is an excellent thing that this struggle has not been passed off in Texas as a trivial squabble among pedagogues but that the state has been deeply shocked and has come to see what is at stake. They have sounded an alert to the Republic, notifying the rest of us to be on our guard.'[1]

---

[1] Bernard De Voto, The Easy Chair, *Harper's Magazine,* August 1945.

This dramatic call to the nation had its beginnings during Moore's term as president of the AMS. One afternoon in 1937, his long-time colleague Harry Yandell Benedict, having just completed ten years as president of The University of Texas, dropped dead from a heart attack. Thus, the campus newspaper the *Daily Texan* had a special edition on the streets by early evening recounting his long career, dating back to his arrival as a student at The University of Texas in 1889. Benedict, a Texan through and through, was a popular president who knew his people, and in spite of his somewhat lax appearance in clothes that looked as if he had just slept the night in them, he had a fast brain and understated stamina in the face of adversity.

His shoes were temporarily filled by a lesser man, John W. Calhoun, whose first action was, however, a brave one. He gave the faculty a pay raise, the first since the Great Depression. This came as something of a surprise since he had been the tightest comptroller the University had ever known. He responded unequivocally and with a smile to anyone who asked: 'My job was to save the University money. Now that I am president, my job is to spend the University's money.'[2] There were a number of people on the Board of Regents who did not agree with that sentiment and began to nurture the thought of using finance and budgetary matters to establish controls of a different sort within the University. A prime-mover in this regard was W. Lee 'Pappy' O'Daniel, the newly-elected Governor of Texas, a maverick businessman who made his name selling Hillbilly Flour and whose only true claim to fame at that point came from his appearance in radio advertising jingles and broadcasts accompanied by the Light Crust Doughboys.

He toured around with his band to deliver a mix of Hillbilly music and old-time religion and, allegedly at the behest of his radio fans, filed for governor on 1 May 1938. He ran on a ticket of Democrat ideals, such as abolishing capital punishment, raising old-age pension and tax reforms (all secretly written by a public relations lobbyist who would be today known as a merchant of spin) and spectacularly reneged on every one of them. In reality, O'Daniel was a conservative of the Jeffersonian Democrats ilk, who was opposed to Roosevelt's New Deal and highly critical of many of his social policies. Jeffersonian Democrats gained support alongside O'Daniel, and with each passing year their disaffection hardened. In 1940 they were appalled when Roosevelt ran for a

---

[2] Frantz, *The Forty-Acre Follies*, p. 139.

third term (and again four years later when he repeated this outrage), which, they believed, was leading the nation toward dictatorship. They also abhorred his support of labor unions, the New Deal spending programs, and, most of all, the replacement of Vice President John Nance Garner of Texas in 1940 with left-winger Henry A. Wallace of Iowa.

The price-fixing of Texas oil also concerned them, as did the actions of the President's wife Eleanor in her outspoken support for black equality under the law. W. Lee O'Daniel, therefore, had plenty of elements to draw upon and he won again in 1940, still popular for his homebody approach and a religious fervor propounded with masterful radio showmanship and aided by the publicity he achieved after wiring President Roosevelt that he had proof of a 'fifth column' operating in Texas. None was ever found. Nor was any evidence made available to the University following a committee of inquiry led by Congressman Martin Dies into alleged Un-American Activities at The University of Texas. Dies claimed there were Stalinist and Marxist 'cells' operating at the University and a covert investigation by the FBI was also in place, but similarly no evidence was produced.

By then, Homer P. Rainey had been appointed to the presidency. He was firm and to the point: 'If anyone has any evidence, it's about time to produce it,' he told several local reporters at a press conference. 'We've done all we can to turn up [un-American] activities on the campus. It's about time someone quoted the old imperative, "Put Up or Shut Up!"'[3] To amplify his statement, Dr. Rainey distributed photostatic copies of seven letters exchanged between former Congressman Martin Dies and his associates and Major J.R. Parten, former chairman of the Board of Regents, by which the University made a determined yet unsuccessful attempt to secure transcripts of testimony taken in a Dies committee investigation. None was forthcoming.

So Rainey, who was appointed in 1939, was already a target and that was a pity, Joe Frantz wrote, because 'few could have brought to The University of Texas more noticeable assets'. Rainey rose from humble beginnings, on through Austin College in Sherman, Texas, where he had been one of the finest athletes in the college's history, won his PhD at the University of Chicago, was for a short while a professional baseball player, then an ordained Baptist minister. He went on to teach at the University of Oregon and then became president at Bucknell University,

---

[3] *Dallas News,* 3 April 1941.

Pennsylvania. In the meantime, Roosevelt had called him to Washington to become head of the American Youth Commission to help children from poorer families get a better chance in life, especially through education. The latter role, of course, brought him into contact with groups working to improve race relations, which provided easy fodder, later, for those who attacked him. But, nevertheless, here he was in Texas and Governor W. Lee O'Daniel did not like him one bit.

O'Daniel, whose predecessor James Allred was a New Dealer, began to place his own conservative sponsors and supporters in positions of power. These included the Board of Regents of The University of Texas, which consisted of nine people appointed directly by the governor. He packed the Board with hard-liners who carried forward his campaign to 'rid the university of subversives, Communists and homosexuals',[4] enforce tighter controls on budgets and academic life in general, and close down some courses. Chief among this invasion of O'Daniel appointees was Orville Burlington,[5] a tough, no-nonsense lawyer who made headlines when he charged that the New Deal was a Communist plot. Another Regent, D.F. Strickland, proposed a patriotism test for all UT professors and then wrote to Rainey stating that if the abolition of tenure made it more difficult to recruit out-of-state professors, Texas would be better off. The same policy was continued by Governor Coke Stevenson, who replaced O'Daniel when he decided to run for the Senate in a special election in 1941.[6]

The scene was set for an ongoing confrontation that would drag on over the ensuing four years, during which time relations between the

---

[4] Bernard De Voto, The Easy Chair, *Harper's Magazine,* August 1945.

[5] He was a delegate to eight Republican national conventions and a powerful figure at the national level. In the 1952 campaign, he was involved in Republican infighting to weaken support for wartime leader Dwight Eisenhower in the state of Texas in favor of Robert Taft. Later that year, he was charged with three others with fraudulently conspiring to defeat Eisenhower.

[6] O'Daniel beat his leading opponent, New Deal congressman Lyndon Baines Johnson, in the race for the Senate after a flurry of late returns. In office, O'Daniel attempted to force through a number of anti-labor bills. They were all thrown out. He ran for re-election the following year, facing former Governor James Allred who was endorsed by a number of leading Conservatives and newspapers, embarrassed by the antics of O'Daniel. Again claiming to be a supporter of Roosevelt, he hung on to enough rural votes to scrape through. Later, however, he became a leading supporter of the Texas Regulars, formed to attract enough Democratic votes to stop Roosevelt being re-elected for a fourth terms. O'Daniel once again became known for his inflammatory radio rantings, stating that America was being sold out to Communists and opposing moves to desegregate the nation's schools.

R.L. Moore walking down Congress Avenue in Austin

Regents and Rainey became ever more difficult. Several Regents began exercising their resolve to abandon faculty tenure and drop some courses. Social science research funds were also cancelled. Rainey resisted but the Regents continued to go over his head on controversial decisions. By a four-to-two vote, the Board themselves fired three economics instructors for allegedly espousing New Deal policies and a fourth who had attempted to defend federal labor laws at an anti-union meeting in Dallas. Various issues arising from these internecine squabbles in turn inspired many hours of debate at general faculty meetings as traditional areas of employment conditions came under scrutiny. Committees were formed to produce proposals for the faculty's own response to the ongoing crisis and the level of concern and anger among faculty could be measured by the turn-out at meetings and by the worry on the faces of those who attended. R.L. Moore's presence at these meetings added both spectacle and drama to the occasion as he made long and explicit speeches, word perfect and grammatically correct to the letter, and meticulously structured as if delivering a precise mathematical treatise.

Many times, he over-stepped the time limit for speeches from the floor, but no one dared bring that to his attention. His views were those of a man who had studied the problems every which-way and came to his own personal conclusions, based on his personal beliefs and attitudes, regardless of whom he might offend or upset. He could be equally dismissive of the case presented by those in authority or by the colleagues around him. He took no sides in that regard, and was capable of being critical of one or the other in equal measure. During a debate in the general faculty meeting on a committee report on the introduction of new criteria by which qualifications and service of teachers might be judged, an amendment was floated which called for an annual report on the personality, character and ability of each member of faculty. Moore said he would vote for it, but 'only for the purpose of making the report by the committee as ridiculous as possible and thus increasing the prospect of its defeat'.[7]

In later discussion over conditions of tenure that the Regents were trying to abolish, a general faculty committee responded with recommendations, which included secure tenure for any instructor with four years experience. They also sought automatic promotion after a certain period of time. Moore did not like the idea at all and his speeches outlining his feelings at these gatherings apparently took on the aura of high drama. In regard to tenure, Moore spoke quietly but forcefully, and it was not what many in the hall wanted to hear:

'I address myself ... to those of you, if there are any, who have some conception of what a university of the first class really is and who are determined to try to ensure that, at least as far as your own departments are concerned, it will really become one, if it is not even now, and that, if it is now then it will remain that way if it is in your power to have it so. I ask you, are you willing to have the Board of Regents requested to adopt regulations requiring that if an assistant professor, perhaps a man who has just received his PhD degree, is added to your department, then, before you have known him for more than about nine months, you must decide definitely to drop him or decide definitely to keep him in your department for four more years? In my opinion, if this legislation is adopted, its adoption will be one more step towards making this university safe for mediocrity. And there is already entirely too much of that sort of security here.'[8]

---

[7] Minutes of the General Faculty, University of Texas, 9 November 1937, pp. 1323–1324.
[8] Minutes of General Faculty, University of Texas, 20 May 1943, pp. 3163–3164.

Homer P. Rainey was already in the firing line when another spectacularly explosive dispute arose with the Board of Regents. He was accused of allowing perverted literature into the university, referring to the third book in the trilogy *USA* by John Dos Passos. The board ordered the book to be withdrawn immediately and instructed Rainey to fire the professor who placed the volume on the English department's sophomore reading list. One of the Regents said it was 'the filthiest book in the English language'.[9] Another distributed a six-page pamphlet throughout the state to 'demonstrate to Texans the vileness their children are exposed to at The University of Texas'. The pamphlet contained quotations taken out of context but, of course, revealing profanity, allusions of sexual experience, mentions of Communism and praise of Russia. All of these aspects were rolled up into a package that was designed to raise the ire of the masses and bring the University and its faculty into disrepute. Since the selection of the book for university reading had been a committee decision no-one was fired, but Rainey himself now became the focus of an unstoppable witch-hunt.

On 12 October 1944, five days after the Board instructed him not to talk in public about University matters, and especially not about any communications made to him by the Regents, Rainey threw caution to the wind and blew the whole situation wide open. He made a dramatic public statement to a packed general faculty meeting of over 400 people. He was not going to cave in to this unprecedented pressure and listed no less than sixteen charges of repression, actual or attempted, by members of the Board of Regents, collectively or as individuals. He still insisted, however, that the fight was not beyond settlement so long as the Regents recognized, guaranteed and protected essential freedoms in the University. 'That is freedom of thought, freedom of research and investigation, and freedom of expression,' Rainey declared. 'Second they must recognize and observe those legitimate functions of administrative authorities which by all human experience and tradition have been assigned to responsible executives in every type of human organization.'[10]

The Regents refused to even consider Rainey's olive branch. Instead they seized upon it as a heaven-sent opportunity to fire him, which they

---

[9] De Voto, *Harper's Magazine,* August 1945.
[10] Homer Price Rainey, *The Tower and the Dome,* Pruett, Boulder, Colorado, 1971; Frantz, *The Forty-Acre Follies*, pp. 71–87.

did on 1 November 1944 for being too liberal with his administration of the University and his attitude towards racial issues, even though he had publicly stated that integration was not practical for UT at that time. Regent Marguerite Gibson Shearer Fairchild cast the sole vote to keep Rainey. Uproar on the campus followed. University students went on strike, and 8,000 marched from the campus to the Capitol and the Governor's Mansion in the form of a funeral cortege, complete with a coffin representing the death of academic freedom. Ahead of it was Black Baptist Minister and early civil rights leader Oscar Blake Smith in black clerical robes.

The faculty met again and drew up a resolution in support of Rainey, although 132 members publicly dissented, which in turn set colleague against colleague with a bitterness which even death did not resolve. Almost a decade later, one professor was asked if he was attending the funeral of a long-time colleague. 'Hell no,' he replied. 'I'll never forgive him for the way he behaved in the Rainey controversy.'[11] This simmering stew lingered for months and as the board cast its net for a new president, the issue of job security arose among the faculty. The general faculty was asked to submit to the Board of Regents proposals whereby any instructor who had completed four years service should be automatically offered either a commitment for promotion to the next rank, or given a year's notice to find alternative employment. But the nervousness and continuing personal antagonism often verging on hatred, was still evident among the faculty.

R.L. Moore, the loner in all this, remained steadfastly true to his own beliefs and continued to apply mathematical logic to whatever arguments he was about to submit. At a meeting where the new tenure recommendations were to be discussed, Moore rose to his feet amid murmurings of expectation, it having been earlier rumored that he was to contribute to the debate. It was likely he would use the issue of promotion to deliver a stinging attack on certain members of faculty who had helped bring the University into disrepute. The rumormongers were right and they were in for a classic performance, as is evident from this extract, in language and sentiment that was as powerful in a packed room as the right hook delivered when he was 25:

'I do not claim to know how to define a university of the first class, but I wonder whether anyone here would challenge the assertion that no

---

[11] Frantz, *The Forty-Acre Follies*, p. 84.

university is of that class unless, (1) a very substantial amount of really fundamental research of a high order is carried on by members of its faculty, and (2) there are some members of its faculty who are intensely on the alert to discover and develop outstanding research ability on the part of their students and who are both capable of recognizing such ability in the early stages of its manifestation and of developing it when it is discovered.

'If no objection is made to this assertion then I wonder on what ground, if any, one could object to the assertion that if a faculty member carries on no fundamental research worthy of the name and neither discovers nor develops the ability of any of his students to do so, then he does not belong in a university of the first class unless perhaps in some sort of secondary sense. Certainly I think it will be admitted that a university cannot possibly be of the first class if *all* of its faculty members are of this secondary type and it cannot be if *too many* of them are of this type unless the remaining members are of such outstanding character as to make up by very extraordinary quality what they lack in numbers in which latter case, unless their *influence*, as well as the quality of their work, is out of proportion to their numbers, there is grave danger that the university will not maintain its position after they have passed from the scene.

'Now let us consider the [probable] effect of this automatic promotion scheme if it is put into operation here at The University of Texas where we *already* have on the faculty entirely too many people who do not belong in a university of the first class. I ask you to picture to yourself a certain type of hypothetical instructor. His personal qualities ensure that he will get along well with students. He meets and dismisses his classes promptly, grades quiz and examination papers carefully and conscientiously and returns them promptly and makes out reports neatly and gets them in on time. If there is a grade curve, he follows it. He attends all meetings of committees of which he is a member and usually, if not always, votes with the majority. His dean receives no complaints from his students or the parents of his students. He gives freely of his time to students who come to him for assistance or consultation. The chairman of his department receives frequent requests from students who wish to transfer into his classes and none from any who wish to transfer out of them. He is a good example of what is ordinarily thought of as a satisfactory teacher of students of average ability. He has a PhD degree from a leading university where his dissertation was supervised by an able investigator who supplied him with enough partly worked out problems to furnish material for the publication of several papers beyond his thesis.

'But he does not have the ability to continue under his own power when this material is exhausted and after five or six years have gone by he will never produce anything that would be accepted for publication in any reputable journal of national standing, though he may keep up some appearance of activity through the medium of newspapers, university bulletins, and other local or semilocal publications and perhaps the radio. I think that after such a person has been an instructor here for four years it would be a very unusual budget council that would refuse to recommend his promotion if the only alternative were to drop him one year later and I believe that, even of those who would otherwise be inclined to do so, a considerable percentage would be deterred by the thought that his being dropped after a trial period of four years would probably make it difficult, if not impossible, for him to secure a suitable position elsewhere. And if he is promoted to an assistant professorship, does anyone think that four years later, after eight years of service in this University, he will be dropped instead of being promoted to an associate professorship?

'And who, if anyone, thinks that if such a man comes to The University of Texas as an instructor at the age of 26 and is made an associate professor at the age of 34, then it is very likely that he will remain at that rank for the remainder of his life? The chances are, I think, that if this rule is adopted, then by the time he is 42, if not before then, such an individual will be a full professor and as a member of a budget council will vote for the promotion of more people of his own sort and later they will vote for still others and so on....

'I do not believe that this University will ever be of the first class (except with reference to a few departments that somehow manage to maintain high standards in spite of the obstacles imposed by such people) if it is dominated by the ideals of those who are more concerned with *uniformity* of standards and "fair" treatment of the mediocre than they are with the establishment and maintenance of *high* standards and the discovery and fostering of outstanding ability.... [Too many] have been promoted from the rank of instructor all the way up to that of full professor but who clearly do not belong in a university of the first class unless in the secondary sense ... [and] too many of them are likely to support legislation calculated to favor their own interests at the expense of the welfare of the institution and to lead to the propagation of more of their own kind to its still further detriment. I am unqualifiedly against this mechanical scheme regardless of whether it is proposed to put it into effect now or a hundred years from now.'[12]

Moore did not convince the majority with his argument. Too many were still nervous and the divisions had widened into a chasm. The Regents invited a faculty committee to search for a successor for Rainey, and subsequently appointed a member of that committee Theophilus S. Painter, an internationally respected geneticist, to the job. He accepted with some reluctance. Outrage was expressed at a general faculty meeting where a vote to show support for his appointment was won only by a handful. The faculty was almost evenly divided, for and against the new boss, and as one faculty member surmised, 'I see little but turmoil for two or more years and of course, it is the University that will suffer.' Outside in the wider world, the impact and ramifications of the sacking of Rainey were still being felt. The Southern Association of Colleges, the accrediting agency for The University of Texas, held its own investigation and placed UT on probation for a year. The American Association of University Professors announced its censure of the University and did not release it from the blacklist for a further nine years. The American Civil Liberties Union, Phi Beta Kappa and other educational and libertarian organizations reprimanded the University. And then came the devastating article by Bernard De Voto in the August 1945 issue of *Harper's Magazine*. It is, however, as well to bear in mind that its appearance came at a time when the horrors of the German concentration camps were being exposed across Europe, and in the same month America dropped two atomic bombs on Japan. De Voto wrote:

'A group of unscrupulous but very clear-minded men have destroyed The University of Texas as an education institution — destroyed it, at least for so long as they or anyone who represents their point of view remain in control.... The service of the regents is to entrenched wealth, privilege, powerful corporations.... But clearly they could neither have won nor maintained their victory if they had not succeeded in getting the support of many Texans who want no truck with fascism and are not enlisted on the side of privilege.'

De Voto went on to state that many thousands of profoundly troubled Texans honestly believed that the Regents had been defending their

---

12 Minutes of General Faculty, University of Texas, 12 May 1945, pp. 3150–3153.

state from outside domination, that they had struck a triumphant blow for individual freedom and saved Texas from terrible evils. But to an outsider, said De Voto, they seemed regressive and anachronistic, and that they had only reared a wall against modern government, modern thinking, and modern literature. The full article, with the authority of a respected scribe and the impact of his skills as a wordsmith, caused reaction right across America. There was a body of opinion that the dispute had been overblown and that Texas would have handled it in its own way, given time. In that long, hot summer of 1945, when the defeat of the Axis fascists was the main story, the educational difficulties in Austin slipped quietly from the national headlines. But they were never forgotten, nor forgiven, whichever side you were on. Indeed, some of the underlying causes remained in place for years afterward. There is evidence in the R.L. Moore papers that he was seething at the De Voto article, but he had already written his own last word on the subject in a letter to the Secretary General of the American Association of University Professors in Washington, DC:

'In my opinion, if the Council of the American Association of University Professors votes to censure the University of Texas on the ground that academic freedom and tenure do not exist here, it will thereby discredit not the University of Texas, but the American Association of University Professors. I do not know a single instance in the last twenty years in which any board of regents of this University has violated what I consider to be sound principles, either of academic freedom or of tenure. I ask that you inform the Council of your association of the contents of this letter.'[13]

Soon after he wrote that letter, another set of figures, produced nationally, demonstrated that things were not all bad at The University of Texas. The *New York Times* reported the annual assessments of the most influential scientists and mathematicians in America in 1945. Out of the eighty mathematicians listed, three were from The University of Texas, all from the Department of Pure Mathematics: the chairman Milton Brockett Porter, Robert Lee Moore and Harry Shultz Vandiver.

---

[13] R.L. Moore to Ralph E. Hinstead, General Secretary of the AAUP, 23 May 1945; R.L. Moore Papers in the AAM.

13

Class of '45
(1945)

As the war ended, Moore's students included a collection of most talented students who went on to make their own impact on mathematics. They included R H Bing, Mary Ellen Rudin, R.D. Anderson, G.S. Young and E.E. Moise whose personal reminiscences provide fascinating insight, as well as some frank and diverse views on Moore's methods.

Moore at the chalkboard

The rush of students returning to universities across the nation quickly replaced the difficult years of World War II. There were students galore, and not enough teachers. For those fortunate enough to survive the hostilities there was a good deal of catching up to do. In R.L. Moore's classes, this presented some immediate problems in regard to the way he taught. Those who had been yanked out of class for war service, such as R.D. Anderson, E.E. Moise and C.E. Burgess, came back to find themselves among another select band of new Moore students who were showing great promise, as they themselves were before they were so rudely interrupted by Hitler.

Before going on to discuss the fortunes of those returning from war service, mention must therefore be made of those other star pupils by then in place and whose discovery was important to Moore and indeed the whole mathematics community of the United States. First among them was one of the most unlikely Moore students of all: R H Bing, who was 31 years old in 1945, the year he gained his PhD under Moore. Bing was a fascinating character, a brilliant student, a fine teacher, a future contributor of substantial mathematics and another of Moore's students who became president of both the American Mathematical Society and the Mathematical Association of America. He also had a characteristic that set him totally apart. He had no names. He was christened just 'R H'. His father, Rupert Henry Bing, was a teacher who became Superintendent of the Oakwood School District in Texas while his mother Lula May, was a primary school teacher at the same school. After their marriage, they decided to give up teaching and move into farming. But Bing senior died when R H was just five years old, leaving his wife to bring up their two children in impecunious circumstances. She had to return to teaching to make a living, and practiced on the children. R H recalled in later years: 'My mother felt that it was an excellent idea for mothers to train their children at an early age. Long before starting school, I was able to read and do arithmetic and regarded these things as great fun. I think I owe a great deal to my mother's early training for my interest and success in school.'[1]

Bing junior took a job in the college cafeteria to help pay his way through Southwest Texas State University (now called Texas State University) at San Marcos, twenty-five miles south of Austin, and

---

[1] S. Singh, R H Bing: A study of his life, in *Collected Papers* Vol. 1 (Providence, RI, 1988), pp. 3–18.

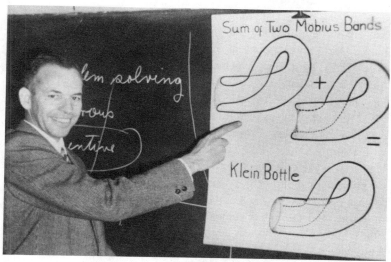

R H Bing

spurred on by the need to earn a living, he completed the course in little over two years. He then taught mathematics for four and a half years at three different high schools in Texas, as well as coaching football teams. For the summer, he enrolled in a mathematics course at The University of Texas with the sole aim of gaining his Master's degree, which he undertook for the simple reason that in order to improve public school education, the Texas state legislature had decided to give a decent pay rise to all teachers who gained their M.Ed.

Bing achieved his ambition in 1938, the year in which he also married Mary Blanche Hobbs whom he had met in the summer school at UT. It was while taking these classes that Bing first encountered R.L. Moore and, like many a summer student, was intrigued by the tales of this famous iconoclast and his renowned method of teaching. However, after obtaining his Master's, Bing took up a position at Palestine High School. The opportunity to study with Moore arrived in the early 1940s when Bing was planning a career move to increase his income and wrote to Moore for a reference.

Moore wrote back suggesting that Bing return to UT in a teaching role, while at the same time continuing his own studies for a PhD. This was unusual, because Moore tended to deprecate older students, and especially Bing who spoke loudly, had a tendency towards boisterousness and was *always* laughing. But clearly, he had decided that the man had a considerable talent that deserved a greater exposure than Bing himself had in mind. Consequently in 1942, Moore secured a teaching position at UT for Bing, which allowed him to earn a salary while continuing his graduate study to work toward a doctorate, as well as trying his hand at research. It was in that year that Bing encountered another new Moore student destined for a bright future, Mary Ellen Estill, product of a middle-class Presbyterian family originally from Tennessee with a strong tradition of education on the female line of both sides of her parentage. Her father Joe brought his family to Texas while engaged on civil engineering projects, and took up residence in the isolated township of Leakey, which at the time could be reached only via a 50-mile dirt road through a canyon, which forded the Frio River seven times. Mary Ellen, who was an only child until the age of ten, attended a tiny high school in a class of five students and eventually won a place at The University of Texas in 1941 with no idea as to the subjects she wanted to study.

She went nervously into the large hall full of students on the day for registration and found herself standing by a table at which sat 'an old white-haired gentleman'. She sat down and they started talking about anything and everything for a long time. Mathematics was barely mentioned. Mary Ellen Estill (and hereafter I shall refer to her by her married name of Rudin by which she became universally known throughout the American mathematical community) signed up for a trigonometry class but was so engrossed in the conversation that she quite forgot to ask the name of the professor who took her signature. When she turned up at class the next day, she was re-acquainted with the dear old man whom she had met at the registration table. He stood there in front of the class to welcome her: R.L. Moore.

It was the beginning of a long association. She would take one of his classes every year until she graduated with her B.A. degree in 1944 and onwards towards her PhD which she gained in 1949. Moore had spotted the talent in her right away, although Mary Ellen had no idea she had been given the 'once over' that sealed her fate. As she was soon to discover, Moore was determined that she would become a mathematician. 'There is no doubt,' she would recall, 'I'm a child of Moore. I was always conscious of being maneuvered by him. I hated being maneuvered. But part of his technique of teaching was to build your ability to withstand pressure from outside.... He built your confidence so that you could do anything. I have that total confidence to this day.'[2] The challenge in the classroom, as Mary Ellen saw it, came as much from the content of the course as from Moore's method of teaching. In an interview with Douglas Forbes for his dissertation[3] in which the merits, or otherwise, of the Moore Method were discussed, she was asked to rate those aspects that she saw as most important: 'I think the challenge of the problems is first. Moore's interest in the course content was certainly a factor. The feeling of really understanding the material would come second [followed by] competition with other students. He certainly showed his pleasure with your proof of a good theorem, and he definitely encouraged students when they began to show some idea of what was going on. There was never any criticism of the method of proof, particularly, no matter how messy and peculiar the proof might be as long as it was logical. Messy proofs if logical, were accepted, but all good proofs were based more on originality of ideas that evolved, and not so much on the correctness of every detail in the steps. The most important, the highest value, was a new idea. He encouraged any sort of positive reaction, even from the poorer student. If they made a poor conjecture, they were probably encouraged to see why it was wrong themselves. I'm sure that he enjoyed a wrong conjecture almost more than a correct conjecture in that it brought out how you can make conjectures. This was a tool.'

Moore made no attempt to segregate his slower students from the fast learners, the high fliers. When he did split a class, it was for reasons that affected the top end of the learning scale rather than the bottom. Rudin candidly recalled her own experiences in graduate school, when she found herself alongside not only R H Bing, but those returning from military service, including R.D. Anderson, E.E. Moise and C.E. Burgess:

'These are all very powerful people, and at the time they were too different in background. He split us up, Bing and Moise in one class, and the other three of us in another class. They had too much information for us. He didn't want us to know the things that Bing and Moise knew, and he split the class for that reason. Bing and Moise were much more sophisticated and much more advanced towards their PhDs [Bing got his that year, 1945, and Moise in 1947].'

---

[2] www-groups.dcs.st-and.ac.uk/-history/Mathematicians/Rudin.html

[3] Interviewed on 7 December 1970, by Douglas R. Forbes for his dissertation *The Texas System: R.L. Moore's Original Edition*; tape/transcript in the R.L. Moore Legacy Collection in the AAM.

E. E. Moise

Rudin found that Moore was continually on the alert for situations where too much information might be imparted by students who had acquired the knowledge by some means or other, and this applied throughout his sequence of classes. It was vital to the system that the knowledge should be self-acquired, and not passed on through any cooperative effort. No such thing existed in Moore classes. Students were forced into getting to grips with their problems through their own devices. It was the only route to success. Rudin went on:

'It was a case of "If you know the answer, shut up!" That was definitely the situation, and you were supposed to keep quiet until you were asked to give the answer. Moore always seemed to know when people had answers. This was again [an example of] his terrific ability as a psychologist. He knew when you walked into the room, almost for sure, what you could do. I'm sure that sometimes he was surprised either that one of the less advanced students had a proof of something or that some good student didn't have, but this always pleased him. To some extent, with his system of calling on the weakest student first, you knew where you stood by where you were called upon, and this again reinforces your position in the group, which is very discouraging to a poor student and very encouraging to a good student. So, this means that it builds the ego of the good student, but might decimate the ego of the poorer student. They're aware of this.

'Some of them were willing to take that, and had a powerful enough already-built ego that they can take this time and again and still fight it. In fact, he liked to have such people in the course and frequently you reached the point where you had very few such students because they simply didn't have what it takes. They couldn't take the competition. He wouldn't necessarily fail them. Grades were sort of irrelevant. The class knew that. Grades didn't have anything to do with it. He generally graded very leniently so that it wasn't the matter of whether you made an A or B, it was whether you succeeded in class. I have a notion that there weren't a lot of failing grades, that most grades were modestly good.

'The administration was not anything to be reckoned with. It was irrelevant so far as he was concerned. The thing that he was doing with grades was strictly a matter between him and the student, and he did not like to discourage even the poor ones if they wanted to continue. The only people he really tried to discourage were people who were much too quick to answer and knew too much, or thought they did.'

In that Class of '45, however, with Bing, Moise, Rudin, Anderson and Burgess — although the latter came late, arriving in the second semester — Moore must have been in his element. Even so, there were still times when no one had anything to show, and as Gail Young indicated in his recollections [in Chapter Eleven], his brother-in-law Ed Moise was still likely to flare-up over Moore's diversions to fill in time. R.D. Anderson had similar memories: 'This was a time when there was pressure on us to come up with something. At least one of us had to get some theorems proved otherwise Moore would start to talk about other things. Any subject. Politics. Newspaper articles. Driving fast to San Antonio. Issues of departmental or University policy.'

Mary Ellen agreed and like the others who shared the experience, accepted in retrospect that this was very much part of the learning curve set up by Moore. But she never did satisfactorily resolve in her own mind whether he was serious or being creatively controversial:

'This was sort of a threat that was held over you. If you didn't have a theorem he would discuss social issues, which would antagonize you. He did so intentionally, knowing that. So did he hold those views? I don't know. It was never entirely obvious to me whether these ideas were things that he had already figured would immediately elicit a negative reaction from everybody present or whether they were just fundamental beliefs. Moise, for example, was already committed to a very different type of understanding of the social issues of his day, and part of this was Moore's anti-Semitism. Moise was a Jew, and was a very good speaker who could express his ideas very well. Moore obviously took pleasure in getting him to speak. He wanted him to argue with him. This gave him pleasure; this was part of the fun of the game. If you didn't have a theorem, you knew [what would happen] and you tried to avoid it like the plague. He felt this was a learning experience for you, too. It was tremendously important what you learned in class, and this is what you proved theorems for, to learn it as you presented it in class. He felt that the presentation in class was a major part of the learning experience, but if you didn't have anything, then you should in a way be punished for this, and this also was part of the learning experience. You should learn to always have something.'

While Moise usually rose strongly to the bait (and the debate that followed), some students might react against certain other aspects of the Moore Method. Rudin, for example, rebelled against the practice of students leaving the room when a proof was about to be presented because they themselves had not yet solved the problem: 'I never left the room, never at any time. I wouldn't do it. I defied him, and that was something he admired in a way, too. I took my punishment and listened to the theorem. I didn't approve of that system, which was clearly built into the process. He didn't ever actually say, "Now, who would like to leave the room, please?" He never presented it as a good thing if you leave the class, but it was presented in a very abstract setting, by saying [something like] a student he'd had ten years ago always left the room if he had not proved the theorem, and you could choose to follow the example of this interesting guy or not. Everything was done in a backhanded kind of way so that you never had the direct attack.

'He teased people with ideas, and how much pleasure you got out of being teased depended upon whether he was teasing you that your idea was a good one or teasing you that your idea was a bad one. There were different ways of handling people, very much so, and the poorer student was always handled with just enough gentleness, shall I say, that he didn't always understand when he was being duped.'

Equally important to the mathematical scenarios and mind games being set up by Moore was the fact that they had to be confined, totally

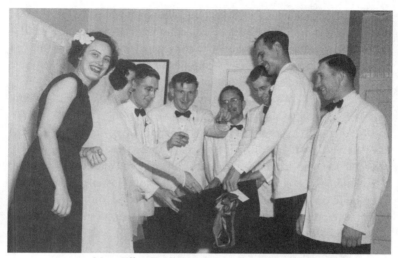

*Mary Ellen Estill, Lida Barrett, John Barrett
and others at Lida and John's wedding reception*

and utterly, to a classroom setting. He would chat in the halls, of course, but not about that day's work, or any specifics. Nor was there any social intercourse, in which Moore might lead out-of-class discussions or off-campus activity. It simply did not happen. Rarely would he engage in a discussion about a theorem or any other issues which he considered should be dealt with in class and. subsequently, Rudin reckoned she visited his offices no more than a couple of times a year.

She also made the same discovery as many of her predecessors that visits to his home or his office were limited. They happened only on rare occasions, in her case something like once every two years. Students could, of course, go to his office if they had anything specific to discuss, such as what courses to take in the future or even matters of a personal nature, in terms of attire, habits, even musical tastes. He was also quite particular about student interaction once they were out of class. Indeed, it went much further. He tried his best to exercise a level of control over student activity outside class to ensure that there was no conferring over difficult matters, as R.D. Anderson[4] discovered: 'There was a fair amount of social life among graduate students, of course, and their spouses. I played chess with Bing but Moore did not think Bing should divert my

---

[4] R.D. Anderson interviewed by Ben Fitzpatrick Jr. for the R.L. Moore Oral History Project, 17 April 1998, R.L. Moore Legacy Collection in the AAM.

attention from mathematics by playing chess with him. So he, I think, discouraged Bing from playing. He liked to keep a tight control on talking outside the class. He would not permit people to say, "I have proved a theorem" or "I've disproved" the conjecture. He wanted people to say, "I've settled it" so that they did not reveal to other students which way you had taken it. We had some occasional social contact with Moore, chiefly Sunday afternoons.

'We might go over to the house with our children, depending on how late that was. We liked Mrs. Moore and, of course, you're in the presence of somebody who is a senior professor, someone you respect. It was never a free and easy, casual acquaintance, but it was enjoyable. I don't think my wife liked it. Of course, we also had the Joneses. Burton Jones was a highly supportive person, individually and otherwise in terms of the students who were currently taking courses with Moore. He had been through the process, and he was, always at Texas and later even after he'd left, the kind of person to whom Moore students could talk freely and easily, and get advice. I've always felt that one very good service he gave was that of being an effective communications bridge between people trained in the Texas tradition and those with more traditional education.

'Bing was the same, to some extent. I developed a close association with Bing while there and later after he went to Wisconsin and at math meetings and so on. My wife and I even went once on a trip to India in the early '70s for a conference and we did some traveling around India with the Bings. This contact between Moore students and former students was, I believe, important and helpful because it provided a kind of sounding board and an interchange of views between former Moore students that was otherwise not available when you were around Moore.

'I'm not certain that the rigidity with which Moore dealt with this interrelationship outside the classroom was entirely justified. Much of what Moore did when he had a group of people in class was go through a kind of group learning exercise with Moore, the teacher, in control. I know that Moore did not believe in cooperative learning[5] but in some sense, his classes involved Moore-guided interaction among students. It wasn't that the students worked cooperatively, definitely not, but the

---

[5] The cooperative (or group) learning 'movement' is credited to Neil Davidson who was a student of Bing at Wisconsin. According to Dancis it was intentionally developed as a variant of the Moore Method, in effect departing from one of its key tenets. See www.discovery.utexas.edu/rlm/reference/dancis_davidson.html.

overall learning experience was very similar to that now thought of as cooperative learning. It was a different motivation, a different format, of course, but I think much of Moore's teaching techniques can be done formally using cooperative learning, and effectively using that as an alternative from the highly competitive nature, which he generated in class.

'My own view is that Moore sort of got hung up on the form as distinct from the substance of what worked for him. He thought that no talking, no exchange of ideas outside of class, was the substance of it. I think that the substance of his successful teaching was largely his interest in the development of ideas, intellectual ideas, mathematical ideas, on the part of his students. His reward system clearly encouraged students, both competitive students and some of the weaker ones, to have confidence they could do things. A large part of his success was in building up the confidence in the individual, and not in the strictly competitive nature of it. The whole spirit of constructivism was part and parcel of what Moore was doing. So in that sense he was a kind of prophet. It wasn't his teaching that made other people change their teaching, in the sense there's a lot of effort now to change the way kids learn elementary mathematics. His approach to doing research mathematics was very compatible with kids learning to solve problems on their own, which is one of the major emphases of the reform movement, but I do not think Moore viewed it that way at all. He was committed to research and committed to the success of his students. Yet, essentially, he never lectured. He led students towards their success by building up their egos.

'When you proved a theorem in his topology class, Moore might make a statement like this: "That's only the third time I have seen anybody prove that theorem in that class and the other two people who proved it are members of the National Academy," or maybe he would say it's only been proved twice and one of those was Burton Jones and one was John Roberts, people whom you thoroughly respected because they were ex-Moore students and Moore would tell you about them. Now that certainly boosts your self-esteem and the belief that you can do mathematics.'

This was certainly the case in regard to that class of '45, as one by one they began to peel off and go their separate ways, although all remained in contact throughout their lives under the umbrella of what Anderson described as the brotherhood of Moore students. There was one other important point Anderson made: that when they were ready to go out into the world, Moore went to great lengths to help his students secure good positions. He was also very concerned with the stature of the places where his students sought employment. Anderson reckoned that Moore had, in his own mind, at least, a mental ranking of institutions and was conscious of these. He was especially happy if the departing student found a place with another of Moore's ex-students. Bing was the first of the wartime group to depart having come through what was undoubtedly a life-changing development of his mathematical talents and inspiration towards what became a renowned teaching ability. It will be recalled that he had no real thoughts of a career in mathematics when he came to UT to study for his M.A.

Bing's key motivating factor was to secure a pay rise in his high school teaching position. That was before he met Moore, and instead of going back for higher pay, selected a lesser income to study for his PhD. The transformation had been remarkable. He completed a thesis entitled *Concerning Simple Plane Webs* and long before his time came to leave, he was casting his net for university positions. A number of possibilities arose and among those who went as far as to seek a reference from R.L. Moore was the chair of mathematics at UCLA, Professor Paul Daus. Moore replied on 18 October 1944[6]:

'I am expecting him to receive a PhD here next June. I am not sure that I ever had a student who, before taking his PhD degree, showed more promise as a research man than does Mr. Bing. This is a pretty strong statement in view of the fact that G.T. Whyburn, W.M. Whyburn, R.L. Wilder, J.H. Roberts, W.L. Ayres, W.T. Reid and others have been students here. Furthermore I think that he is one of the best teachers in our department, that he is genuinely interested in teaching and that he would welcome the opportunity to show that he can lead others in research work.... He has submitted, as a thesis, a long manuscript ... I don't think it at all possible that this piece of work could have been done by anyone who did not have [a] really rare research ability.' Nor was this recommendation lightly given. An unofficial rating scheme sometimes used by Moore and his colleagues went something like this: 'You could expect a student with Brown's talents and abilities

---

6 R.L. Moore to Professor Paul H. Daus, 18 October 1944, from the R.L. Moore Papers in the AAM.

every year; you could expect a student with Lewis's talents and abilities once every four years; but a student with Smith's talents and abilities came along only once in 12 years.'

Bing's talents and abilities threw him in the 12-year class, or in an even higher class, since he became one of the most distinguished mathematicians ever to have received his degree from The University of Texas at Austin. Several of Moore's later graduate students have written that in the days after Bing, Moore used to judge his students by comparing them with Bing, probably not to their advantage.[7] Bing received his PhD degree in May 1945, and in June 1945, he proved a famous, long-standing problem of the day known as the Kline Sphere Characterization Problem.

The news was treated with skepticism in some quarters. It was Moore's policy to cease to review the work of his students after they finished their degrees. Moore believed that such a review might tend to show a lack of confidence in their ability to check the work themselves. When a famous professor telegraphed Moore asking whether any first-class mathematician had checked the proof, Moore replied, 'Yes, Bing has.'

In any event, Moore made it clear that while acknowledging the value of the Kline Sphere Characterization Problem, which was a much better known topic, he felt that Bing's work on planar webs demonstrated his true mathematical ability. Apart from the interest of UCLA and other universities, Bing was now formally offered positions at Princeton and at the University of Wisconsin, Madison. He chose to go to the latter, joining the faculty there in 1946. Some suspected that he did not wish to play second or third fiddle to Lefschetz at Princeton, nor did he enjoy the prospect of serving under another topological giant who, like Moore, dominated his home turf. He was to remain at Wisconsin for 26 years except for leaves. He returned to The University of Texas at Austin in 1973, by which time Moore had finally retired. It was, however, at Wisconsin that his most important mathematical work was undertaken to bring him international repute. He began with research concerned with general topology and con-

---

[7] R H Bing, Memorial Resolution, Documents and Minutes of the General Faculty, The University of Texas at Austin, pp. 19185–19202a, prepared by a special committee consisting of Professors Michael Starbird (Chairman), William T. Eaton, Cameron Gordon, and Robert Greenwood with the assistance of Professor S. Singh, Southwest Texas State University, 1986.

tinua theory. He proved theorems about continua that were to remain central to the field, among them his characterization of the pseudo arc as a homogeneous indecomposable, chainable continuum. The period from 1950 until the mid-60's was Bing's most productive period of research. He published 115 papers in his lifetime, most during this period at the University of Wisconsin. His research success brought him many honors, awards, and responsibilities. These included the presidency of both the Mathematical Association of America and the American Mathematical Society and election to the National Academy of Sciences.[8]

Like so many other aspects of work for which Moore students became well known, Bing's interest in the pseudo arc began in a Moore class. An example is recalled by R.D. Anderson, concerning work being done at the time by Edwin 'Ed' Moise, who was in the final stages of his PhD thesis. Moise had also created a good degree of interest in what Anderson described as a 'beautiful characterization of the pseudo arc'. Bing came to discuss it with him. Moise was convinced at the time that the pseudo arc wasn't homogeneous. Bing became very interested, and a few weeks later he came back to Moise and began picking up on what Moise had done, eventually producing his characterization of the pseudo arc as a homogeneous indecomposable, chainable continuum. This result contradicts most people's intuition about the pseudo arc, including, at the time, Moise's, and directly contradicted a published, but erroneous, 'proof' to the contrary. 'What I found particularly interesting at

---

[8] Bing was quickly promoted through the ranks at the University of Wisconsin, becoming a Rudoph E. Langer Research Professor there in 1964. He served as a Visiting Lecturer of the Mathematical Association of America (1952–53, 1961–62) and the Hedrick Lecturer for the Mathematical Association of America (1961). He chaired the Wisconsin Mathematics Department from 1958 to 1960, but administrative work was not his favorite. He was President of the Mathematical Association of America (1963–1964). In 1965, he was elected to membership in the National Academy of Sciences. He was Chairman of the Conference Board of Mathematical Sciences (1966–1967) and a U.S. Delegate to the International Mathematical Union (1966, 1978). Bing was on the President's Committee on the National Medal of Science (1966–1967, 1974–1976), Chairman of the Division of Mathematics of the National Research Council (1967–1969), Member of the National Science Board (1968–1975), Chairman of the Mathematics Section of the National Academy of Sciences (1970–1973), on the Council of the National Academy of Sciences (1977–80), and on the Governing Board of the National Research Council (1977–1980). He was a Colloquium Lecturer of the American Mathematical Society in 1970. In 1974 Bing received the Distinguished Service to Mathematics Award from the Mathematical Association of America. He became President of the American Mathematical Society in 1977–1978. He retired from UT in 1985, having lectured in more than 200 colleges and universities in 49 states and 17 countries.

the time was this interchange of ideas between two young and really impressive geometers,' Anderson noted.

Moise then followed Bing in leaving UT having secured his doctorate with a thesis entitled *An indecomposable continuum which is homeomorphic to each of its nondegenerate subcontinua*. He obtained an immediate position at the University of Michigan, where he joined his brother-in-law Gail Young who, since he left Texas in 1942, had first taken a post at Purdue before moving on. Almost immediately, Moise began to attract recognition for his work on 3-manifolds which culminated in his proof, completed at the Institute for Advanced Study at Princeton, that every 3-manifold can be triangulated. He subsequently went to Harvard as a James B. Conant Professor and like everyone in Moore's Class of '45, held high office, first as vice president of the AMS and later as president of the MAA.

R.D. Anderson was next to be dispatched from Texas. In the fall of 1947, Professor J.R. Kline was looking for a good man to join his department at the University of Pennsylvania where Moore himself, of course, had originally been given the freedom to experiment fully with his method of teaching. Kline had tried to hire both Bing and Moise and had later written to Moore to inquire if he had any other up-and-coming young students whom he might take on: 'I am anxious to get men with the mathematical power shown by these two latest members of your long group of fine students.'[9]

Moore did not reply to the letter for almost seven months, when he wrote back to recommend Anderson, demonstrating that he 'has shown that if M is a compact continuous curve in a metric space there exists, in Euclidean space of three dimensions, a compact continuous curve K such that there is an upper semicontinuous collection of mutually exclusive continua which fills up K and is, with respect to its elements regarded as points, topologically equivalent to M. I regard this as a really fine piece of work requiring unusual ability and imagination of a high order.... I feel that where Mr. Anderson locates may have to do with his further progress and I hope you may have a place for him at Pennsylvania. PS I don't like this picture of him very much.'[10] Bad photograph or not, J.R. Kline immediately took to the personable young man and Anderson was to remain at Penn for eight busy years, encom-

---

[9] J.R. Kline to R.L. Moore, 10 October 1947; R.L. Moore Papers in the AAM.

[10] Moore to J.R. Kline, 7 May 1948, copy of letter in the R.L. Moore Papers in the AAM.

passing his deep interest in teaching on the one hand and ongoing research work which initially centered on the geometric topology of continua. In later years, he was largely responsible for the development of infinite-dimensional topology, along with his students to whom, like Moore, he always gave credit and recognition. He also tells an interesting anecdote which occurred some eight years after his move to Penn, when the opportunity arose to join the faculty at Louisiana State University. It was an interesting offer, with an encouraging hike in salary. His wife had never entirely settled in the east, and coming from Tennessee herself, and with four young children and a fifth on the way, was happy to move south. On learning of this intended move Moore wrote Anderson a letter, which he subsequently treasured, concerning his move to LSU. It said simply: 'If it is true, commiserations. If it is not true, congratulations!'[11]

LSU at that time did not fall into Moore's category of places of higher learning that suited his students. Anderson saw it quite differently, principally as a challenge to himself to help put mathematics back into a respectable place at LSU. Over time, this was achieved and with it came numerous recognitions of Anderson's contribution to LSU and the mathematical community at large.[12] It is also worth noting that when Anderson concluded his graduate studies at Texas, moves were being discussed to restrict the use of graduate students as full-time instructors, a system that Moore had utilized to the full for the benefit of his doctor-

---

[11] R.L. Moore to R.D. Anderson, 29 February 1956; R.L. Moore Papers in the AAM.

[12] R.D. Anderson became a strong participant in the affairs of American mathematics in general but especially those of the Mathematical Association of America, emphasizing the teaching component of the higher education community, over a period spanning four decades. He was noted for his committee work for both the MAA and the AMS, the research component, and became something of a beacon for cooperation between the two organizations whose joint activities he supported as being complementary in spite of their separate focus. He served as vice president of the AMS and president of the MAA. He held appointments at the Institute for Advanced Study and received the MAA's Distinguished Service Award. As these words are being written, he remains an active contributor to mathematical meetings and in his retirement, devotes much time and energy towards reforms in American education in mathematics and the sciences, both at high school and university levels. His own list of student successes is evident from the distinguished careers achieved by a number of his students, ten of whom received their PhDs at LSU under his direction. At Penn, he also contributed to the supervision of Lida Barrett in her PhD studies, which she began with Moore at Texas. She moved to Penn when her husband accepted a position at the University of Delaware. She went on to a distinguished career, which like Anderson included the presidency of the MAA.

al students thus far. As we have seen, many of his students were able to continue on with his courses by virtue of their being hired as instructors, usually teaching four courses in each academic year while simultaneously pursuing their own degree work. In 1946, it was proposed that this should end and that graduate students should be placed on halftime, teaching two courses and reducing their salary accordingly.

When the Faculty Council met to consider the proposal, Moore was there to make a stand and he did so with what many regarded as an unwarranted attack on some of his faculty colleagues. But as we will see in the next chapter, Moore was never one to worry about his popularity rating, at faculty meetings or elsewhere, nor indeed did he go out of his way to engage in unnecessary discourse with his fellows without real purpose. There was, as one of his former colleagues once put it, little point in saying 'Good morning' unless you were prepared to back up that statement with an accurate description of what, exactly, was good about it!

14
Clash of the Titans (1944–1950)

Intriguing developments arising from an ongoing internal feud between Moore and colleague Harry S. Vandiver, best known for his work on Fermat's Last Theorem, resulting in their never exchanging a single word again; far reaching implications; also a closer examination of Moore's 'ruthless' exclusion of students whom he either disliked or considered had 'too much information' to participate in his courses.

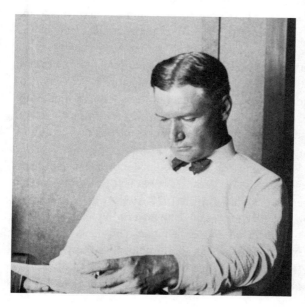

R.L. Moore

The post-war review of faculty requirements in the Department of Pure Mathematics at The University of Texas concerned an issue that had been bubbling uncomfortably in the background for some years. It finally came to a head in 1946, resulting in Professor Harry Schultz Vandiver, whose own research work had become internationally recognized, transferring out of Pure Mathematics and into Applied Mathematics in which he had little interest. But the cause of this curious move had nothing to do with a change in his allegiance in terms of his work, but simply that the ongoing feud between himself and R.L. Moore had finally made his position untenable.

The two men were giants in their profession and dedicated to their work, but unable to settle personal differences, the cause of which few could even remember. At that point, they had worked in the same department for 24 years. Early correspondence between the two indicated an amiable exchange of views and opinions and there were notes referring to generally humorous inside matters of the sort that two colleagues, who at least got along, might exchange. This fell apart sometime in the late 1930s and soon they had reached the point of no contact. Their joint efforts to avoid ever having to converse with each other, let alone being in the same room if at all possible, became the talk of the university. The animosity between them was the butt of anecdotes and jokes that ultimately became an embarrassment to both men, and the University itself. The unseemly rift seems to have arisen over a culmination of events and attitudes. Some identified the original arguments as arising from Vandiver's 'wanderlust' tendency which led him to accept appointments elsewhere and thus considerably reduced his contribution to the overall teaching effort in the Pure Mathematics Department of The University of Texas. Even the committee that drew up an In Memoriam resolution for the faculty council after his death in 1973 wrote that following his arrival in Texas in 1924, there were 'those who bemoan the fact that he did not *stay*'. They continued:

'He was always receiving grants and senior fellowships, and therefore moving about in an age when most people tended to stay put. He received a Guggenheim Fellowship in 1927–1928, was a lecturer at Princeton in 1934, and a lecturer at the University of Indiana and Notre Dame in 1947. He received numerous research grants from The University of Texas, the American Philosophical Society, and the National Science Foundation.... One of the coveted prizes offered by the American Mathematical Society is the Frank Nelson Cole Prize in the Theory of Numbers. Vandiver was awarded this prize in 1931. Also he was elected a member of the National Academy of Sciences, and thus was entitled to put the initials M.N.A.S. after his name in the faculty directories.... In the thirties, The University of Texas had a new distinguished professorship, with a new distinguished professor named each year. The holder was relieved of half of his teaching duties, and was expected to give a series of lectures each spring. His series of lectures [in 1953] was on ... Fermat's Last Theorem (or conjecture).'[1]

In addition, Vandiver had been hugely productive in the publication of authoritative expository, historical and survey articles. But as Ben Fitzpatrick states: Vandiver 'once told me that he was the only distinguished professor of applied mathematics and astronomy in the world who knew not a damn thing about either one.'[2] He only achieved his first degree by default when, in 1946, the University of Pennsylvania where he had taken some graduate courses in 1905 conferred an Honorary Doctorate of Science upon him. This, some believed, was anathema to Moore, first that Vandiver should receive such an honor and second that it came from a place that was one of his own beloved places of higher education.

The fault-line in the Moore-Vandiver relationship, however, went much deeper, involving practices within the mathematics department of The University of Texas that many of his colleagues, Vandiver among them, in both Pure and Applied Mathematics believed were unfairly giving favor to Moore students. The issue came to a head at the end of World War II when several UT professors began openly complaining that financial support made available to students of Moore and his senior colleague in Pure Mathematics, Ettlinger (in the way of teaching appointments), was denied to students who chose to work under other professors in the same department, namely Dodd, Beckenbach, Lubben, Batchelder and Vandiver. It was seemingly over this issue and Moore's predilection towards vying for graduate students, regardless of the designs of his colleagues that led to the 'fisticuffs' between himself and Associate Professor Edwin Beckenbach in 1944. One of the most

---

[1] *Harry Schultz Vandiver (1882–1973), Memorial Resolution*, Documents and Minutes of the General Faculty, pp. 10926–10940, prepared by R.E. Greenwood, Anne Barnes, Roger Osborn and Milo Weaver, University of Texas, 1973.

[2] Ben Fitzpatrick in an interview with Frank Vandiver for the R.L. Moore Oral History Project, June 1999; R.L. Moore Legacy Collection in the AAM.

Harry Schultz Vandiver

informative insights into this curious state of affairs comes from Vandiver's son, Frank, who provides remembrances of the feud. Frank Vandiver was based for most of his first 20 years in Austin, although it was an erratic upbringing in terms of his home life. His early education before graduate school was through private tutoring because his father did not think highly of the general quality of public schools. He traveled often as his father went on his jaunts to other universities, including Europe. Once, when Vandiver was to become a visiting lecturer at Göttingen in 1934, Frank and his mother were dropped off in England because Vandiver didn't want them to go into Germany. He was afraid that it wasn't safe because of the rise of the Nazi party.

Frank's classes at The University of Texas were in history and philosophy. Martin Ettlinger, the son of his father's colleague Professor H.J. Ettlinger, was one of his best friends, although as he pointed out the wanderlust of his father meant they never had a permanent home during his childhood, or at least never as a family of owner-occupiers. They lived in rented accommodation or hotels, the latter for quite extended periods. For a time, when Vandiver was at Princeton, they lived next door to Einstein and became close friends. Frank has good memories of those days but could not identify either the time or reason for the animosity between his father and R.L. Moore. He remembered some of the associated anecdotes, such as the day he was getting his mail, and a colleague, Walter Prescott Webb, looking out the window saw the flag was at half-mast. He said, 'Harry, why is the flag at half mast?' Vandiver senior didn't even look up but replied, 'It's never for the right one.'

There was also an incident involving Frank Vandiver when Moore produced a handgun. It was well known within the university that Moore owned at least one revolver and had been involved with target shooting and gun club activities. His papers include a receipt for an Officer's Colt pistol bought through the University's cooperative society in 1934 for $34. There was also a pamphlet entitled 'How to organize and conduct a gun club' and it was a subject that he regularly introduced during his discussion in class. He was a vociferous defender of the right to bear arms.

Frank Vandiver recalled[3] the aspect of Moore's interest that concerned himself and confirmed rumours that Moore had pointed a pistol at him. 'That really happened. I was on the way home from school, walking home from school one day. I don't know what year or where I was in school, but I was walking home carrying my books and this car pulled up by me on the curb, and Dr. Moore was in it. I thought he was going to offer me a ride home which I was willing happily to accept. Instead of that, he pointed this pistol at me, and said, "Ah ha, what do you think of this?" I was absolutely terrified. I don't remember what I said. Moore pointed it at me, and I realized that Moore and Daddy were not friends, and I had the feeling that maybe he was going to kill me, but I think it was sort of a grim joke he was playing. The gun was loaded, that I could tell, so I was not enamored of that moment. I went quickly home and told my parents.'

As it happened, the Moores and the Vandivers were due at a dinner party with the Ettlingers later that same day. The whole atmosphere was a little chilly, as Frank Vandiver recalls: 'Mother was not good at keeping her emotions under control. She would have let them know what she thought. The animosity between Moore and my father had been going on for some time, in fact I can't remember a time when there wasn't animosity. Daddy never said anything about it, but Mother would let me know that they were not friends, and that we were not associating with them.' By the mid-1940s, however, Vandiver was tiring of this situation

---

[3] Frank Vandiver, reminiscences of his father, quoted from a tape-recorded interview conducted by Ben Fitzpatrick Jr. and Albert C. Lewis, 30 June 1999 for the R.L. Moore Oral History Project, tapes now lodged in the R.L. Moore Legacy Collection in the AAM.

at a time when his own intense concentration on mathematical research, in addition to completing a prodigious array of reviewing and refereeing assignments from journal editors around the world, threatened to bring on a nervous breakdown. Quite often, said Frank Vandiver, he might check into a hotel by himself to do work and he kept a permanently packed suitcase in his office if the need to do so arose. At home, it was not uncommon for him to lock himself into his room while engaged on some research project. 'Mother, who understood all this, would just leave a tray outside the door, wouldn't even bother him. For two and three days at a time he would be incommunicado working on some paper, I guess, some theorem. And we all walked very gingerly around the house so as not to disturb him.'

Eventually, however, he began to shows signs of a weakening resolve, which subsequently required him to cut back on his workload. But it was the ongoing antagonism between himself and Moore that, in the end, compelled him to threaten to quit. He was told that his resignation would not be accepted[4] and instead a compromise solution was arrived at, which meant he was transferred immediately to Applied Mathematics. Nor did this resolve the situation because, in any event, by the end of that year of 1946, plans were already being mooted to merge the two departments, which infuriated Moore to an even greater degree for reasons that had nothing to do with Vandiver. So here, and for years afterwards, The University of Texas harbored two of the most outstanding mathematicians in the United States, of great international repute, who still diligently avoided any contact with each other whatsoever and studiously and meticulously plotted their movements, such as the time of their arrival and the entrance they used to enter their building, so that they did not even have to endure eye contact. The divergence in their objectives was also already apparent. Vandiver focused increasingly on research, as the In Memoriam resolution at The University of Texas demonstrates:

'Vandiver's intense, single-minded concentration brought him to the brink of physical collapse. He would forget to eat, and he would work far into the night. Eventually he found it necessary to withdraw temporarily from this schedule. By following a regular and less-demanding schedule, his health returned. In a few years he was working away on his research with renewed vigor and enthusiasm.'

---

[4] R.E. Greenwood, *History of the Various Departments of Mathematics at the University of Texas at Austin, 1883–1983.*

Much of this time was taken up on Fermat's Last Theorem and number theory. In 1952, for example, Vandiver used a computer to study the Theorem and was able to prove it for all primes less than 2000 and it was for this work that he became best known in the post-war era. His output continued for many years and when he retired, he had written some 175 research papers and historical notes.[5] Moore, on the other hand, had more or less given up his own research work and publishing to devote himself entirely to his students and their mathematics. His research output, once so prolific, diminished to a trickle in the first half of the 1950s, and apart from a major revision of his book in 1962, he published only two further papers,[6] the last in 1953.

His switch from research to devote more time to teaching became evident not only in the number of doctoral students he supervised, but also in the influence of those students as they progressed in their own careers. To some extent, the latter involved the practice Moore had adopted some years earlier in Pure Mathematics, that of hiring graduate students to teach four courses a year while simultaneously pursuing their degrees. This was one of the areas that came under attack from his colleagues in the difficult times after the war. As R.E. Greenwood points out, it was seen by some mathematics professors as providing Moore students with an advantage over others. The issue was raised at a General Faculty meeting when it was proposed that graduate students

---

[5] In 1961, when he was in his 79th year, the Texas Section of the Mathematical Association of America honored Vandiver by inviting him to deliver the principal address at the annual meeting of the Section. It was his last public professional appearance, and some two hundred of his colleagues, friends and ex-students attended. His lecture "On Developments in an Arithmetical Theory of Bernoulli Numbers" included recent advances and simplifications, which Vandiver had made. Work in Bernoulli numbers was second only to Fermat's Last Theorem in Vandiver's studies, and indeed certain properties of Bernoulli numbers were used to establish some of Vandiver's results on Fermat's Last Theorem. In 1965, when Vandiver was 83, the editors of the *Journal of Mathematical Analysis and Applications* decided to dedicate one issue of the 1966 volume to him. A number of papers contributed by former students and by friends from over the globe appeared in this volume. Vandiver was mentally active until late in life. Professor Alfred Brauer of the University of North Carolina wrote to Vandiver, saying 'In my opinion there does not exist another mathematician in the history of mathematics who has done research so successfully as you after attaining the age of 70.' Finally, at age 80, he resigned his Modified Service appointment and became Professor Emeritus. Although he lived on for six and a half more years, his poor health made rest home care desirable for him. He died on 4 January 1973.

[6] A characterization of a simple plane, *Proc. Nat. Acad. Sci. USA* 32 (1946), 311–316 and Spirals in the plane, *Proc. Nat. Acad. Sci. USA* 39 (1953), 207–213.

should be restricted to two courses a year and that their pay should be graded according to the degrees held. This, in turn, would have had serious repercussions on payment because as the opposers well knew, Moore did not encourage his doctoral students to take an M.A.

He said as much when he rose to address the General Faculty[7] on the subject. Once again, he pulled no punches in his assessment of the situation and stated that in his view graduate students were often better teachers than some full-time instructors. They ought not to be discriminated against by an inferior title and lower pay. He was particularly outspoken on the proposal to include an M.A. degree as a requirement for graduate teachers, which in his opinion constituted a sheer waste of time in many cases. He concluded: 'If a graduate student displayed real scholarship, he ought to go right on to the PhD degree and if he did not give evidence of sound scholarship, he ought to be sent elsewhere.'

Greenwood suggested that people who did not suit Moore's course, or didn't suit him, for whatever reason, were indeed invited to leave, or not taken on in the first place. He cited one young man in 1948/49 who had a particular interest in the modern dance craze and went around with an arty crowd, spoke in the modern vernacular and dressed accordingly. Moore encouraged him to leave. He subsequently did well at the University of Virginia.

The accusation of elitism crops up from time to time in any discussion about R.L. Moore but a closer survey of those who came into his sphere through their participation in his courses or to work directly for the PhD reveals more about Moore, to coin a phrase, than the students themselves. The selection process was, and always had been, purposeful and deliberate and no less so as he entered that postwar period when, on passing the age of 65, the pace seemed to quicken. Few would have guessed that at that stage he still had in him two further decades of delivering the Moore Method message to a succession of talented young people. This whole postwar era of Moore-supervised PhDs was, as R.D. Anderson points out, to have a substantial effect both locally and nationally, and indeed reflected both Moore's own history as a research mathematician and that of those inspired by him. The fact is Moore was by then rather uninterested in anything but the training of mathematicians, preferably research mathematicians and possibly mathematics

---

[7] Minutes of General Faculty, The University of Texas, Austin, 9 April 1947.

teachers who would use their own modified Moore Method. That is to say, he did not spend much time or effort on the Mathematical Association of America.[8] He was more interested in the research organization, the American Mathematical Society. Someone once heard him describe the Association as the Salvation Army of Mathematics. Anderson said: 'That sort of represents his attitude. I was told that he joined the MAA three times, which meant that he must have resigned at least twice. But as it turned out, while his students and those influenced by his students were highly active in the AMS, particularly those like Wilder, Whyburn and Bing who all became president, he had an even greater influence through his students on the MAA.

From 1950 onwards, five of his students and a sixth who studied with him but gained her degree under Kline, became presidents of the Association which meant that for almost a quarter of the time in the remaining half of the twentieth century, the person leading the MAA was a student of R.L. Moore. Further, a good many more of his students took other influential roles and it is remarkable, considering the number of mathematicians there are, that one man could produce so many students who would play such a sizeable role in the development in modern mathematics and the teaching thereof in the United States.

There is no doubt that this had a tremendous effect on issues of collegiate education. Perhaps more interesting is the effect he has had on K–12 education, which again was passed down through his students. Ed Moise, for example, after completing some great work on 3 manifolds, became involved in education issues and he wrote a number of successful textbooks, and a treatise on *Geometric Topology in Dimensions 2 and 3*. Later, one of R.L. Wilder's students, and thus a second generation Moore student, named Ed Begle was among the leaders of an early reform movement known as the School Mathematics Study Group (SMSG) and, in turn, his own students continued this trend in a major way. This is just one example of many in which the spread of Moore's influence across mathematical research and mathematical education began to expand across the whole community, first through his own personal efforts and then through those of his mathematical descendants. The network over the coming decades would become vast and significant.'

---

[8] Interestingly, however, it was this organization that sponsored and promoted the film *Challenge in the Classroom* on Moore's teaching method.

Anderson's sentiments are entirely borne out by his former classmate Mary Ellen Rudin, both in terms of Moore's personal motives and the influence he exerted, without planning it, across the entire American educational system. Rudin, who went on to a career in which she was rewarded with many accolades and honors (see Chapter 15), described precisely the experience of herself and many others who came into Moore's realm, which shows that quite a percentage of them originally had no intention of being so intensely trained in a particular field of mathematics: 'Moore made me a mathematician. I would never have normally been a mathematician, I suspect, if I had gone to Wisconsin and taken the trigonometry, the probability that I would have been a mathematician is practically zero. But I went to Texas, and because Moore dragged me through by the hair, I'm a mathematician and it is something that gives me tremendous pleasure and something that I think I have talent for. He was efficient in recognizing talent.'

Conversely, when his students moved on to create their own domains, as so many did in substantial manner, Moore did not attempt to exert undue influence on the direction they took in their work, or the methods they used in their teaching. He was not averse to showing his displeasure when one of them went to a position he believed was below his station, but seemingly made no effort whatsoever to see that his style of teaching, his method, was perpetuated through those students who had learned through it. Rudin explained:

'This is a very personal thing that has nothing to do with the method, I think. It was very much his religion that no one should ever tell anyone how to teach, and the idea that someone would go to a class and check up or that there should be common examinations for two sections of a course was something sinful. He was never curious whether you were using his method. Absolutely not. It would be against the rules of the game to ask. He really stuck to the rules. My methods were very different. I believe that it is very nice to have a course where you are allowed to prove theorems because I did enjoy these courses very much. I'm certainly a mathematician because of this. The method has two virtues. It gives a student a chance to really see what a mathematician does and gives him an opportunity to do some of these things, and it gives you an opportunity to see who the really talented ones are in a group. So, I think it's interesting to teach at least one or two courses in a curriculum like this. Realizing how much you have to break down a theorem is one of the most difficult parts of teaching in this way. You have to really break down the theorems into very, very tiny statements. It's fun to have some courses in the curriculum of this type.

'There are people who are interested in this technique, who are very good at it. I did it only in a very limited sort of way. If you just present the theorems and the definitions, it's a very slow process, and it requires a tremendous amount of work from the students to get it growing. I confess that I both hate to demand quite that devotion to the subject and I haven't the patience to wait on it. So, I end up talking about the theorems and describing the ideas involved and giving too many hints, and end up more or less proving a lot of the theorems, and they prove some, particularly if I'm teaching the undergraduate course which I think is really built for it terrifically. I try not to do much of that, and let them really have a chance to prove the theorems.'

There was, however, another teacher who was about to become a convert to the Moore Method at The University of Texas, and he did so at a time of turmoil in the two mathematics departments when the localized war between Vandiver and Moore spread to cause a considerable rift between the two departments. It was also a time when the administrators were attempting to combine the two departments and move them into a new building, Benedict Hall, which was due for completion in the early 1950s. Factions were being formed and bitterness was breaking out when into this difficult situation stepped Vandiver's replacement in Pure Mathematics, Professor Hubert Stanley Wall (1902–1971) a calm and considerate man who had much to offer the troubled mathematics arena at Austin. In his previous professorship at Northwestern, Wall had supervised 25 Master's candidates and five doctoral students. Following Northwestern, Wall was professor at the Illinois Institute of Technology. It was at IIT that Wall had Pasquale Porcelli as an undergraduate student. At both Northwestern and the Illinois Institute of Technology, Wall was associated with Ernst Hellinger, a student of Hilbert. Wall related that he and Hellinger conducted seminars in Chicago in which a main feature was students' presentation of their work. Peter Lax once indicated to John Neuberger that Wall saved Hellinger's life by helping to get Hellinger out of Nazi Germany to come to Northwestern. In the years ahead, Wall would tell anyone prepared to listen that his move to Texas and the resulting metamorphosis pulled him out of the doldrums and gave him a new outlook.

This he demonstrated in results over the coming decades, during which he joined and then overtook R. L. Moore in the Texas production

Hubert Stanley Wall

line of PhD mathematicians. He and Moore got along well and Wall was fascinated by the Moore Method, with which he soon began experimenting, and then adopted for some of his courses. Thus with the expansion of teachers utilizing the method, which now included Wall, Burton Jones and, from time to time, Ettlinger, the Moore Method was subsequently referred to thereafter by some commentators as The Texas Method. Wall and Moore had much in common, apart from the fact that they had both undertaken appointments at Northwestern. Wall came to Texas ready to try new techniques and, teaching the Moore way, became an unqualified success.

He, like Moore, also began to downgrade his own research and to curtail his publishing and the refereeing of papers for the various mathematical journals to concentrate on his students. Although in the early days at Austin, his name was associated with that of the still-dominant Moore, he very quickly established a reputation for his own excellence as a teacher. He became a firm proponent of the view that students should depend less on textbooks and monographs and devote more of their time to constructing mathematical proofs of theorems on their own account.

On many occasions Wall spoke of how fortunate he was to have been placed in a situation where he could see the advantages of this mathematical philosophy and where he was given the freedom to pursue it. Many would testify that Wall also seemed able to add a unique humanitarian side to the method, and instilled in individual students a creative spirit along with a challenge to create new mathematics. He did so, in the early years at least, against the unhappy and difficult situation that existed among the faculty at UT. Enrollment had rocketed to an unprecedented 16,000 in 1946 and the population of the University was clearly set to remain at these high levels for some years to come.

It was evident to everyone in UT planning and administration that more mathematics was needed for upper level students in Chemistry, Geology, Physics, Engineering, Biology and Economics but Moore remained firm about the focus and direction of his own courses. A measure of the dissent that arose across the spectrum of mathematics may be judged from the somewhat indignant recollections of Professor R.E. Greenwood, who had taken courses with Moore as a student himself and was now a member of faculty in Applied Mathematics: 'It has been said that the intransigence of R.L. Moore prevented corrective measures being taken to care for these new needs. Some have said that R.L. Moore had blinders on and did not realize the new conditions. It is true that as he grew older, Moore did not "mellow".'[9]

As will be seen, this accusation is vehemently dismissed by many of his students, who in turn point to his achievements up to that point, and certainly in the ensuing years, for which they became the living proof. Mellowing was not an attribute to which Moore subscribed. He was clearly motivated by his continuing capacity to discover students of great ability and point them towards success through their own stamina and effort. It may well have been that while the need for modernization was fully recognized, The University of Texas was slow in establishing new offerings, methods and attitudes. It is often the case that those institutions that have enjoyed considerable success over an elongated period are loth to introduce drastic change through fear of ruining a perfectly good system. Notwithstanding the difficulties encountered during political shenanigans, the university had historically produced its fair share of the nation's most prolific mathematicians, and continued to do so.

---

[9] R.E. Greenwood, *History of the Various Departments of Mathematics at the University of Texas at Austin, 1883–1983*, p. 52.

Their work in terms of achievement and publishing had placed Texas consistently among the top echelon of the nation's universities for its ability to turn out the highest quality PhD students.

Various charts and assessments produced over the next decade or so reflected the tremendous impact of Texas-trained mathematicians on American academic life. It is worth noting that in 1955 the National Science Foundation sponsored the setting up of a Survey of Research Potential and Training in the Mathematical Sciences. It was a substantial undertaking, based on data collected from a lengthy questionnaire sent out to a very large number of people who received their PhD in mathematics between 1915 and 1954. The data gathered became known as the Albert Report, compiled under the chairmanship of Professor A. Adrian Albert of the University of Chicago.[10] Among its many wide-ranging conclusions, the survey would produce some startling results. They included collating the work of PhD students after graduation— their output was examined and placed into one of four categories. The highest category represented an assessment of the top quality PhD graduates as judged by the amount of published mathematical research. The results showed that of the top 15 per cent, 25 per cent were graduates from Texas,[11] 20 per cent from Princeton, 16 per cent from Harvard and 8 per cent from Chicago. Although those three universities were of course smaller than Texas, it was a creditable achievement given that Texas did not begin producing PhD students until 1923, eight years after the commencement of the survey period.

Results of that survey were, of course, unknown at the time that the administrators of The University of Texas began what turned out to be painful machinations as they prepared to meet the demands of the second half of the twentieth century. Combining Applied and Pure Mathematics under one departmental roof was a process that was initiated by the administrators in 1947 but dragged on for six turbulent years before a resolution was finally found to the problems and issues that surfaced during endless debates, surveys and petitions. Moore was deeply opposed to the move and, according to R.E. Greenwood, offered

---

[10] *A Survey of Research Potential and Training in the Mathematical Sciences; Final Report*, The University of Chicago, March 1957.

[11] The 'highest quality' group from Texas was made up almost entirely of Moore students, and included R.L. Wilder (who graduated in 1923), G.T. Whyburn (1927), J.H. Roberts (1929), R H Bing (1945), E.E. Moise (1947), R.D. Anderson (1948) and M.E. Rudin (1949).

another solution, to close down Applied Mathematics altogether. This idea attracted support and hostility in equal measure. The passion with which both sides presented their arguments is demonstrated in the letter of resignation of F. Burton Jones who had decided to leave Texas for his first professorship of mathematics, at the University of North Carolina. He wrote to UT president Dr. T.S. Painter:

'In leaving, I will not become unconcerned about the welfare of the Department of Pure Mathematics. This welfare and the welfare of the University will not be served by a combination of the mathematics departments, either now or in the future, and no steps whatsoever should be taken in that direction. The Department of Applied Mathematics should be a "service department for engineering" and steps should be taken to insure that in performing this function the Department of Pure Mathematics is not adversely affected. If this policy is not followed, then what *is* to be the reward for distinguished service to the University? Harassment and extermination? I hope not.'[12]

The advice of Burton Jones, and the insistence of Moore, went unheeded and the dash toward combining the two departments progressed into the term of Painter's successor as president, Logan Wilson, who approved the plan advanced by Dr. C.P. Boner, professor of physics and Dean of the College of Arts and Sciences. A stormy period ensued with numerous long and boisterous faculty meetings, which did not always do justice to the issues at hand; indeed, they often resulted in an artificially induced result. The faculty meetings were full of people who waited to find out what stance R.L. Moore would take on an issue before the faculty. As soon as his position became known, X per cent would decide to vote against him as a matter of course while Y per cent would always vote with him.

The upshot was that the unification was achieved in 1952 when the two mathematical staffs were moved into a new campus building, Benedict Hall, the Pure Mathematics Department housed on the third floor and Applied on the second floor. But as R.E. Greenwood records, 'this did not develop into an "Era of Good Feeling." A spirit of antagonism developed in the minds of young graduate students in the old Pure Mathematics Department and R.L. Moore was unrelenting in keeping up the pressure on former Applied Mathematics members.' Undoubtedly,

---

[12] 'Copy' to R.L. Moore: F. Burton Jones to Dr. T.S. Painter, president, The University of Texas at Austin, 28 January 1950.

there were at least two philosophies working toward their goals, and not simplistically as a Pure versus Applied issue.

On the one hand, there were those who wanted to carry on what they had been doing very successfully for some time, and those who had decided to meet the changing circumstances of the world about them in a manner that to some extent discounted past achievement. Whether they could come together in peace and harmony was unlikely. In fact, further seeds of discontent were already being sown.

15

His Female Students

Given that in some quarters he has been regarded as a misogynist, oral history interviews with a number of Moore's female students prove the contrary, including observations from Mary Ellen Rudin and Mary-Elizabeth Hamstrom. To the latter, when only a prospective student, he wrote an incredible letter outlining his method and his expectations of her, given here in full.

Moore in his office

*M*ary Ellen Rudin was one of Moore's most outstanding students and she scored a number of firsts for her gender in an era when the rights and recognition of women in professional life were still barely a conversation topic, let alone a movement towards change. She received her PhD in 1949, having also had the benefit of working with F. Burton Jones, much loved and respected at Texas, and he was on the committee of signers for her thesis, *Concerning abstract spaces*. She, like many of Moore's top students from that era, took advice and guidance from Burton Jones in his less emphatic application of the Moore Method, with which she herself also experimented. When she graduated, Moore arranged an instructorship for her at Duke University in Durham, North Carolina where she began teaching in 1949. Her move also led her toward a disappointing development, as far as Moore was concerned, that is. She got married! Little more than three years after her departure, Mary Ellen Estill became Mary Ellen Rudin.

'Whatever did she want to do that for?'[1] was Moore's reaction when he heard the news. It was a statement that leads us to another aspect of Moore's personal feelings as a teacher, that of his attitude towards female students. There were suggestions in later years that he was somewhat ambivalent about having women in his classroom, and that he was never keen on accepting them into his courses because they were likely to go off and get married and have babies. One even accused him of misogynous tendencies. At first glance, statistics might demonstrate some evidence towards that view: there were just six women among his fifty doctoral students between 1914 and 1970. To this might be added the testimony that he was 'heartbroken' when his first woman PhD, Anna Mullikin, chose to 'waste her talents' by becoming a high school teacher. Twenty-one years would pass after Mullikin before he supervised another woman who completed her PhD, a mature student named Harlan C. Miller, who had already taught at the Hockaday School in Dallas, Texas, prior to her graduate program. Indeed, she was 45 when she gained her doctorate in 1941.

Moore, however, would not have it said that he showed any bias against women in his selection of students and in fact during a lively Faculty debate in the late 1950s on the issue of segregation, he had accused both Harvard and Princeton of discrimination against women[2],

---

[1] Recalled by R.D. Anderson, in an interview with the author, January 2001.
[2] Traylor, p. 179.

which was presumably meant to show that women were welcome at The University of Texas, and in his classroom. In fact, over the years there were quite a number of female students taking his courses who did not go on to their degree, given that throughout his time as a university professor, the emphasis on women choosing a career in higher mathematics was far less prevalent than after he had retired. If he objected to anything in relation to female presence, it might well have been that at times his classroom took on the air of a dating agency.

At least six women students who took courses with him became the wives of other Moore students. Among the couples who married as a result of their meeting in Moore classes was Joe and Gayle Ball. Gayle was quite open about her intentions as soon as she spotted B.J. Ball back in 1946. She first saw him when she was visiting a Moore geometry class and, she said, 'that settled his fate ... I decided right away that Dr. Moore had a pecking order in asking questions and figured out what it was and noticed that after a sequence of questions to X, Y, and Z he would always end up asking Mr. Ball. And I admired the way in which Mr. Ball invariably gave the correct answers. I turned around and looked at him and decided that, well, at least I would make an attempt in that direction. So I took the same class as Joe and we married early in 1947, at the semester break.'[3] Gayle, who was the only woman in her class at the time, did not pursue a mathematics career and gave up Moore classes when she became pregnant, but Joe gained his doctorate in 1952, with a thesis entitled *Concerning continuous and equicontinuous collections of arcs* and subsequently went on to major academic appointments at Virginia and Georgia. His own work remained in the area of continuum theory, general topology and, later, shape theory. In their remaining graduate years at Texas, Gayle and Joe Ball enjoyed the social scene and occasional contact with Dr. and Mrs. Moore. Gayle's reminiscences of that time provide some insight into Moore that might only come from a woman involved in the day-to-day classroom activity:

'I liked Mrs. Moore very much and after Dr. Moore had students from his class there, she invited me back for further visits, which I did. I grew quite fond of her. What struck me, however, was that Mrs. Moore was indeed resentful of the adulation that was bestowed upon him and the

---

[3] Mrs. Gayle Ball, interviewed by Ben Fitzptrick Jr. for the R.L. Moore Oral History Project, in Oxford, England, February, 1998; tape in the R.L. Moore Legacy Collection in the AAM.

Mary-Elizabeth Hamstrom

manner in which this put her well into the background. I remember once when I went back to Texas for a visit to the University, the first day I was there I went over to call on Mrs. Moore and she asked me if I had seen her husband yet. I said No, and she then said, "This is very unusual. They all want to see him first, and then they see me incidentally." She was quite pleased that someone wanted to see her before seeing him.'

Gayle also mentioned Moore's concerns at the time that Joe was planning to go to a northern university after graduation. Moore was against it, and spoke to her specifically on the subject: 'Dr. Moore was sensitive to the needs and feelings of women. He felt it would be very hard on me to have to, perhaps, get on a bus and sit down next to a person of color. He thought this would be very offensive.' Asked in her interview with Ben Fitzpatrick if she felt that Moore was 'trying to get a rise out of you', she replied: 'I think he was trying to get my reaction. He said, "What would you do, Mrs. Ball, if this happened to you?" [I replied that] I would say: "How do you do?"' Moore's reaction to her reply is not recorded.

Gayle and Joe had a daughter, Margaret, while at Texas and, incidentally, she followed in the footsteps of her parents and took courses from Moore 19 years later when he was in the twilight of his career. She had spoken of becoming a mathematics major. Moore was very unhappy when she changed her mind. So women were not second-class citizens in Moore classes by any means but there was, however, a hint of reluctance on Moore's part or, at the very least, an attempt to establish the talent, suitability and seriousness of the candidate before taking them on for the long haul to a PhD. One of the most famous of his students to be subjected to this scrutiny was Mary-Elizabeth Hamstrom.

In 1948, at the age of 21, she was awarded her BA at the University of Pennsylvania and in the course of her move towards higher education, Ms Hamstrom wrote to Moore in April 1948, having already been accepted as a graduate student at The University of Texas and given a scholarship. She said[4] that at the time she knew nothing of Moore's teaching methods but when the decision was made to go to graduate school, she wrote to seek guidance as to what she should be reading during those summer months of 1948, prior to her arrival. The letter she received in reply was, in her view, astounding.

It was indicative of the extent of his assessment of potential students and demonstrated the amount of time he must have spent thinking seriously about each one of them. The letter, extracts of which are given below, was classic Moore, in that it provides a fascinating insight into the general attitude he adopted towards his fresh students as he prepared to accept them into courses.[5] The detail is precise, and in many ways historical:

> Dear Miss Hamstrom:
> I was glad to have your letter of April 27. You say, 'I have had no formal work in point set theory other than that included in the Real Variable course I've been taking. This, and the little reading I have done, have appealed to me very strongly.' I wish you had never taken a course in Real Variable Theory and that you had read *even less* about point set theory than I imagine you have. You indicate that you would like to have some detailed information concerning courses to be offered next year and add 'This is necessary so that I can plan my summer reading.' Whatever else you read about this summer, do not read any point set theory if you can help it.

Moore went on to quote case histories of former students to demonstrate the success of students who came to him without having previously taken courses in point set theory elsewhere or even having read

---

[4] In an interview with Douglas Forbes for his dissertation research, 13 February 1971; tape/transcript is in the R.L. Moore Legacy Collection in the AAM.

[5] Moore to Hamstrom, 7 May 1948; R.L. Moore Legacy Collection in the AAM.

up on the subject and while he expressed difficulty in presenting a satisfactory explanation in regard to that stipulation in a letter, he assured her that after she had been at Texas a while, she would understand — 'and I *hope* you will agree'— and he then went on to explain the Texas scenario:

I expect to give, next year, P.M. 24 and P.M. 88. I am afraid that, in the course you are taking on Real Variable Theory, you have had a good deal about theory of measure of point sets and Riemann and Lebesgue integration. Please let me know whether or not this is the case. If it *is*, then I imagine, you had *read* or *listened* to statements and proofs (or approximations to proofs) of many theorems concerning these matters *instead of working out proof of them yourself* and thus if you took 24 you would, I think, in many cases know in advance the answers to certain questions I would ask (and would have already seen approximations to *proofs* of many of them) while certain other members of the class (I have in mind, in particular a Mr. Pearson who is now taking a first course in calculus and showing unusual ability) would not know in *advance* what the answers are or, if they should have enough insight (as contrasted with information gathered from reading etc.) to have some idea of the answer, at least they would not have been deprived of the opportunity of working out proofs for themselves by having had someone tell them proofs either verbally or in books in a course previously taken.

To *read* a proof of a theorem or to *listen* to some professor prove it is very different from proving it yourself without any such assistance. Did it never occur to you that when you read or listen to a proof of a theorem that you had never heard of till someone stated it (perhaps a few minutes before, without even giving you enough time to first realize just what it means) you are thereby acquiring *information* (of a sort) but depriving yourself of the opportunity to work it out for yourself and thereby, perhaps, to develop that much more *power* instead of just acquiring that much more *information*?

What does *information* amount to compared to *power*? Suppose you and Mr. Pearson (and perhaps some others) should take 24 next year and I should define the word 'countable' and say 'See if you can prove next time we meet that the set of all points on the *X*-axis is countable'[6] and Mr. Pearson will work hard trying to show that it is countable and you will say to yourself 'Oh that question was settled long ago by Cantor. I

---

[6] On the face of it, Moore was asking his students to prove something that wasn't true and his students knew full well that that is exactly what he intended. He used the English language with an extraordinary precision, states John Worrell, and likely hoped that the way he posed the problem would have a result of enhancing the mathematical experience of one to whom it was thus presented and who subsequently, independently, might solve the problem.

Page 5 of the letter from R.L. Moore to Mary-Elizabeth Hamstrom

can give you his argument right now. I don't need to wait till next time.' Suppose Mr. Pearson however had never heard of this word countable before and started out with no hint from anyone he would come in next time and say 'I can settle that question' and would do so in a manner very different from any that you had ever seen before. And suppose at the next meeting I would say 'Prove that if a set of segments covers a segment some finite subset of it does so' and you would say to yourself 'Oh, that is the Heine-Borel Theorem (or is it?) and that was one of the theorems in Chapter I of that book we used last year, let's see how was it proved?'

And suppose Mr. Pearson with no such previous knowledge would say to himself 'That is an interesting question. I am going to get to work on that as soon as I can' and suppose at the next meeting he would come in with some wrong ideas and next time after that with some other wrong ones, but finally after a week or two would obtain a conclusive argument of his own. And suppose, you having told me that you had seen a proof of the Heine-Borel Theorem long ago, I had said 'Well don't say anything to Mr. Pearson about it, let him come to his *own* conclusions' and accordingly at every meeting for say about two weeks you watched Mr. Pearson try time after time, making mistake after mistake perhaps, and finally arrive at a correct conclusion and in doing so make some real progress towards developing power in the subject, would you or not be saying to yourself 'I wish I didn't know so much and that I could be working these things out for myself like Mr. Pearson is doing instead of knowing, in advance, answers that have been told me or that I have read somewhere?'

Do you see why I want to know something about what you have had in the Real Variable Theory course?

Suppose you have had so much Real Variable Theory that it would be inadvisable in certain ways for you to take 24, that your *previous knowledge* of the subject would make you too much of a spectator in that course, an onlooker observing others working to get answers that you already know. Then what? Perhaps take 88? I think that in 88 next year I will probably have four or five students who are now taking 24 and who have been developing power working on questions that I have raised, somehow trying to prove something is true that is really false and finally discovering that it is really false, sometimes coming in with arguments and having it pointed out that they are wrong, trying again, again getting something wrong and finally either settling the question correctly or having the experience of listening to some other member of the class finally straighten it out, etc. I remember that in one class I had a student who for a while, made mistake after mistake. For I don't know how long I was inclined to think, if he said he had a theorem proved, 'he thinks he does, but he probably *doesn't*'. But after a certain number of months had passed if he said, 'I have proved this theorem' I was inclined to think, 'If he says he does he probably *does*.' Do you think such a change could have taken place in him if he had been given no opportunity to work things out for himself, no opportunity to make mistake after mistake and have them corrected instead of listening to (or reading) *other people's* arguments?

While I feel that I would have preferred that you come without a previous course in Real Variable Theory and start here in 24, I am inclined to think that, with the situation as it now is, it might be preferable for you to start in 88 in spite of the fact that if you do, you will probably be competing with some students who have had 24 and who have thus been accustomed to working things out for themselves for quite a while. Would it not be better to do that than to take 24 and be, for a very considerable part, if not most, of the time, an onlooker watching others work on theorems which are new and interesting to them but which you have already heard or read about?

As to the question of a minor subject for the PhD degree, in the past it has for sometime been possible for a student, with pure mathematics as a major subject, to absolve the requiring of a course in a minor subject by taking a course in Differential Equations given by a member of our department but counting as Applied Mathematics 22 instead of Pure Mathematics 22 if the student desires to count it that way. I think that Mr. Anderson, a candidate for a PhD degree next month, will do this.

If you are going on to a PhD degree and are going to become a research worker in mathematics and it is not necessary that you first get an MA degree for financial reasons then I don't think you should divert, to the fulfillment of the additional requirements for an MA degree, time and energy that might otherwise be devoted to the developing of *power* to do mathematical *research*. Of the last three people who took PhD degrees here in Mathematics, only one had a Master's degree....

Don't hesitate to write me if there is anything I have written which you would like to have explained in more detail or there is anything else you would like to ask about.

Please remember me to Miss Mullikin,
R.L. Moore

The upshot of that extraordinarily long letter was that Moore took on Hamstrom in 1948 and, on reflection, it was something that even Hamstrom herself could not quite fathom: 'Dr. Moore wouldn't ever let you in a class with the preparation that you would have had from Pennsylvania today. I'm not quite sure why he did it even then. He may not have thought I would do well or something, but I think that he let me come there with some trepidation.'[7] Hamstrom's own intentions, ambitions and aspirations were clear, and she soon discovered they were shared by most of those around her in Moore's courses. She explained this in a 1971 interview with Douglas Forbes:

'I went there *wanting* to be a mathematician. I guess I thought I was becoming one sooner than I might otherwise have because I was sitting there doing it. I just liked thinking about the mathematics and doing it myself, and we all had the impression that we were really pretty serious mathematicians already, not that we were going to get a degree and then be a mathematician. We thought we were hot shots right there and then.

---

[7] From a tape-recorded interview with Douglas Forbes in 1971, for his PhD Thesis: *The Texas System: R.L. Moore's Original Edition*; transcript is in the R.L. Moore Legacy Collection in the AAM.

It's just that his idea was that the business of getting a degree is doing something original in mathematics and we realized that we were getting training in doing something original in mathematics. He certainly played down all of the other requirements for the degree, course requirements and things like that. His main interest was in having us do the mathematics, and so we did it, and we were very serious about our work. This was our work that we were doing, but it gave you a very good feeling.'

Like others who have, some years down the line, expressed reservations about the intensity of the Moore Method, Hamstrom was left with what she described as 'very mixed emotions about the whole experience'. There were many aspects of Moore's teaching method that she felt could have been done differently and many years later she admitted to herself that there were some things she disliked immensely. She said she probably realized these dislikes at the time but sublimated them. On one hand, the Moore experience opened up incredible vistas in mathematics, but at the same time she could not fully warm to to the man himself and these memories jarred when she thought about them in later years:

'He had a great many personal failings, characteristics that I just don't care for one little bit. I didn't worry about it then, and I don't know why I should worry about it now. One wonders whether we ought to have, or somebody ought to have given him hell now and then. He may have liked that; I don't know. Sometimes, there were some very tense situations. Dyer[8] once stayed up all night working on a theorem and he got it, but the way in which he talked about it did not please Moore, and Moore exploded. He was very serious about that. There were situations of that nature. I certainly didn't care for the way Moore was acting. He got after me once because I had given a proof in class that this fellow also had and then I discovered after class, although I didn't know it during class, that my proof had been wrong. Moore had never said anything. I told Moore that, and I made the mistake of telling this guy that I had made a mistake. I told him that my proof was wrong and Moore got terribly upset about that. "*Miss Hamstrom, never in a hundred years*

---

[8] Eldon Dyer, who joined the same year as Hamstrom and stayed the course to gain his doctorate in 1952 with a thesis, entitled *Certain Conditions under Which the Sum of the Elements of a Continuous Collection of Continua is an Arc*. He went on to a distinguished career that included appointments at Georgia, Johns Hopkins, Chicago, Rice and CUNY, where he became chair. He followed Wilder and Bing into the role of a consulting editor of the *Encyclopedia Britannica* and played a very active part in AMS affairs, serving as editor of the *Proceedings* and associate editor of the *Transactions*.

Lida Barrett at a UT Roundup Dance. (She began her studies with Moore but finished up at Penn.)

would you understand this method of teaching." He was absolutely furious. He had a tendency to be very polite to females. And I think that, had I not been female, I would have gotten a lot more from him.'

As with most former students who have, over the years, contributed recollections of their time working with Moore, Hamstrom also came to the point in the course where no proofs had been delivered for some time. She too, circa 1948–1951, had the experience of many of those before her of discovering Moore's penchant for arousing political ire among his students by raising issues from the Roosevelt era. She remembers one student who was the only person in Moore's class at the time and suffered Moore's meanderings on a one-to-one basis. 'That student once went six weeks without getting a theorem and he heard an awful lot about the New Deal during that time. We all got a good bit of it because there were days when you didn't have anything, particularly when you got to 690, so you would hear from him about all sorts of subjects, whatever it was that was exciting people at the time. There was a time when they were building the new math building, and so there was

Mary Ellen and Walter Rudin

talk about that. I certainly didn't want to get into any arguments with him about this kind of thing. Whatever we discussed, however, he was the one that decided what was going to go on in that room.'

One of Moore's objectives in selecting outside subjects for discussion was to make students analyze the use of everyday language by asking them to apply correct English usage to articles and quotations taken from daily newspapers. But this elongated period of discussing matters that, in theory, had not very much to do with mathematics, has puzzled many who have studied the Moore Method, and even those who studied under it. The most common challenge arose from the possibility that Moore students were not given a sufficiently broad canvas on which to base their futures. While that does not appear to have been borne out by the experiences of the majority, it is clear also that many of them took remedial action to avoid it. Hamstrom was certain that there were limitations:

'One has some kind of knowledge about how to go about doing mathematics. As Dick Anderson once said to me, you get to know a lot about what's not going to work. When you follow the conventional way of doing mathematics, you don't get to bark up wrong trees, and there certainly is something useful in having the experience of barking up the wrong tree. You certainly get good research attitudes. You get the idea that research is important and interesting. You get certain capabilities at performing it. You do not get any training in learning the kind of mathematics other people do. It's something that you need; you need to learn to read the literature. You just can't sit down and do it. If you're going to be able to do it efficiently it requires a certain amount of training. As far as teaching attitudes and capabilities are concerned, this is something, again, that's unusual. You don't get any capabilities out of it because we all taught. You certainly get experience at talking in front of people, but one of the things you get, not out of his method, but out of Moore, is an interest in teaching. All of his students came out of there wanting to teach and feeling that it's important. They don't get the feeling that doing mathematics is the only important thing. He believed in teaching and his students, at least when they get out, want to be good at it.'[9]

It certainly worked for Hamstrom, who established herself in major academic appointments, principally at the University of Illinois, and directed nine PhD students. She also continued her own research in the field of geometric topology. She and Mary Ellen Estill thus presented a strong case to demolish any thoughts Moore may have had about not teaching women. In 1953, Mary Ellen Estill married Walter Rudin, who was a mathematics research student at Duke when they met. The first of their four children was born in 1954 by which time Walter had been appointed to the University of Rochester where Mary Ellen got a part-time appointment. The difference was that she did it for the love of mathematics, rather than the money. With that attitude from the start, Mary Ellen Rudin's career as a mathematician took a unique course, which, as it advanced beyond the early years of the demands of her young family, became outstanding in every regard: as a teacher, a researcher and as a mathematician of substantial repute. When once asked how she managed to combine being a full-time mother with being a full-time mathematician, she replied: 'I have never minded doing mathematics lying on the sofa in the middle of the living room with the children climbing all over me. I feel more comfortable and confident when I'm in the middle of things, and to do mathematics you have to feel comfortable and confident.'[10]

---

[9] From a tape-recorded interview with Douglas Forbes in 1970, for his Ph.D Thesis: *The Texas System: R.L. Moore's Original Edition*; transcript is in the R.L. Moore Legacy Collection in the AAM.

[10] Donald J. Albers, G.L. Alexanderson and Constance Reid, *More Mathematical People*, Harcourt Brace Jovanovich, 1990, p. 298.

She and Walter moved to the University of Wisconsin in 1959, and Mary Ellen became a part-time lecturer, a role in which she remained until 1971 when in one leap she was promoted from lecturer to full professor. Mary Ellen herself commented: 'The guilt feelings in the mathematics department were such that nobody even asked me if I wanted to be a professor. I was simply presented with the job.' In fact, the University of Wisconsin had been most remiss in not recognizing her accomplishments earlier.[11] From the beginning she had shown a considerable talent in pursuing a number of aspects relative to the work of R.L. Moore and R.L. Wilder, notably in set-theoretic aspects of topology, and finite and infinite combinatorics. Her continuation of this theme had been recognized as early as 1963 by the Mathematical Society of the Netherlands when she was awarded its Prize of Nieuw Archief voor Wiskunde. It was the first of many such awards, culminating in further recognition at Wisconsin in 1981 when she became the first holder of the Grace Chisholm Young Professorship at Wisconsin where she remained for the rest of her career. In 1980, she was elected vice-president of the AMS, served as Governor of the MAA, and was elected a Fellow of the American Academy of Arts and Sciences and a member of the Hungarian Academy of Science. Even as the twentieth century drew to its close, she was still publishing, this time four papers in sequence aimed at characterizing the Hausdorff continuous images of compact linearly ordered spaces, confirming a conjecture by J. Nikiel that they are precisely the compact Hausdorff monotonically normal spaces. One of her own proudest moments was an invitation to be the Emmy Noether Lecturer for the Association for Women in Mathematics. There were other female students who came to study with Moore who did not make such a mark, but were nonetheless glad to have had the opportunity. One example was Mary Spencer (née Foster)

---

[11] Having produced a dissertation that looked at the implications, and relations to various alternatives, of an axiom system for point set theory proposed by R.L. Moore in 1932, she went on to look at spaces satisfying a subset of R.L. Moore's axioms in 'Separation in non-separable spaces' published in 1951. Her next paper 'A primitive dispersion set of the plane' (1952) provided a positive solution to an unsolved problem contained in R.L. Wilder's book *Topology of Manifolds* (1949). Also in 1952 the paper 'Concerning a problem of Souslin's' continued her examination of R.L. Moore's axiom systems, this time motivated by a 1920 problem due to Souslin. The above papers were all published under her maiden name of Mary Ellen Estill, but beginning with 'Countable paracompactness and Souslin's problem' in 1955, she published under her married name. She became particularly known for her ability to construct counterexamples.

Harlan Cross Miller

who, for whatever reason, felt compelled 21 years later to write to her former teacher to express her thanks:

'I took your beginning calculus course at The University of Texas in 1946. I had done well in the usual sorts of public school mathematics courses, mostly because symbolic manipulation was natural and easy, and I had enjoyed them, but when I entered your classroom I had no real appreciation of the beauty and significance of mathematics. Your very first lecture became, now that I look back, one of those turning points in my intellectual development that stand as major events in the memory. You will not remember, perhaps; you had given us a set of simple equations to graph, and we had proceeded merrily along as we had been taught, evaluating great large conglomerations of points and then trying to connect them with some sort of a smooth line. And then you showed us how, by factoring the equation and considering what happens between the root points, to develop a picture of the essential shape and placement of the curve, almost just by looking at it. I don't know whether you realized what a tremendous impact that simple notion, presented as it was, had. Lights began to dawn all over the room—my lord, there was a connection, a real meaning, here! The idea that broke over us there, and was developed and reinforced throughout the rest of the course, was that mathematics was not just fun and games but an inquiry into fundamental meanings. It is an idea that has informed practically all the thinking I have done since. I did not become a professional mathematician. As with most women, most of my energies have been directed towards home and children. But I think it is true that I see farther into the world, and understand better what I see, because of the work I was privileged to do for you. I thought that perhaps, as one of the few real teachers this country contains, you might be pleased to know.'[12]

Such letters were not uncommon, and most written much later in life, when they were reflecting on the experience, but as Hamstrom, Rudin and others confirmed, the experience was not for everyone.

---

[12] Mrs. Mary Spencer to R.L. Moore, 19 July 1965; R.L. Moore Papers in the AAM.

16

Moore's Calculus
(1945–1969)

Teaching reforms, especially in mathematics, begin to take on special resonance in accommodating the needs of postwar industries and new technologies, with a fully compatible range of courses, and there was a particular study into teaching calculus to engineers. This chapter, then, focuses upon his calculus courses from that period with input from students who experienced them.

Moore in his office

The busy post-war years of high enrollment and the demands for more and better qualified engineers and technicians for burgeoning new industries brought fresh pressures on the educational system of America. Various committees were formed across the land to examine the need to accommodate these students with a fully compatible range of courses, and there was a particular study into teaching calculus to engineers.[1] Moore made few concessions to the demands of the modern age, and indeed seems to have made a determined effort to reinforce the content and methodology of his own calculus courses to fit more precisely his own agenda of teaching pure mathematics with the principal aim of producing researchers and teachers of mathematics rather than engineers and scientists. It is a noticeable fact that from the 1940s onwards, he relied on his calculus more heavily to discover talent for his higher level courses. This continued even after he had — officially at least — been restricted to modified service and reduced working time after he reached the age of 70, restrictions incidentally that he wholeheartedly opposed and totally ignored. Calculus opened the door to his own mathematical world, and for many of the students who now came into it. In the 1966 film *Challenge in the Classroom*, he made a point of mentioning what may have been the start of the evolution of the Moore Method: his experience as a youth when calculus became a beacon in his own life. It will be recalled that at the age of 15 he taught himself calculus by first reading a statement of a theorem while holding a sheet of paper under the statement to hide the proof. He would reveal a line to give himself a clue, and repeat the process until he was able to obtain a proof, classic inquiry-based learning that is not uncommon among the great names of mathematics. Out of that youthful exercise in self-instruction came two paths of discovery, as a teacher and as a researcher. Those two paths crossed back and forth continually throughout his life as he produced groundbreaking mathematics that won him international acclaim and a succession of world-class students whose progress we have noted in these pages.

We have seen that in the early years of his career, when Moore was experimenting with his Method and, by the very nature of his lowly position did not have the freedom to experiment with his teaching, he gave courses that were generally posted by his departmental head. His special-

---

[1] Committee for Teaching of Calculus to Engineers, mentioned by E.E. Moise in his paper Activity and Motivation in Mathematics, *Amer. Math. Monthly* 72 (1965), 407–412.

ties were Foundations of Mathematics and Point Set Theory. But in the early 1920s, when he had returned to his alma mater, the Moore Method began to achieve even greater promise. Another turning point came in the postwar years when his advanced mathematics courses were full of likely candidates owing to the demand for calculus, but his objective remained the same: to discover research mathematicians, regardless of whether the student in question had even thought of doing such a thing.

The content, the class size, the way he taught calculus, depended on his assessment of the students in a particular class and he could alter any one or all of those conditions to suit them as a group and, more importantly, as individuals. This innate ingredient is mentioned time and time again. It shows itself, in particular, when two people who took the same course, or were perhaps even in the same class, get together to talk about old times. Their reflections, their memories often do not concur in the slightest. It is a recurring and fundamental element in Moore's whole approach and demonstrates that he went well beyond merely providing the student with an insight into what mathematics was all about.

Moore guided his undergraduates through his carefully planned courses in which they could discover the power and beauty of mathematics in the way that it had so totally consumed himself. He began with his inspirational calculus classes for undergraduates and, with a minimum of personal contribution by way of lectures, turned each one into an eye-opening experience that also encompassed 'real life' problems and topical issues, as well as essential elements such as time and motion, energy and matter and logic.

He used calculus to unlock his students' thought processes to solving problems, as opposed to the serious business of proving theorems, and it was here among the virgin minds that Moore went fishing for students he considered had that special talent he could release and exploit, much to the chagrin of members of faculty who may have already had designs on them. As we have seen, some were so inspired by him that they moved completely away from an intended direction. Among them was one on course to become a physician who ended up as a theologian. Another, John Worrell, already on the road to becoming a doctor of medicine (see below) became a superior mathematician involved in the Space program. Both cited the experience with Moore for the change of direction, and there were many like them.

Over time, the popularity of his courses came in part due to his reputation as a teacher and as an irascible character whose personal beliefs

Moore in his office with Michael Proffitt

and attitudes remained steeped in traditions of manners and a view of social order that was fast being overtaken. He intrigued and attracted students, but undoubtedly the main topic of discussion was the style and content of his presentation. Word got around, and by its very nature, calculus lent itself to an ever-searching talent-spotter like Moore. He was not everyone's choice, both for reasons of a personal nature and the distinct differences in the way that he taught calculus compared with other more conventional teachers. As his long-running feud with Applied Mathematics continued, he deliberately magnified these differences to highlight his calculus offerings under the Pure Mathematics banner as compared with the courses posted by rival faculty. He varied his courses each year, partly to fool the opposition but mainly to keep the calculus fresh and challenging to new students.

The author has chosen a small selection of views that reflect the last memories of students who came into his calculus during the latter stages of his career when, as one put it, he should have been sitting in a rocking chair on the front porch. Their views of this period are important, first to provide insight into Moore's unique approach, and then to show, as he neared the end of a career in which he had produced some of the most outstanding mathematicians in America, that he remained highly motivated and unflinchingly dedicated to the goals he had set himself so many years earlier. What may also be seen here is Moore's own assessment of his calculus and his maintenance of the clearly defined ground rules that continue to evoke fascination in the twenty-first century. Bearing in mind that he gave only the briefest of talks and banned any conferring between students, the following description is itself a lesson in the inquiry-based learning techniques that he pioneered:

'I do not, *at the outset*,[2] give my calculus class either a definition of a function as a collection of ordered pairs or a definition in terms of ε and δ of the statement that as $p \to a, f(p) \to b$. I proceed on the supposition that since some or all of them have rough rudimentary ideas concerning these notions, and I prefer to try to refine and develop these ideas of theirs as we go along instead of beginning with definitions which may seem to be arbitrary and artificial and which, if introduced too soon and too abruptly, may tend to inhibit the development of their spatial intuition.... If someone should say, "Don't you want your students to know what a function is?" I think I might be inclined to ask, "Do you know what a function is?" If so, who told you and why did you believe him? Because you thought it was the proper thing to do? And I might be inclined to ask, "Do you know what a set is and will you define the meaning of: the and that and if?"'[3]

Other notes[4] illustrated his determination that his calculus classes were recognizably different from any offered elsewhere. The courses were first and foremost aligned to his own teaching style but also, as a deliberate act, to ensure that there was absolutely no mistaking his classes from those offered by those he saw as his adversaries. He stated that a course in Advanced Calculus when given by him was usually so different from courses with the same title given by others that he decided to use a different title and a different number. About 1941 the title was changed to Introduction to the Foundations of Analysis and the number was changed to 24 and later to 624. He did not usually allow in 624 any student who had taken or was taking Advanced Calculus or Differential Equations. A student who had studied one of these courses

---

[2] The italics, here and in subsequent notes, denote Moore's own emphasis.

[3] Handwritten notes by R.L. Moore, for the film *Challenge in the Classroom*, R.L. Moore Papers in the AAM.

[4] Relating to a draft script for his appearance in the 1966 MAA film about him, *Challenge in the Classroom*.

was likely to know in advance the answers to so many of the questions raised in 624 that he would be inclined to call on him in that course either seldom or not at all. Moore went on: 'One of the questions that I ask, usually near the beginning, in 624 is whether or not there exists, on the x-axis, a closed and bounded point set M such that each point of M is a limit point of M but M contains no interval. On one occasion, sometime after I had raised this question, a student indicated that he could answer it and went to the board. I don't think he had written more than two or three sentences before I became suspicious, stopped him and asked him whether he had read anything on this subject. He replied that he had not but that he had talked to someone about it. I said "Well, that's *enough*! You have *spoiled* this question for this class" and he sat down. It was a long time before I called on him again about *anything*. In the hall after this class period another student said, "He certainly *did* spoil this question. After he said what he did it was easy to see the *answer* to the question."'

An example of what *not* to do, in Moore's view, would be to tell the class at the *outset* to prove that if M is a simple graph whose projection onto the x-axis is closed and bounded then M itself is closed and bounded if and only if it is continuous — to tell them that at the outset and thereby deprive *them* of the opportunity to both *think* of it *and* prove it. He often told a class to prove something that he knew was not true, for example to prove that if a point set is closed so is its projection onto the x-axis. Moore asked: Isn't this much better than to tell them to prove that the projection onto the x-axis of a closed and bounded point set is closed? Why should any teacher want to follow the latter procedure and therefore deprive a student of the opportunity to discover independently that one of these propositions is true and the other one is false? But propositions, true or false, are not the only things to be considered. If it is granted that it is better not to prove or disprove a proposition for a student without at least giving him an opportunity to prove or disprove it for himself, then what about *concepts* and *definitions*? He went on:

'Suppose you would like to know whether there is anyone in a certain class who is capable of thinking of a certain *concept*, thinking of it without the help of any *hint whatsoever*.... Suppose you know of a problem that you believe no one could solve without thinking of that concept. Then how about first carefully avoiding any reference to the concept or anything too closely related to it until you are ready to propose the problem and *then* proposing it.... I would much rather have a concept introduced ... by a student naturally in the course of an investigation even if (I am tempted to say *especially* if) it is not perfectly stated at first.... Sometimes when a student gives a long argument [to prove] a theorem and I know of a much shorter one I do not tell him there is a shorter one. A mathematician from Europe once said to me, "Oh if you know a shorter proof you should tell him. If you don't you are not teaching." I am tempted to paraphrase an often-quoted saying about governments by saying "that student is *taught* the best who is *told* the least."'

This statement was indeed the nub of Moore's style. But how did his students react to it, and especially those for whom mathematics was not their major? How did he leave an impact on their lives at such a vital time in their education? The author has selected Oral History[5] and written testimony of four students to amplify this theme as a pointer towards all that follows in this account, and given the importance to those who were involved, first person recollections are used here to demonstrate their views in their own words.

William S. Mahavier had no intention of pursing a career in mathematics; in fact he hated mathematics all the way through high school. He entered The University of Texas in 1947, and by accident finished up in Moore's class and thus, unbeknown to him at the time, his destiny was settled, and another major figure in the R.L. Moore story was to undergo a life-changing experience. He was undecided on his future career, but had aspirations towards either an engineering or physics major and when he went to discuss registering for suitable courses, he was advised to take applied mathematics because he'd get to work out a lot of problems. Mahavier was not keen, but there was no going back. To major in physics, he would have to take mathematics and so registered for a course he found in the catalog, and it wasn't under the Applied Mathematics banner:

'I did not know who was teaching it and didn't care. I just didn't want to take all those applied classes. So, I walked into class and there was this old man who wrote this strange equation on the board. I had no idea who he was but I remember that first day more than I remember any of the subsequent days in the class. He started by putting the second degree polynomial equation in two variables, $Ax^2 + Bxy + Cy^2 + Dx$

---

[5] Compiled for the R.L. Moore Oral History Project, by Ben Fitzpatrick Jr. for which original tapes and transcripts are maintained in the R.L. Moore Legacy Collection in the AAM.

$+ Ey + F = 0$, on the blackboard in huge letters. It started at one end and went all the way to the other end. Then he went over and sat down in his usual chair to the side and front of the class and said he wanted us to graph that equation. Somebody said "I don't understand what you mean. There isn't any equation; there are no numbers." Moore said, "Oh, well what would the graph be if each of *A, B, C, D, E* and F were 0?"

'I think I was the one that said, "Well, it wouldn't be any graph then; it wouldn't be an equation." Then Moore got up on the blackboard and said, "There wouldn't be any equation?" And he wrote $0x^2 + 0y^2 + \cdots$ and so on ... equals 0 and said, or maybe he put in a 1 and a 2. But he said, "Isn't (1, 2) on the graph?" And then he carefully said what he meant by a graph, and I remember being impressed by that. That was the first time anybody had actually told me what they meant when they wanted me to do a problem. Then he asked if it wasn't true that every point in the plane satisfied the equation and began shading in the entire blackboard with his chalk while saying, "Isn't this what the graph looked like?" And that was the beginning. I don't remember what went on the rest of the day, but I do know that by the time I came to class the next time, I think I knew what that graph looked like in every case where, at most, two of the numbers were not 0, and maybe some of the cases where three of them were not 0. I know I worked a lot, and it was the beginning, and I think of it as amusing now, that I got into that class because I didn't like working problems. In retrospect, I was hooked at that moment.

'I didn't know that for probably four more years, but for the first time I had found something that was challenging and fun and I could do it. I knew what the rules were and I knew how to play the game. I enjoyed that course more than any course I had ever had. The rest of that semester went to plane analytic geometry; there were some very hard problems, and lots of them I didn't get, period. But, he had a real knack for presenting problems that were almost beyond your reach. If you just worked hard enough you could get them, and that was how he wound up developing people. He was a master at deception in a sense, in making you think you were doing things right, and only later, when you look back on it did you realize ... you didn't really know what a limit was, and didn't really know this, and didn't really know that. But the whole scheme of his teaching amounted to gearing his courses to the level of the students. So, what happened was a gradual, almost unobserved, increase in the rigor that Moore demanded in his class. Everything he did was geared to individual students.

'He seemed to know instinctively what was going on in their brains and he would often give problems that were geared to some particular persons in that class, who were perhaps having trouble with some concept. There were often times when I was stuck on something and he would give a hint, which I didn't realize at the time, and got it when I started to think about the problem. Also, it is important to point out that we were having fun. At least I was, certainly as an undergraduate. It had never occurred to me that a person could make a living doing this kind of stuff. I'd never had so much fun in my life. It was more serious, of course in graduate school but through it all, one fact stands out: Moore was totally dedicated to his students and devoted enormous amounts of time to the development of their mathematical abilities. This required that he treated different students in different ways, being harsh with some, gentle with others. But in essence, he gave help when it was needed and refused when it was not, and then stood back to allow the student proudly to present his accomplishments to his peers.'

Mahavier was able to relate an anecdote that provided evidence of patience and, on occasions, humor, from his time with Moore. Mahavier was known to be meticulous about his work and only went to the board if he felt he had a correct proof and, like others, he would be 'devastated' if he made a mistake. One day, about fifteen minutes before the period ended, he had gone to the board to deliver his opus and as he progressed, became annoyed that Moore kept interrupting him, talking about a squirrel outside the window. He kept on talking about how bad it was that some people on the campus chased squirrels away, and some even shot at them.

'I was flabbergasted,' said Mahavier. 'I couldn't imagine what he was going on about, and I just wanted to finish my proof. But he kept talking, until finally the bell went and Moore said, "You can finish it next time." Of course, there was a reason. When I got home and checked the proof, I found a hole big enough to drive a truck through. I spent two nights trying to patch it up and couldn't fix it, so I started over. When we went back to class, Moore asked me if I'd like to continue. I said I would like to start afresh. He just grinned and I went to the board and gave the proof.'

After obtaining his bachelor's degree in physics in 1951, William Mahavier interrupted his studies at The University of Texas and traveled to California, working as a physicist for a year. He returned to UT to teach and begin his PhD studies in mathematics; supervised by

Moore. He was awarded his doctorate in 1957 for a thesis entitled *A theorem on spirals in the plane*. His long academic career thereafter included appointments at the Illinois Institute of Techology, the University of Tennessee and Emory University. His wife Jean also took classes with Moore and, in fact, teaching became the family business, with their two children following in their footsteps.

Another student from the same era was John Worrell, who also had no intention of becoming a mathematician when he went to The University of Texas. Born in El Paso, he went to school in Colorado City Elementary School through the eleventh grade but moved to Odessa to study violin. When he graduated from Odessa High School he had to make a decision whether to seek a career in music or in science. At the time, his ambitions in music were matched by a desire to study medicine, and he eventually decided on the latter route. He enrolled in The University of Texas as a pre-medical student in the Plan II Honors Program and from the beginning, committed himself to taking the fundamental courses in science and not, as he put it, 'the ones that were diluted for advantages for one sort or another, for persons that weren't planning on a career in science.' Worrell decided to take engineering chemistry in the second quarter rather than go with the routine chemistry course. In general, efforts were made in all areas to give Plan II students an opportunity to study with especially distinguished professors. Professor Norman Hackerman[6] was one of them. Worrell had very little background in mathematics. To his great surprise, it was in Hackerman's course that he began to appreciate that he had mathematical talent. He realized in this course that he would need to have more advanced mathematics, and that included differential and integral calculus:

'In the summer, one of the most important things happened that was decisive for my whole life. I went into an algebra course, freshman algebra, and the teaching assistant was away ill and in [his place] walks a rather distinguished-looking individual. His hair was already white, and immediately there was something about him that attracted my attention. It turned out this was Professor R.L. Moore. Instead of simply getting

---

[6] Norman Hackerman, a native of Baltimore and a graduate of The Johns Hopkins University, was in the mid-term of his career as a renowned chemist and academic administrator. He first came to The University of Texas in 1945, fresh from working on a method of producing Uranium-235 for the Manhattan Project. He subsequently held numerous positions at UT, including Chairman of the Department of Chemistry, Dean of Research and Sponsored Programs, and President of the University from 1967 to 1970.

into the class and starting with some lectures, he asked certain questions and he put certain simple things on the board and asked students to give answers to the problems. These were actually rather simple problems in fractions as it turns out, and I was impressed that he would approach the thing from that standpoint. More importantly, I realized instantly (at that time I think I was still 17 years old) that this was a man of extraordinary characteristics. I did not regard him as a person of the ordinary ilk, even comparing him with someone such as Professor Hackerman, I realized there was something extraordinary here.

'After the class session I went up and spoke with him and inquired how I might study with him. I wanted to know what I had to do to get into Professor Moore's sessions. I was able to enter Professor Moore's calculus course in my junior year and I asked him what calculus book I might read in preparation. He made it clear to me, for certain, that if I wanted to study calculus from him I should not read any calculus book at all. So I had the problem of guarding myself against getting premature information about calculus and nevertheless, proceeding ahead with my other science courses.

'I took calculus under Professor Moore's instruction when I was in my last year on campus at The University of Texas. He actually posed problems and expected the students to work them out, and he gave them time to do it in a reasonable sense. He would give you no reward for rote memorization. And in fact, he gave the veiled threat that if you try to memorize from the calculus *I will detect it, and I will excommunicate you. You will not be allowed in my class.* That carries a very powerful message. It was very clear then that he wanted students to work independently. He would raise questions on derivatives; he would raise questions in the first session and ask students to derive representations of a derivative of a certain function. He also introduced, and I think this is an important thing here, definitions of derivatives from a geometric standpoint. He didn't simply give a definition of what a derivative to a function was. In fact, I'm not sure that he even used the word derivative that I recall in the class session. I don't think that he even brought up the topic of the nomenclature derivative. He would ask us to find the slope of a graph at a point, and he used the geometric concept and the geometric approach. I can say, and I think this is important to realize, that Professor Moore had enormous patience. But, he had very little patience with impertinence, and he had very little patience with certain kinds of things that I think a person should not have patience with. But, when it came to matters of intellectual development among students who were earnestly seeking to achieve that, he had an incredible degree of patience. He introduced concepts using geometric formulations, and at first some of what he was trying to get across was obscure. It took me a significant amount of time to begin to comprehend and understand what he was doing, even with his geometric definitions of slope, for example. As a teenager in Professor Moore's class, I found it a challenge to understand *even his* approach to the *definition* and concept of the slope of the graph of a function.'

And now John Worrell was hooked. He had already decided by then that he would continue his studies in medicine, but having had what he described as 'the extraordinary experience' of meeting Moore, he wanted above all else at that time to study with him further. There was a great temptation simply to change direction completely, not go to medical school, and continue working with Moore. In the end he went ahead to medical school and graduated with his M.D. but, as we will see, Moore had not seen the last of Worrell. The experience with Moore remained with him, and indeed developed into a passion for mathematics that brought him back into Moore's classes. Others, however, were not to be diverted from their chosen course, although they found their experiences with Moore an inspiration to their studies. David Briles, for example, went on to become Professor of Microbiology and Pediatrics at the University of Alabama at Birmingham. Although calculus was by then long ago washed from his brain, he was able to look back on his time with Moore as an experience in which he learned how to think and use his mind:

'[He] was someone who has positively affected my life in many ways. I took two semesters of calculus from [him] during the 1964–1965 school year.... I majored in Zoology and during my freshman year had registered as a pre-med student. It was rare that a Zoology major would take classes from Dr. Moore. As I recall, there were two types of serious mathematics courses at UT in the 1960s. Those for the math and physics majors were in a theoretical mathematics tract where the emphasis was on how mathematics worked and how the theorems were developed. Those for other natural science students and engineering majors were of a more practical sort that emphasized working problems rather than them understanding mathematics.

'I came to take courses from Dr. Moore because of the influence of Dr. Wilson Stone who had become my unofficial advisor at UT. Dr. Stone was a member of the National Academy of Sciences, Head of the

A still from Challenge in the Classroom, *a film about the Moore Method*

UT Genetics Foundation and a member of the Zoology faculty. Because of his senior status within the Zoology Department he was not generally assigned to advise students but my father had studied with Dr. Stone as a UT undergraduate before the Second World War. When I decided to go to The University of Texas, my father (by then a successful avian geneticist) called Dr. Stone and asked if he would watch out for me ... and help me pick out my courses.

'In my second year at UT I was ready to take calculus and Dr. Stone told me to sign up for the course with Dr. Moore ... because his major focus was not just on the mathematics but on teaching students to think. By 1964 Dr. Moore must have already been far past retirement. His hair was completely white except for faint permanent yellow stains in the hair along both sides of his head that came from the sweatband of his hat. He wore high-top leather shoes of a type I had only seen in museums or in movies.... Because of his age, formal dress, and his dark, wide-brimmed hat Dr. Moore was easy to recognize even from a considerable distance.... I would often notice him on campus. I might see him walking in to work in the morning or back home in the evening. I would also see him in the library. He frequently came into the library to look up people listed in *American Men of Science*.

'His teaching style was outstanding.... I had heard that he had been a very famous young mathematician, but at the time I never knew exactly what he had been famous for. I respected him for his patience in teaching us very young students a topic that for him must have been very elementary, even when he had originally learned it as a student himself.... I also had a tremendous respect for his dedication to teaching young students. He was still molding young minds when he could have been rocking on a porch somewhere.... His lecture manner was always very casual and open [although] it was also a bit confrontational. He talked to us as though we were not far from his level. His lack of formality was very much like that of the graduate students that taught me algebra, English, or German. Occasionally he would talk to us about things that were not totally related to the proofs at hand. One day I remember him commenting on a view by some psychologists that tremendous leaps of thought and ideas often occurred while one was asleep. A student had raised a question in class on this issue. Dr. Moore said that on originally hearing this claim he had placed a writing pad on his night stand and had written down any thoughts he had on mathematics in the middle of the night that were so intense that they woke him. "Invariably," he said, "the notes on the pad were nonsense when examined in the light of day." He was certain that serious progress in mathematics did not occur in the subconscious, especially not while sleeping.

'Dr. Moore was one of my heroes at UT and ranked with the best of my favorite genetics and zoology professors. [I can't] remember as much about any of my other professors. I like to think that some of what I may have learned about how to think and use my mind from Dr. Moore has helped me all of these years.... The spirit of thinking and learning that [he] tried to impart to us is something that I try to convey to our graduate students.'

As will be evident, whenever Moore is discussed, the praise is invariably tinged with controversy. Don E. Cowley of Austin, Texas, for example, was not alone in the views he expressed about Moore's calculus courses during the 1960s. 'I absolutely agree on the importance of inspirational mentors in young students' lives,' he wrote. 'However, personally I cannot imagine Dr. Moore in that role.... To me he simply seemed to be a self-centered, opinionated, crusty, and very old man, possibly senile. Perhaps, he was no worse than any other of my professors, but as far as I'm concerned, he is a perfect example of everything that is wrong with modern higher education in general. The biggest impression he left on me were of those shiny, black, old fashioned, high-top shoes he always wore.... Maybe Dr. Moore was a brilliant mathematician, but there was no way I could ever have known that from his classroom manner.'

Mary Ellen Rudin was, as we have seen from her earlier recollections, never one to shun the alternative view of Moore's style. She makes no bones about the disadvantages to some students: 'There was no distinction that I could tell between the teaching of trigonometry and teaching of the most advanced course in graduate school, regardless of the students in the class. And [he used] exactly the same technique for calculus. You didn't end up knowing much calculus as a result. You had no encouragement or training in doing the standard type problems or certainly none in being fast using the definitions to construct a proof of a theorem. For this reason the student who was interested in physics or engineering and wanted calculus as a tool, did not find this course useful.

'We would prove theorems about being able to find a minimum and you'd prove theorems about being able to find a maximum and the circumstances under which that is true, but you did not develop any techniques or word problems for handling these. As for the poorer students, they understood more of the theorems than you would think because they essentially, by going to class and listening to this time after time, learned something about it. But most of them dreaded and hated class for the simple reason that it was a painful experience for them. Because they had difficulty presenting things and because the things that they conjectured were often wrong, they were used as an example of how you could be wrong. It wasn't necessarily a pleasant experience for them. In fact, I think that even for the good students, it wasn't necessarily a pleasant experience altogether. You didn't enjoy seeing people fail. You have sympathy for the people who are not doing it properly and to some extent this is a painful process.

'You become antagonistic towards Moore himself, as a person. He elicits this reaction, a very negative one, almost on purpose. I'm sure it was a deliberate act. I still have a very negative reaction, frankly. I have a tremendously negative reaction combined with a very positive one too. That is, there is a combination of these things. I recognize that he was a real artist at his technique. He knew how to elicit your reactions. He was a brilliant psychologist, and he did give you a tremendous amount of training in proving theorems. He built your ego very solidly and very effectively. But [middling students] ended up with mostly an antagonistic feeling and a feeling of defeat. Over and over again.'[7]

The baldness of Rudin's summary is seldom spoken with such firmness, nor one that can be ignored. Indeed, whenever Moore students, or those interested in his work, get together it is not long before the meeting erupts into lively discussion. Counterexamples, as with proofs, abound to provide a balancing view that in fact what Rudin described was all part of the Moore experience, and that those who came through it were just as likely to be prepared for life in so many other aspects than mathematics alone. As to the question of whether his calculus course had value for studies in physics and applied mathematics, there are numerous positive examples, including the work of Dr. John Neuberger. He and John Worrell were in the same calculus class. Both enrolled in the class strongly motivated by their views of a need for understanding calculus in their studies in science. Neuberger subsequently took his doctorate in analysis under the direction of H. S. Wall. He continued with a distinguished career as a research analyst. He has expressed strongly the importance of Moore's calculus course for his subsequent mathematical development. At Legacy of R. L. Moore Conferences at The University of Texas, and elsewhere, he has remarked not only on a progressive awakening to a critical understanding of the foundations of the subject that took place as a consequence of his experience of this course but also on the technical power he gained through it for work in the applications of mathematics to physics.

He compared his ability subsequently in physics courses with that of others who had not had the benefit of having had calculus under Moore and had just had other "applied" courses. As he has described it, there was essentially no contest by them against his academic performance. In a retrospective view spanning over forty years of his work, still intensely active in mathematical research, consultation, and teaching, Neuberger[8] has said that the basis for his accomplishments in mathematics over these years derived chiefly from the grounding he had received in Moore's calculus.

---

[7] From a tape-recorded interview with Douglas Forbes in 1970 for his PhD thesis: *The Texas System: R.L. Moore's Original Edition*; transcript is in the R.L. Moore Legacy Collection in the AAM.

[8] At the time of this writing, he is a mathematical consultant at Los Alamos National Laboratory, among other places. He is Regents Professor at the University of North Texas.

Similar testimony is afforded by John Green, who received two PhD degrees, the first with R.L. Moore, and a number of years later, the second from Texas A&M in Statistics. He subsequently became Principle Research Biostatistician at DuPont's Haskell Laboratory for Toxicology and Industrial Medicine where he has been described as 'in the firing line of some of the most important questions facing civilization'. He gives Moore much credit for his ability to withstand the pressures of such an environment, and actually found value in what he described as a relatively narrow field of mathematics within which Moore worked:

'One of the keys to the success of the Moore method of teaching is I think going from strength to strength. That is we develop an extraordinary depth of understanding in a relatively small area of mathematics. The strength this gave us can later be used as a stepping-stone to building strength in other areas. Many of Dr. Moore's most successful mathematics students went on to make significant contributions in subjects in which they had no formal training. This is the hallmark of a great education.... The Moore method is not for the faint hearted. Having said that I would also point out that I was excruciatingly shy as a boy and as a young man. I would never volunteer to go in front of class and had a horror of presenting to an audience, and rarely spoke out in groups of more than three people. Dr. Moore did not ask for volunteers. Indeed, he ignored volunteers; he called on whomever he wanted. Had it been otherwise, I wouldn't have survived in his class.... [The] self-reliance that was developed and continually strengthened in Dr. Moore's classes has proved invaluable in my industrial career.... We learned that Dr. Moore was never going to tell us how to prove a theorem or construct a counterexample. If we didn't do it, it wouldn't be done.... One benefit of working with Dr. Moore was learning how to work with [the intimidation] factor. First, there is the man himself. [He] was a very imposing figure, he dominated his environment. We were all in awe of him on many levels.... [And] having worked under [him], no one else can intimidate me. [Equally humor was] an important part of Dr. Moore's classes. He had a keen sense of humor and shared it with his classes. He encouraged us to laugh at mistakes. But, did so in a way that even the person making the mistake did not feel he or she was being laughed at, just the mistake. Everyone was treated with respect and dignity, but statements and ideas could be laughed at, torn apart, discarded....

'He wanted us to want to follow his approach and to do that, we had to understand it.... Indeed I had to go through a period of anger towards

Dr. Moore, to stop working on the problems that he gave us and start working on what became my dissertation. The one time I told him what I was doing, he discouraged me from doing it. As angry as I was at the time, I ignored that advice. If anything, it encouraged me to continue in order to demonstrate to him that I had the insight that he might not credit.... By the time my dissertation was finished, I was over my anger.... Whether Dr. Moore deliberately made me angry at the time of my dissertation I don't know. I am sure he did not take a simple approach to teaching. He looked for and used many tools to develop students.

'He did not employ the same approach with everyone;... [nor was] Moore's influence upon his students restricted to the subject matter of his classes. It was an entire way of life. It permeated everything that we do.... What made a big impression on me was that he had taken such an interest in my future. This is characteristic of his approach to teaching. He did take a very keen interest in who took his classes, what they got out of them, how best to develop them, what to develop in them. He told us early on that he had no use for the university guidelines stating that we should expect three hours of outside class work for each hour in the classroom. He said he wanted us to think about his class all day, every day, to go to bed thinking about it, to wake up in the night thinking about it, to get up the next morning thinking about it, to think about it walking to class, to think about it while we were eating. If we weren't prepared to do that, he didn't want us in his class. It was also quickly evident that he meant exactly what he said.... He approached the teaching of mathematics with his entire being. He conveyed a deep love of the subject and all aspects. He saw mathematics as an activity. A way of life.... I truly experienced the love of learning under Dr. Moore's influence. He took me to another level of happiness that I had not experienced before, and I have carried with me ever since.'[9]

Thus speaks the evidence of a variety of students exposed to Moore's ways early in their academic careers, and the diversity of opinion generated in their comments is interesting and reflective of the man himself.

---

[9] Extracts from a presentation by John Green at The Fourth Legacy of R.L. Moore Conference, 3 May 2001.

17

Changing Times (1953–1960)

Moves are afoot to weaken Moore's domination of the Pure Mathematics Department by forcing him on to modified service in advance of merging Pure and Applied Mathematics which Moore had maintained would never be a single unit. Some were campaigning to have him retired altogether. Moore was having none of it; altercation and drama ensued...

Moore in his office, ca. 1963

*H*aving lost the battle to keep the Department of Applied Mathematics separate from his own Pure Mathematics in 1953, R.L. Moore now faced a secondary and more personal confrontation in regard to his continuing service at The University of Texas. It was one that developed eventually into bitterness and a prevailing mean spirited atmosphere which undoubtedly left him with the feeling that elements among the hierarchy wanted him to retire regardless of his outstanding contribution, and no one could argue against that, to the reputation of the university and to mathematics at large.

His steadfastness, combined with his popularity among his students, was such that it would require a devious ploy to depose him, and an attempt to hatch one began as he headed toward his seventieth birthday in 1952. For some time, correspondence between administrators and senior faculty focused on a single department that would, in the fullness of time, either cancel out Moore's dominance by sheer weight of numbers or bring the man himself to discover other passions in life that he might now like to address. Typical was a memo from Dean C.P. Boner to President T.S. Painter stating that effectively it would be impossible to combine the two departments until Dr. Moore became inactive. In other words, he should be encouraged to retire. The Dean was right about the fate of the Mathematics Department. Although in theory the two departments were merged under one roof, they continued to operate on a self-contained basis in the new campus building, Benedict Hall, with Moore and his colleagues on the third floor and Applied Mathematics on the second floor. The antagonism between them permeated down through faculty and students alike, and the friction manifested itself through the emergence of two groups: those who were 'third floor' and those who were not.[1] This de facto separation continued for some years, and was recalled by a student of the era, Linda Cheatham: 'Some students know it as a kind of Cold War between the third floor and second floor faculty members of Benedict Hall.... Faculty members call it a "difference between two points of view." Whatever the identifying tag, math faculty as well as almost any student who has taken a math course on this campus are well aware of the active, long-standing disagreement between two factions of the University mathematics

---

[1] R.E. Greenwood, *History of the Various Departments of Mathematics at the University of Texas at Austin, 1883–1983*, p. 58.

department: the proponents of the "theoretical" or pure school of math versus the champions of the "applied" school.'[2]

As to the possibility of his retirement, Moore would not hear of it. He was determined to remain at his post until he dropped, or was sacked. This resolve was supported by a strong rumor that he was subscribing to an annuity that would not begin to pay out until he was 120! He certainly gave the impression his personal goals were far from maturity. Further public indications of his determination to remain in situ arose when a new battlefront opened up, coincidentally at exactly the same time that Pure and Applied supposedly became one.

The move into Benedict Hall coincided with Robert Lee Moore reaching the age of 70 and, according to university rules, his terms of employment would be reduced to modified service. It was a day for which certain members of the administrative body of the university had long been waiting. Those who felt that mathematics at Texas should not be so strongly influenced by Moore were taking steps to diminish that influence, and behind the scenes there had long been close study of the rules under which modified service operated, especially those that governed the amount of time a professor on part-time duties would be allowed to spend at the university. The situation, as the administration people perceived it, was clear: modified service meant half time and half pay. But was there any rule that specifically barred the modified professor from working full-time for half pay? The thought may not have occurred to those pondering Moore's situation, but it would soon be brought to their attention in fulsome manner.

During the fall of 1953, a petition from the Faculty Council to the Board of Regents was made, asking for firm clarification of the rules and regulations of the Board in regard to modified service. This inquiry was placed before an October meeting of the Faculty Council when Dean Boner 'sought the advice of the Faculty in regard to questions which had arisen in the administration of the modified service regulations'.[3] It was the beginning of a long, drawn-out saga which ran for more than fifteen years before a dramatic climax that would bring Moore's teaching career to an end.

Some months earlier, University President Logan Wilson had appointed a committee to study the issue of modified service, resulting

---

[2] Linda Cheatham, Cold War Going in Math Department?, *The Daily Texan,* April 16, 1969, Vol. 68, No. 149, p. 4.

[3] Minutes of the General Faculty, University of Texas, 9 March 1954, p. 6825.

ultimately in recommendations by the Faculty Council to inject precision into the rules under which members of faculty who had reached the age of 70 were allowed to continue teaching. As well as Moore, these included Ettlinger and Vandiver although it was assumed that both of these would 'go quietly'. Proposals were to be submitted to the General Faculty in formal session after a ten-day period to allow protests to be lodged. A minimum of eleven protests were required, otherwise the recommendations of the Faculty Council would become final.

The principal element of their deliberations, listed as Item I, called for a firm ruling that no person on modified service should carry out more than one half the regular duty of a full-time employee. Thus, Moore's presence would be halved if the rule were adhered to. Eleven faculty members duly raised their protests and not unexpectedly, they were led by Professors Moore and Ettlinger, both of whom would, ere long, become candidates for the new modified service directive. They were supported by a number of their colleagues from the third floor and Moore entered a two-sentence objection in which he stated: 'I protest the action of the Faculty Council taken on January 18, 1954, concerning modified service, and I request that this matter be referred to the General Faculty. I regard Item I as absolutely inexcusable and indefensible.'[4]

The General Faculty was therefore convened on 9 March and the proceedings opened with the secretary calling for the adoption of the recommendations of the Faculty Council, supported by David Miller, chairman of the Special Committee that had presented the original report to the Faculty Council. Miller pointed out that the original intention was to have everybody completely retired at 70 but experience had shown that it might be wise to retain members over that age provided they were capable of performing their duties. Therefore, the committee had proposed that the budget councils of the departments should recommend to the administration whether or not they thought that the person in question was capable of continuing to teach. These were merely recommendations to the administration and to the Regents so that they would have a basis for determining whether a particular individual should be completely retired or retained on modified service. Mr. Miller thereupon moved the adoption of Item I, the principal bone of contention as far as R.L. Moore was concerned. It read: 'No person on

---

[4] Contained in Faculty Council minutes and documents, The University of Texas, 24 February 1954, p. 6255. All other excerpts from the meetings quoted within this chapter are taken from the same source, between pages 6286–6290.

modified service shall carry more than one-half the regular duty of a full-time employee. This work may consist of classroom teaching, conference instruction, research, directing research projects, or other departmental or university duties, but in no case shall the sum of these responsibilities exceed one-half of regular full-time duty.'

Not for the first time, and certainly not the last, there was an air of tense, almost electric, expectancy as Moore rose immediately to his feet and turning half towards the chair and half towards his fellows, pronounced his verdict of the committee's deliberations:

'Is this a university or is it a union shop, its so-called faculty consisting of mere employees, each doing a specified amount of work for a specified wage, each fearful that if one man does more than he is paid to do there will not be enough work left for the others to do, a union shop where, according to the Faculty Council minutes, one member says "the question is whether you could put more than one-half time work on a man who got only one-half time pay, even if he wanted [to do] more," as though teaching a class that he is not specifically paid to teach is bound to be a burden and if a man asks that such a burden be "put" on him then either he is just trying to appear to be accommodating, but is secretly hoping that his request will not be granted, or he is so far gone mentally that he does not realize what he is doing and therefore should not be allowed to teach at all? A union shop where the chairman of a committee brings in a report containing an item requiring that no one on modified service shall be allowed to teach more than half time and apparently is strongly in favor of that requirement but, when asked by the president whether the committee had given any attention to the minimum load of persons on modified service, shows little or no concern as to whether or not these people should be required, or even expected, to do anything at all worthwhile, saying that if a person is "not capable of teaching he could be assigned to research or something of that nature?" If a person on modified service who had formerly taught full time is no longer capable of teaching at all, I wonder what sort of research, if any, he would be capable of carrying on?

'I hope that most members of this faculty do not have this union shop attitude. If a man on modified service does more than he is paid for and his work is sufficiently good and he is doing it with no expectation whatsoever of any extra pay then, instead of condemning, should not the administration approve of the extra work he is doing on the ground that, by doing more than he is paid for, he compensates to some extent for those who are doing less than they are paid for, and in that way makes some contribution toward making the modified service system pay its own way?'

Moore was under no illusions as to whom the new rules were aimed at. He said he had made investigations among other departments and there were perhaps only two, at most, on modified service who were working longer than the hours expected of them and then went on to make what one close to him at the time described as a 'remarkable and quite moving summary of his career, tinged with irony, but arrow-sharp in its point' as he outlined the impact of the new rules on one man in particular, himself, which he delivered entirely in the third person:

'... the only present target [of the intended alteration to the rules] is a man who is giving five courses. I have full authority to speak for him. I will call him Professor X. He has felt in the past, and he still feels, that he can contribute much more to the advancement of his subject through his own researches and those of his students ... [and] with this in mind he has been giving a sequence of five courses. The first one, 613, is one section of a sophomore course which he has been giving partly, but not wholly, because he felt that he might discover some new material there. The next course, 624, is only one of a number of sequels to 613, the other possible sequels being given by others. The next one, 688, is only one of a number of possible sequels to 624, and so forth through 689 and 690. No course in this sequence beyond the first one has ever been given here by anyone else. At the present time, in each of these courses except, of course, the last one there is prospective material for the next one and thus there is a reason for continuing to give all of them ... it would be unsatisfactory to have one man give one or more of them and have another man give the others.

'The first PhD with the dissertation under the supervision of Professor X was conferred in 1916. The last one so far was conferred in 1953. The interval from May or June, 1916, to May or June, 1924, will be referred to as an eight year period. So will the one from May or June, 1917 to May or June, 1925 and so forth.... If Professor X were forced to pick out one of these groups of students and say that it was the best one I think he would say the last one, the 1945–53 group.[5] He would be

---

[5] The group referred to consisted of the following: R H Bing (1945), E.E. Moise (1947), R.D. Anderson (1948), M.E. (Estill) Rudin (1949), C.E. Burgess (1951), B.J. Ball (1952), E. Dyer (1952), M.E. Hamstrom (1952), J.M. Slye (1953).

inclined to say that in spite of the fact that [an earlier group] included, among others, both the man who is now serving as President of the American Mathematical Society for the two years 1953 and 1954[6] and the one who has been elected to serve in that capacity during the two years 1955 and 1956.[7]

'The total number of people who have received PhD degrees with dissertations under the supervision of Professor X is 26. Nine of these belong to the 1945–53 group. Seven of the 26 are now members of the Council of the American Mathematical Society, a body which in 1953 had a total membership of 48.... There are five of the 26 who are either starred in *American Men of Science* or listed in the Distinguished Group in the recently published *Leaders in American Science*.... So far as I have discovered, the total number of all people who have received PhDs at this University with dissertations under the supervision of any present member of this faculty outside of the Department of Mathematics and who are either so starred or so listed is two. One of these two received his degree in Zoology in 1915 and the other one received his in Organic Chemistry in 1941....'

At this point, the President interrupted Moore to say that the secretary had called his attention to the faculty rule limiting a speaker to ten minutes on any one proposition and that this period had been exceeded by Moore. From the floor, it was moved that the speaker be given the privilege of extending his remarks, and the motion was seconded and adopted without dissent. Moore continued:

'Making use of *American Men of Science* and other sources of information, I have consumed much time trying to find out whether or not any one of the approximately 65 people who have taken PhD degrees in Physics at this University is starred in *American Men of Science* or is on the Distinguished List in *Leaders in American Science*, or is a member of the Council of the American Physical Society or is on the faculty of any member of the Association of American Universities other than The University of Texas. So far I have found no-one fulfilling any one of these conditions except the last one and the only one I have discovered who fulfills that is a professor at the University of Illinois and he is a professor of Electrical Engineering, not of Physics.'

---

[6] G.T. Whyburn.
[7] R.L. Wilder.

'In 1948–49, in 613, the first course in the above mentioned sequence of five, there were four undergraduate students who were majoring in Physics. In 1949–50 all of them were in 624 and they were joined in that course by a student also majoring in Physics who had graduated at the California Institute of Technology. Of these four undergraduates, one graduated in 1951 and continued in Physics. The three others graduated in 1951 (one with highest honors) and are now all enrolled in 690, the next most advanced course in that sequence of five. The graduate major in Physics,[8] who was with them in 624 in 1949–50, took no more courses in Physics after that year though he wrote a thesis for a Master's degree in Physics and received that degree in 1951–52. In 1953 he received a PhD degree with Pure Mathematics as major and Physics as minor subject. He went to the University of Minnesota as an instructor in mathematics.... I consider him one of the very best of the entire 26. Professor X would like very much to have the opportunity to try to make the eight year period from 1954 to 1962 at least as good as the one ending in 1953. But I think that it certainly will not be at all possible for him to do it if Item I is put into effect....

'There is a principle involved here. Should anyone have to ask the permission of the administration every time he wishes to do something that is not prescribed?... I hope that this faculty does not approve of any motion which even suggests that doing more than one is required, or expected, to do is irregular and that anyone who wishes to do it must first get the permission of the administration. Has uniformity become a virtue and has every departure from it become an offense to be tolerated only by special permission in each individual instance?'

Now it was the turn of Ettlinger to lambaste the perpetrators of the recommendations before the faculty, although his contribution was barely necessary. From the murmurs around the hall it appeared that Moore might well have won the day, but as has been previously noted, there were those present who made a point of voting against anything suggested by R.L. Moore, regardless of their views on the motion. Ettlinger, therefore, continued the attack:

'The principle of "Thou shalt not render more than you are paid for", is extremely undesirable. Who will hold the stopwatch on a faculty member engaged in his own research, or in conference with a graduate or undergraduate student, and tell him, "Just seven more minutes and

---

[8] J. Slye.

you must cease and desist or you will be violating the work rules of The University of Texas...." In Item I, the department is adjudged incompetent to exercise good judgment to protect the younger faculty members in their desire to offer courses in their fields, and also to protect students from incompetent teaching. But who will install the time clock in the department, who will patrol the laboratories, who will take the reading of the light meters? Will quarterly reports be made and to whom as to the precise expenditure of time?... How will it be determined whether "classroom teaching, conference instruction, research, directing research projects, or other departmental or University duties" add up to two-thirds or five-fourths of regular full-time duty and hence constitute a violation of the work code for modified service? This item just does not add up and this General Faculty should not recommend a regulation which is unenforceable or without meaning. This regulation is not compatible with the ideals of a university of the first class.... Alfred North Whitehead served brilliantly as a faculty member of Harvard University from the age of sixty until close to eighty-five, without any regulation circumscribing his intellectual activities. Cannot The University of Texas afford the luxury of an Alfred North Whitehead?

'For these three reasons, that the regulation will prove to be unenforceable, that it is undesirable since the department can best protect the interest of its own faculty members as well as its students, and because it will handicap unnecessarily The University of Texas, I favor the deletion of Item I of the report.'

Put to the vote, Item I was defeated by a margin of 82 to 62. But Moore was not let off the hook. A secondary mantrap had been prepared in the guise of Item V which, if adopted, would decree that all employees would be retired automatically on their seventy-fifth birthday. Moore was not in favor of that rule either:

'If a man over 75 shows by his actual performance that he is still outstandingly capable of continuing to teach and inspire and contribute to the advancement of his subject, he shall nevertheless be replaced either gradually or otherwise by some younger man who no one has any good reason to think could come anywhere near matching either his past or his future performance.... Now if this Committee has planned its program without reference to the possible existence of such outstanding men, then it has been planning for a mediocre university, perhaps for a university in which there is such a dead level of mediocrity that whenever a member of its faculty nears the age of seventy then, no matter in

R.L. and Margaret Moore in front of their home, June 1954

what subdivision of what field he is working, there is always a younger man who is just as good (and probably better since he is younger) and who is available and procurable and easily made ready to gradually take over and continue with his work. Does anyone think that really first class men are that plentiful?'[9]

Once again, he carried the day, leaving the hierarchy to ponder their next move. They were not long in making it, this time attempting to swamp the Moore camp by doubling the number of protest votes required to ensure that new legislation could be put before the General Faculty. When that, too, failed to win support the attempts to reduce Moore's influence and to halt his avowed intention to teach as he wished when he wished, were temporarily halted, but not for long. Although the whole issue flared with predictable regularity over the coming decade or so Moore continued on regardless for years hence, disregarding the rules and regulations, which were supposed to restrict his activity.

Indeed, it may be seen that in the year of 1953, when this whole unseemly business forced itself into public debate, Moore had already set himself on a course that would steer him into his most productive era, when measured by the number of students he supervised towards their PhD. In addition, his close associates H.J. Ettlinger and H.S. Wall added substantially to that roll call and together the three men had established a hold on Pure Mathematics that was formidable in every sense. Wall had long ago thrown himself wholeheartedly into the Moore tradition, with his own interpretation of the Moore Method, and there was a good deal of cross pollination of students through their courses, some steered to their PhD by Moore and others by Wall and Ettlinger. Between them, they continued to dominate PhD guidance in Pure Mathematics throughout the 1950s and 1960s. The respect and affection demonstrated by their students remains to this day a memorial to their efforts. The recollections of those under Moore abound on these pages and there is ample testimony for Ettlinger and Wall whose pedagogical techniques, inspired by Moore, also placed them at odds with the University administration. One student wrote: 'Hubert Wall was completely devoted to mathematics and to his students. However contentious he may have seemed to administrators and colleagues, only this devotion was evident to his students.'[10]

---

[9] Minutes of the General Faculty, 23 March 1954, p. 6297.
[10] H.S. Wall, Memorial Resolution, Documents and Minutes of the General Faculty, The University of Texas at Austin, 1971, pp. 10433–10438.

There is no doubt that the outstanding record of Moore, combined with that of Wall and Ettlinger, and other faculty on the third floor, contributed to what, in the end, was a very substantial outreach of students exposed to the discovery learning techniques embodied in the Texas courses in Pure Mathematics into the major centers of higher education across the United States. By the 1950s, that effect was already being hugely demonstrated in the work, publications and status of their students in the international fields of mathematical research and teaching. Nor, by any means, was this influence on the wane. Quite the reverse, and the Pure Mathematics department continued on regardless of the detractors who felt students should be brought toward more modern frontiers of mathematics through the conventional route of lectures and reading assignments. As Traylor wrote[11]: 'Clearly, Moore's age was becoming a disadvantage to him. It could be suggested, or stated, that his mathematics was old and outdated and his own age would only tend to support such a claim might well be true. What of the problem of hiring new faculty? Who would wish to come to Texas and compete with R.L. Moore for students? Might it not be true that the better student would gravitate toward Moore and away from other faculty, or if not true, it might well be thought to be true. Thus some faculty would choose to reject an offer from Texas, some even stating that Moore's presence would keep them from coming. To an administrator who was trying to hire new faculty, it surely would occur to him that Moore's presence was an obstacle.'

All of these aspects were a matter of concern internally as the 1950s turned into an era of great change, but far more vital issues were emerging across the whole landscape of American social and educational activity, changes in which The University of Texas was well to the fore, sometimes for the wrong reasons and not least over the issue of segregation. Texas was not a prompt participant in the moves toward integration in schools and public facilities, in spite of the progress made elsewhere in outlawing discrimination, notably in the defense industries to which UT was a supplier of manpower.

---

11 Reginald Traylor, *Creative Teaching: Heritage of R.L. Moore*, pp. 180–181. In spite of those reservations, however, it will be seen in Chapter Eighteen that Traylor was responsible for attempting to persuade Moore, by then 85, to teach at the University of Houston when, in 1969, it seemed clear that Moore's long association with The University of Texas was about to be severed.

Certainly no softening of attitudes existed within the educational hierarchies of the South until the issue was forced upon them. As far as R.L. Moore is concerned, his position is well summed up by the recollections of a student from that era, J.C. Davis: 'I treasure the experience of Moore as one of true excellence in liberal education. Few professors at UT came up to his standard.... Alas there was one downside. He was an unreconstructed segregationalist just as that attitude was going very far out of fashion.'[12]

Some would argue that Moore was against any reforms to the Texas state law, which actively encouraged African-Americans seeking higher educational degrees than were available at the black-only establishments to go elsewhere. One who experienced it ironically became a doctoral descendant of R.L. Moore. Beauregard Stubblefield,[13] the son of a Houston watchmaker, showed an aptitude for mathematics by assisting with his father's bookkeeping, but his attempts to turn that aptitude into a career were met by obstacles typical of the times. He was educated at the Booker T. Washington High School and subsequently obtained his Bachelor's degree in mathematics after three years at the Prairie View A&M College for Black Americans in 1940. The establishment was the second oldest institution of higher education in Texas, and had managed to survive difficult circumstances, which reflected the struggle of blacks in Texas for educational opportunities. The Texas constitution, in separate articles, established an *"agricultural and mechanical college"* and pledged that *"separate schools shall be provided for the white and colored children, and impartial provisions shall be made for both."*

Stubblefield took time out working as a teacher for two years and returned to take his Master's when Prairie View became a university in 1943–1944. He was unable to pursue his ambitions for a doctorate at Prairie View because they did not offer that degree, nor could he transfer to The University of Texas: 'As blacks in Texas, we knew what we could do and what we couldn't do and when you're trying to get an education you don't buck the system. You do what you can to beat the system. I knew I could not go to The University of Texas at that time even

---

[12] J.C. Davis, 28 May 2000 in a letter to Howard Cook, in response to a survey of R.L. Moore students for The Legacy of R.L. Moore Project, R.L. Moore Legacy Collection in the AAM.

[13] Beauregard Stubblefield, interviewed by Albert Lewis for the R.L. Moore Oral History Project, 8 April 1999, R.L. Moore Legacy Collection in the AAM.

if I wanted to, and under Texas law, the state would pay my way to any school in the country that I chose. My teacher at Prairie View, Dr. Clarence Stevens, suggested the University of Michigan. I wrote to them, along with three others, but only received one reply from the University of California, stating that they only took their own students into the doctorate program. So I just drove up to Michigan to enroll.'

He subsequently gained a place at the University, but then even more obstacles presented themselves. Michigan ruled that he could not be classed as a resident of that state because he received a student's allowance from Texas. It was necessary for him to drop out of education for almost six years to save enough money and qualify for a doctoral program. In the meantime he took work at a jeweler's shop close to the university campus, repairing watches. It was while working in the jewelry store that he came into contact with a regular customer, R.L. Wilder, by then a professor at Michigan. Chatting over the counter, Stubblefield related his interest in mathematics but did not mention at that time he came from Texas. The upshot was that the watch repairer with a Masters in math gained admission to Michigan in preparation for his PhD. He took courses from Wilder, one of the finest exponents of the Moore Method, and from Gail Young, the latter becoming his PhD adviser. Ed Moise was also at Michigan at the time, and engaged Stubblefield in many mathematical discussions.

Stubblefield said he felt entirely comfortable with the Moore Method and carried much of it into his career in mathematics, principally at Appalachian State University and, after his retirement, as a research mathematician concentrating on number theory. Although he was heavily exposed to the Moore Method en route to his doctorate, and thus became a mathematical descendant of its creator, Stubblefield never met the man himself, even after The University of Texas made its move towards desegregation. Even then, it was always doubtful that he ever would: 'I heard stories [about Moore] over the years, and that one time a black got into his class and sat for a while. R.L. Moore said "I am not going to teach any of this course until a certain person leaves my class." Moore, as I understand it, refused to teach blacks.'

In fact, although Moore had a tendency to open up controversial discussions on social issues during a blank day in class, the possibility of him having to take on a black student in his courses did not arise until the late 1950s. For some years, lawyers for the National Association for the Advancement of Colored People had instigated a series of important legal challenges nationally, in which they argued that segregation meant unequal, and thus inadequate, educational and other public facilities for blacks. Their landmark victory came in the case of *Brown v. Board of Education of Topeka, Kansas* in 1954, when the Supreme Court ruled that racial segregation was in breach of the Fourteenth Amendment to the Constitution, that no State may deny equal protection of the laws to any person within its jurisdiction. The decision declared that separate educational facilities were inherently unequal and the culmination of this series of cases ended in a total reversal of an earlier Supreme Court ruling (*Plessy v. Ferguson, 1896*) that permitted 'separate but equal' public facilities.

The University of Texas, meanwhile, was confronted by its own civil rights case, that of *Sweatt v. Painter*. The plaintiff, Herman M. Sweatt, was a black postal worker from Houston who filed his suit after being denied admission into The University of Texas School of Law. The Texas State Board of Regents responded by creating a "separate but equal" law school for blacks at Texas State University in 1947. The law school opened in Austin and moved to Houston a year later. The Sweatt case, nonetheless, made its way to the U.S. Supreme Court and ultimately the Board of Regents declared in June 1956 that hereinafter The University of Texas would strive toward reaching total desegregation by the mid-1960s. It would, of course, take much longer, especially in regard to facilities around the campus that were outside of university control.

Walker Hunt, whose parents were sharecroppers, was among the earliest to benefit. He enrolled in 1956 and had as fellow students A.N. Stewart and L.L. Clarkson, who became the first African-Americans to gain their PhDs in mathematics at The University of Texas. All three studied with Ettlinger and Wall. Ettlinger was Hunt's thesis director for his M.A. gained in 1962. Hunt reports: 'I also wanted to take Robert Lee Moore's famous Foundations of Point Set Topology [course]. However, that was not to be. The reason: I was black! I used to feel that I was short changed! Over the years, I have come to the realization that it was for the best.'[14] Hunt then spent time teaching, eventually at Prairie View A&M College where A.N. Stewart was appointed department chair. He was subsequently encouraged to return to UT to study

---

[14] Walker E. Hunt, Personal History of Walker E. Hunt, www.math.umd.edu/users/rlj/whunt.html.

for his PhD, but with Ettlinger's health by then failing, his dissertation director was Professor Don E. Edmonson. He had acquired a position at San Antonio College where, at the time of writing, he continues what has been a long and fruitful academic career.

Edmonson was also dissertation director of Vivienne M. Mayes when she became the first female African-American student at The University of Texas to gain her PhD in mathematics, although it was an uncomfortable beginning in that she was the only African American in her class, and the only woman. She too was denied access to a particular course she wished to take with R.L. Moore. The civil rights movement was at its height during the years she was in graduate school, 1962–1966, and she joined picket lines to force restaurants and movie theaters to admit blacks. She later wrote: "[At the time,] I could not join my advisor and other classmates to discuss mathematics over coffee at Hilsberg's cafe ... [they] would not serve Blacks.... It took a faith in scholarship almost beyond measure to endure the stress of earning a PhD degree as a black, female graduate student.'[15] Dr. Mayes was hired at Baylor and in 1971 the Baylor Student Congress elected her Outstanding Faculty Member of the Year.

Although she was denied a teaching assistantship while at UT, she did act as a grader for a linear algebra class taken by Raymond Johnson who earned his B.S. in Mathematics from The University of Texas in 1963. He too points out that when he arrived, dorms were still segregated, as were sports and most other aspects of campus life. He writes: 'I decided to major in math because it was one of the things I had enjoyed most in high school.... At Texas, you were either [Applied] or [Pure]; there was no mixing.... The real head of the Pure math department was Robert Lee Moore.... I learned of his prestige (before I became identified as a third floor student) and I thought of taking one of his courses. [My friend] Walker Hunt [advised me against it]. The image of R.L. Moore in my eyes is of a mathematician who went to a topology lecture given by a student of R H Bing. The speaker was to be one we refer to as a mathematical grandson [of Moore]. When Moore discovered that the student was black, he walked out of the lecture. (Parenthetically..., Bing was by then a topologist of world-renowned stature and Texas desperately wanted to attract him back from Wisconsin. Word was that

---

[15] V. Mayes, Black and Female, *Association for Women in Mathematics Newsletter* 5(6) (1975), 4–6.

Bing had said he would never return to Texas while Moore was there. Moore died, a year or so later [and] Bing returned to Texas. I have a very different image of R H Bing.)'

Raymond Johnson went on to graduate school at Rice University in Houston, which had also experienced difficulties over desegregation. Rice foundation rules stipulated it should educate only white citizens of Texas. The university changed its admissions policy and Johnson was accepted. Two alumni sued and he was forced to spend his first year as a research assistant, without attending classes. Rice won the suit and Johnson was fully admitted. He was the first African-American to graduate from Rice University.[16] In a round-about way, therefore, all of the early African-American mathematics students accepted into The University of Texas were exposed to Moore Method teaching, although not necessarily with the blessing of its founder.

---

[16] From Professor Raymond Johnson's web page at www.math.buffalo.edu/mad/PEEPS/johnson_raymond1.html.

18
Axiomatics Continued (1953–1965)

Moore's advanced courses are examined at a time when issues such as Civil Rights, the Cold War, Vietnam and higher educational needs filled a major part of political campaigns. What of the Moore Method as classes grew? Recollections from exceptional students from this era, along with discussion as to how the Moore Method could fit the modern age.

Moore on the UT campus, clock tower in the background

Attempts by the hierarchy to curtail R.L. Moore's influence and domination of Pure Mathematics at The University of Texas by forcing him into modified service had little effect on his productivity, and certainly none at all when the output of graduates under the ruling triumvirate of himself, Wall and Ettlinger was tallied as they moved out of the 1950s and into a new decade of dramatically changing scenarios, which would in time have a profound impact on educational institutions across the country. The prospect of nuclear annihilation suddenly looked a possibility, Civil Rights issues moved into an even higher arena of public strife, the Soviets had made early advances into Space prior to the American Moon landings, and ahead loomed Vietnam.

Education, too, filled a major part of political campaigns as classes grew larger and the need arose for an ever growing and varied syllabus that fully catered to students bound for industries pioneering new concepts, especially in engineering, electronics and the technologies that gave birth to the computer age. Coincidentally occurring as a parallel issue was the recognition of the legal requirement to accommodate African-Americans and other ethnic groups into mainstream university life. That transformation went at a slow pace in Texas. R.L. Moore, meanwhile, continued to use calculus as a place for spotting talent and, once discovered, it was nurtured in his higher level graduate courses through the use of his axiomatic systems for plane analysis situs. It was a procedure that had progressed through his teaching career and in so-called semi-retirement, which as far as he was concerned had not happened. He continued to show the kind of results that appeared only in the dreams of many professors of pure mathematics. Some considered, and stated openly, that Moore was living in the past, obsessed with a bygone age. Did he not, after all, touch his hat to every female he passed on the sidewalk, wear a straw hat on a hot day, never walk out without jacket and tie, and never be seen without his specially made shiny, black high-topped shoes that belonged in the last century? But many of his old Southern attitudes were still not as out of place in Texas as his shoes, observed as ever by the campus statue of Robert E. Lee.

Simply from the perspective of being a teacher, however, most would say that it was a grotesque misrepresentation to label him a has-been. He possessed decades of experience that he was very capable of transforming daily into a vehicle for imparting knowledge and ideas — not just his own, but those of the many students who had worked with him. This was demonstrated in a letter he wrote to the Office of Naval Research in December 1960 to recommend his former student G.W. Henderson for a post-doctoral fellowship in Naval research.

'In 1921, S. Marzurkiewicz raised the question whether or not it is true that if a continuum in the space of m dimensions has the property of being topologically equivalent to every one of its nondegenerate subcontinua then it is an arc. In 1949, in his doctoral dissertation, E.E. Moise showed that there exists a plane indecomposable continuum which has this property but which is not an arc. That left open the question whether every decomposable one is an arc and I believe I have raised this question with one or more of my classes certainly every year since 1946, until 1958, and probably for each of many years before then, but it remained for Henderson to answer it. In his 1959 dissertation[1] he answered it in the affirmative. I have reason to believe that some very able mathematicians whose names I would prefer not to make public, worked long and hard in a vain attempt to settle this question and I think it was quite an achievement for Henderson to settle it.'[2]

There was no compromise by Moore, either, in dealing with the various external pressures that, in his view, were imposing themselves on university life. Ettlinger and Wall, and the rest of the Pure Mathematics faculty, were happy to accept black students into their classes. In theory everyone was legally bound to do so other than on the grounds of lack of ability, although the commitment of The University of Texas to its desegregation program was tardy and, as we have seen, was not necessarily enjoined by off-campus establishments in the early days of that era of bitter social strife as the civil rights movement pressed its claims for status and recognition.

Moore made few concessions to the changing times, and allowed himself no public utterances on these issues, other than asides and remarks in the confines of his classroom. There was no sign of any mellowing in his personal attitudes or general demeanor as he grew older and his ongoing feud with the second floor had resulted in something of

---

[1] G.W. Henderson, thesis title *Proof that every compact continuum which is topologically equivalent to each of its nondegenerate subcontinua is an arc* for his PhD awarded in 1959 [Due to the inadvertency explained under Henderson in Appendix Two, the thesis should read '*every decomposable compact continuum*']. He later taught at the University of Virginia, Rutgers and the University of Wisconsin, Milwaukee.

[2] R.L. Moore to Dr. Robert Osserman, Mathematics Branch, Office of Naval Research, Washington, DC, 31 December 1960, from the R.L. Moore Papers in the AAM.

a whispering campaign against him, in which words like 'old and bigoted' were used and had grown louder in recent times. Even so, he was generally well supported from within his own area of operations, although his colleagues had long been seeking the appointment of at least one other senior, but younger, professor to the third floor, and by implication they sought a strong incoming professor who had the strength to rise above the existing disharmony. They were evidently at pains to point out that although the third floor was not on the brink of falling apart, it did require new blood as a matter of urgency. This was demonstrated in a petition signed by members of the Pure Mathematics faculty to the university president soon after Moore went on to modified service, urging the hiring of a senior professor to complement their ranks but insisting that Moore himself planned to remain active:

'Professor Moore intends to carry on his work of developing and turning out students and mathematicians of the highest quality, just as in the past. Let us remind you that his students during the past few years have been outstanding and there is no reason to think that he will not continue to bring renown to Texas and its University for many years to come.... There seems to be a notion in some quarters that the Department of Pure Mathematics *is* R.L. Moore and his "retirement" means the "end".... [In fact] Moore's contribution continues as before. However, we maintain that the teamwork in the Department of Pure Mathematics, the common high ideals and the methods of teaching and of selecting teachers are to a great extent responsible for the great record of achievement in that department. It should be emphasized that Professor Ettlinger has developed and is developing outstanding students, and Professor Wall has ... enthusiastically adopted the methods of teaching for which this department is famous.'[3]

One of the most notable aspects of that note was the reference to teaching methods for which *the department* had become famous. Clearly, there was an effort on the part of his colleagues to take the spotlight off Moore and allow him to carry on while enhancing the department's activities to meet modern needs and overcome the increasing strength of their colleagues on the second floor. The dissent between faculty in the opposing camps of Applied and Pure Mathematics contin-

---

[3] Petition to President Logan Wilson, The University of Texas, from members of the Faculty of the Pure Mathematics Department, 4 May 1953, the R.L. Moore Papers in the AAM.

ued as they vied for students by offering courses clearly in competition with each other. Applied Mathematics faculty members believed that they were essentially the key providers of courses geared to modern demands while Pure Mathematics and R.L. Moore in particular would have none of it.

There was no doubting the continued and remarkable productivity of Moore, Ettlinger and Wall and there is ample evidence that they retained extraordinary loyalty and respect from their students. The mesmeric teaching powers of Moore remained as strong as ever, evidenced by recollections of many students specifically from that era, including another exceptional student, John Worrell, who returned to the fold during the period under review, gaining his PhD in 1961 having already qualified at medical school. His story is one of particular note, if for one conclusion alone that according to Worrell, R.L. Moore empowered him toward his contribution to America's mission to Mars. It will be recalled from earlier mentions that Worrell[4] had taken Moore's calculus in preparation for his medical studies and asked Professor Moore if he could take his Theory of Sets in the summer and Introduction to the Foundations of Geometry which was approached from the standpoint of Hilbert's axioms:

'He agreed to let me do that. Because I was going to go to medical school, I didn't know whether I could continue the whole summer, so I asked if I could perhaps sit in the session for three weeks. He said I could. But, what guile, if one can use that in connection with him. He, of course, gave the problem in his point set course [that is, this Theory of Sets course] about the Vitali coverings. This theory of sets course was directed toward measure theory and concepts in measure theory, and he raised an interesting question. That is to say, is it true that if a point set bounded on the real line has measure 0 and G is a collection[5] of segments covering that point set such that for every point in it, and for every positive number ε, there is a segment in the collection G

---

[4] Taped for the R.L. Moore Oral History Project, 16 April, 1998; interviewed by Ben Fitzpatrick Jr. and Harry Lucas Jr., in the R.L. Moore Legacy Collection in the AAM.

[5] Such a collection G is called a *Vitali covering* of the point set. A subset M of the real line L was defined to have property V if and only if it is true that if G is a collection of segments of L covering M, then, for every positive number ε, M is covered by a subcollection G' of G such that the sum of the lengths of the segments belonging to G exists and is less than ε. [Segment of a line in Moore's definition is equivalent to open interval of a line in current terminology.]

which has length less than ε and contains that point, then there is a subcollection of G consisting of segments that cover the[6] point set such that the sum of the lengths of these segments is less than ε?

'He asked whether that's true, and whether the converse[7] holds true ... whether if a point set which is a bounded subset of a line in this context, does have that property [property V of footnote 5] then it can be expressed as a sum of countably many closed sets ... The way he expressed this converse problem, it seemed like such an intriguing problem, and I wanted to solve it, since he said it had been outstanding a long time, and he hadn't had anybody to come up with a solution for it. So, the anticipated three weeks went on, and the first thing I knew I'd attended the whole session of his courses, and this intensified my desire to study with Moore. So, I went to medical school, wanting to solve the problem about the Vitali coverings, wanting to know more about measure theory, and wanting to do work with Professor Moore in point set theory. After the first year of medical school I did come back the next summer and had some more work at The University of Texas in mathematics. Professor Moore did not have any additional sessions, and he would not give me a conference course in his foundations of point set theory. I did study with Professor Wall during the summer, and also I took a course under the direction of Professor W.T. Guy. Then I returned to medical school for the next session. I maintained close contact with Professor Moore over all the years that I was in medical school, and I continued to have the very strong desire to return to study with him. He was always in the background. There was always that question, what would it have been like to be studying with him? It was not simply the man, but also his ideas, but you can't separate the two. I finally reached the conclusion that I simply couldn't handle this any more. I had to come back to Texas. So, I wrote to Moore wondering whether he would regard me as a renegade at that point. I'm saying this in a quasi-facetious form, but still there is that element because he had very distinct predilections. He did not like to be offended. I wrote

---

[6] In *Fund. Math.* 5 (1924), 328–330, J. Splawa-Neyman showed that every closed and bounded subset K of the real line has property V. Moore noted that the hypothesis of closure on K can be relaxed to that of K being the union of countably many closed sets. In the *Proceedings of the National Academy of Sciences U.S.A.* 10 (11) (1924), 464–467, R.L. Moore showed that there exists a bounded subset K of the real line that does not have property V.

[7] There is implicit reference to the extension noted by Moore of the Splawa-Neyman result of footnote 5 with the hypothesis of closure on K relaxed to K being the union of countably many closed sets.

H.J. Ettlinger, a long-time colleague of Moore

him a letter indicating that I simply had found nothing else in my career that competed with the adventure in ideas I had found in his classroom, wondering whether it would be possible for me to come back to The University of Texas and study with him. Why? Moore brought forth the view that mathematics was an area for innovative thought and the thing that was most important, once you place a premium on it, was the ability to work these problems out.

'So, from the very first few minutes of stepping in his class I was engaged in solving a problem he put on the board, and it was not a situation where he was giving the answers. It was his method of presenting questions in class and then forcing us to think on our own initiative. But it was more than that ... bound up with his personality, bound up with his approach to the world of ideas, his emphasis on developing in individuals their ability to think things through for themselves, not to be dependent on external authority. We're not talking about ridiculous lack of acknowledgement of the work of important mathematicians and scientists; that would, of course, not be the case. He placed a premium on the ability to think things through, rather than on the ability to memorize things, and that was evident almost instantly. Moore gave problems for his students to work out that were research level problems in terms of their quality.'

Moore arranged for Worrell to be offered a teaching assistantship during the first year he was back at The University of Texas and in the second year he was taken on as a physician at the UT Student Health Center and Hospital. He was therefore able to practice medicine while studying with Moore. He also solved the converse problem, on the assumption of the continuum hypothesis, about the Vitali coverings, which became the first abstract he had published, in the *Notices of the American Mathematical Society*. He was subsequently offered a research fellowship, again through the intervention of Moore, in his third year. After completing his PhD[8] Worrell was granted a post-doctoral fellowship through the National Science Foundation, which he conducted at UT with Moore as his mentor. For Moore personally, this was something of a crowning glory. For years, he had been turning into brilliant mathematicians young men and women who initially had no intention whatsoever of going down that route, and most of whom never imagined they were even capable of it. With Worrell, there was an additional buzz to the capture: he had turned a man who had already proved himself in music and medicine to topology and mathematical research, into which Moore had led him step by step until the final act of becoming his mentor. Had Worrell really chosen the path himself? Or had Moore led him down it, step by step until he was hooked? He had performed this maneuver so many times in the past that it occurred now without great planning, so that the talent was snared and suitably indoctrinated. Worrell recalls: 'That year was very important for me. The Fellowship allowed me to concentrate intensively on certain problems that I had been developing with Professor Moore, and it was during that National Science Post-doctoral Fellowship that I solved another problem that Professor Moore had placed great emphasis on, that is whether the property of being a Moore space is preserved under the actions of upper semi-continuous decompositions into compact sets. I got a much more general result, but the fact is that I solved that problem during that fellowship.'

Worrell was subsequently contacted by Sandia Laboratories, prime contractor to the United States Atomic Energy Commission, with an offer to join them under Howard Wicke, a former student of Chittenden, who was head of the Applied Mathematics Division at Sandia. It was a remarkable opportunity, and as Worrell describes it, one that 'gave me absolute freedom to pursue whatever course of research in mathematics

---

[8] His thesis title was *Concerning scattered point sets*.

that I wanted to. At least that's as close as one can get to an accurate statement in terms of all the nuance that applies in the real world.' He discussed the offer with Moore, who wrote a letter of recommendation and Worrell joined the research team at Sandia:

'We began a progressive working relationship in which we worked together on certain problems, but it was more in the spirit of Moore's classroom, and this continued for a decade there. Moore's objective was to empower individuals to work and think independently and this power was applicable in industry and fields other than mathematics. This is an important and central issue here. A time came at Sandia Laboratories when a problem was raised in applications and had been worked on for several years. I was asked if I would consider consulting on this problem. It turned out to be one where they were trying to make certain applications to medicine of a technique that had been worked out at Sandia Laboratories. I looked at it and realized, and correctly so, that if they tried to use the sort of approach that we're doing, it was like playing against the house at Las Vegas. That simply was not likely to yield the results they wanted. My surgical experience, my experience in medicine, was very convincing to me on this. But, I saw another opportunity on that, which led me to having contact with Randy Lovelace, M.D., who was then head of space medicine with NASA.[9] He and I had some discussions together.

'The technique we're talking about was of sufficient importance that when President John F. Kennedy visited the laboratory, Dr. Lovelace wanted him to see this particular technique and what might be done with it. However, I realized they were making a mistake. Authority or no authority, they were wrong! That simply was not the way to go with this thing. So, after conferring with Dr. Lovelace, we saw that this had a tremendous application to a very important NASA problem that was then pending, getting this technology properly integrated into the NASA program. I feel and still feel to this day, that really Moore is one of the persons who helped get us to Mars.[10] Why? Because there's more to it than simply the knowledge.

---

[9] He founded the Lovelace Clinic in Albuquerque.

[10] U.S. Mariner program of interplanetary probes became operational in 1961 to fly by Mars, Venus, and Mercury. In 1969 Mariners VI and VII obtained photographs of the Martian surface and made significant analyses of the atmosphere. Later at Sandia, work was done on the NASA Viking project which centered on the objective of landing a spacecraft on the surface of Mars capable of initiating biological experimentation seeking to answer the question of whether life exists there.

'I wanted to see us get to Mars, and I put tremendous effort behind that, and Moore gave me the confidence that I could trust my own mind to see what I thought and not simply be swayed by what was popular or by what somehow or other would be prestigious, or what would give you the immediate acclaim. The power to see the things that were and then to follow through, to have confidence that one had arrived at a correct answer. One had the self-confidence in knowing that one had a solution and knowing one did not have to check any book to see whether it was there. That kind of confidence had a great deal of application to something that was a very complex engineering project. Also, it saved Sandia and the United States a lot of money.'

While the focus above on the work at Sandia is on the impact of the Moore Method on the capacity for work in applied science, the period at Sandia was very productive in fundamental mathematical research, intensified and catalyzed by Worrell's association with Howard Wicke. Individually and jointly, in this period they made over 60 research reports in the *Notices of the American Mathematical Society* and programs of the International Congress of Mathematicians, a Prague Topology Conference, and annual National Science Foundation sponsored topology conferences held at various universities.

In informal language this work may be said to suggest with strong technical support that certain classical forms of convergence, grounded in Cauchy-like conditions and finite intersection type properties, have significant limitations in scope of application. In the work of Worrell and Wicke these classical concepts are generalized to obtain convergence criteria founded in deeper finite intersection properties removing these limitations. R H Bing remarked on the surprising character of some of their research results, some extensions of which they reported on at the International Congress of Mathematicians meeting in Moscow in 1966. There, the distinguished topologist P.S. Alexandroff publicly commended the presentation. A productive and sustained interaction followed with Russian mathematicians and later Worrell and Wicke were invited to join an International Topology Symposium honoring Alexandroff at Moscow State University and Steklov Mathematics Institute in 1979.[11]

After almost a decade with Sandia, Worrell moved on to give further application to his combined skills of medicine and mathematics, and

---

[11] For a fuller discussion, see John M. Worrell, Jr., Some remarks on the career of Howard H. Wicke and our work together, *Topology and Its Applications*, 100 (2000), 3–22.

none could have been more appropriate than an offer from Ohio University, where he was given the opportunity of founding an Institute for Medicine and Mathematics. This was a unique development at a time when there was a critical shortage of funding in mathematics and people from even the most prestigious universities were happy to accept positions in academia that were probably below their capability level. Worrell considered that, again using his Moore training, it was propitious to introduce a program aimed at helping to prepare young students with mathematical background who subsequently could apply this in the field of medicine and medical research along with other objectives. He recruited young topologist and Moore-method enthusiast G.M. 'Mike' Reed, who had gained his PhD under the direction of Ben Fitzpatrick[12] at Auburn to take a leading role in the development of advanced research in mathematics. Reed had also received considerable exposure to Moore's ideas and teaching. He recalls: 'Auburn, at the time I was a student,[13] was a unique Moore-method experience. Distinguished visitors were commonplace. Sharing Tex-Mex food and beer with Mary Ellen Rudin, R H Bing, F.B. Jones and even Kuratowski was a normal part of graduate life in topology. We had the best of two worlds: inheriting a world-renowned mathematical tradition and creating our own identity.'[14]

Looking back on those years, Worrell came to the conclusion that the way Moore unmasked latent ability and developed it caused things to take place in the student's brain, by inducing a neuro-physiological transformation. Worrell concluded:

'There is a lot of latent ability that is overlooked, and it is sometimes hard to know whether the ability was there latently and brought out, or whether the ability was actually developed. But, I venture to speculate on the basis of my experience that Moore developed ability that was not there, and I'm not trying to make a simplistic statement that you could take persons of grossly compromised intelligence and bring them up to the level of someone like Mary Ellen Rudin. I think he developed it. I think there was something in Moore and the way he approached things

---

[12] Fitzpatrick was a Moore student who gained his PhD under Professor Ettlinger.

[13] A number of students of R.L. Moore migrated en masse to Auburn when he was eventually retired in 1969, as described in Chapter Nineteen.

[14] Mike Reed, oral history interview by Harry Lucas, Jr. and Albert Lewis, 21 July 1998; R.L. Moore Legacy Collection in the AAM.

Nell Elizabeth Kroeger (née Stevenson) at the Spring Topology Conference, Houston, TX, 1971

that caused things to take place in the brain. I'm talking about neurophysiologically in persons who were attracted to him that actually induced a neuro-physiological transformation. I'm virtually positive that in principle what I'm saying is true. I think it's a very important point for education.

'From my experiences over the years since I left Dr. Moore's formal instruction in his courses, I've gotten an impression that it's not appreciated widely that Moore not only unmasked ability or freed it up and allowed it to be expressed, like letting someone out of prison who shouldn't have been there in the first place, taking the key and opening the door and letting them out. He was also one of the few people that I had direct experience with who had the insight to realize that it could be much more important to ask the right question than to give a solution to a point. In fact, if one wanted to have sort of an epigrammatic summary of a part of the secret of the power of what Moore did, it's this: he realized that once the right question is asked anybody, as it were, can answer the question. The real challenge is to find the question to ask. Moore understood that.'

A similar theme was taken up by R.L. Wilder years earlier, in a topical paper he prepared for publication in 1959 for a symposium on the axiomatic method in relation to the development of creative talent.[15] Wilder struck a comparison between Moore's method and Socrates, citing the manner in which Moore exploited competition between students, and the way in which he would encourage a student who seemed to have the germ of an idea, or put to silence one who loudly proclaimed the possession of an idea which upon examination proved vacuous. Wilder, like Worrell, considered a deeper theory by suggesting that these were matters closely related to Moore's personality and capability as a teacher. This, he suggested, was the often unstated but fundamental ingredient of his pedagogic system that formed the platform on which he built his axiomatic approach in the development of creative talent.

That development was launched initially by two crucial elements: his ability to select students mature enough to cope with the type of material to be studied and the unique demand of his teaching, and second his powers of persuasion that captured the imagination of unwary, and perhaps uncommitted or even unwilling, students to join him. This power seemed to rise to greater intensity as he grew older and competition for candidates grew stronger. Wilder highlighted, as have others in these pages, Moore's ability for detecting talent among undergraduates, and he often set his sights on a man long before he was ready for graduate work, believing that this creativity should be unleashed and harnessed as soon as possible, and the younger the better. He reasoned that one could always pick up 'breadth' as he progressed. Wilder concluded that once he had spotted a potentially mathematical mind, he marked that man for the rest of the course as one with whom he would 'cross his foils, so to speak'. There were times when a student of lesser talent did slip by and that student was doomed to a semester of sitting and listening, often with little comprehension, or hurriedly taking notes in the usually vain hope of being able to understand by reading outside of class. They often ended up as casualties of the course, a system Wilder described as 'humane, as well as good strategy' because only the fittest survived.

However, the system itself often inspired and invigorated those in his classes who, initially, found difficulty in following even the basics of what was going on. One such student was Dr. Sam W. Young whose own

---

[15] R.L. Wilder, Axiomatics and the development of creative talent, in *The Axiomatic Method with Special Reference to Geometry and Physics*, L. Henkin, P. Suppes and A. Tarski, editors, North-Holland, 1959, pp. 474–488.

experiences in Moore's 688 class were at first frustrating, but eventually exhilarating. His time began in the Spring of 1959. He recalls that he had struggled that school year to catch on to the theorem proving courses of Ben Fitzpatrick and H.S. Wall and he concluded he was not doing nearly as well as many of the other students. However, he and other classmates had heard about Moore and his famous 688 course and were curious to find out if they could or should sign up for it the following Fall semester.

'I remember that Ben warned us that it would be very different from anything we had experienced before [but we went ahead and signed up].... Looking back on it now, I think that I failed to understand that the theorems were independent of the meaning of the undefined terms. I was laboring under the assumption that the students who went to the blackboard to present a proof knew more about "regions" than I did.... I was simply baffled and flustered by the entire process. I continued to take notes meticulously and this habit turned out to be a contributor to my eventual salvation.... To me, they were *all* impossible ... [but] I continued to copy the proofs, which were just so much gibberish to me.... "Mister Young, can you prove the next theorem?" It was a bit terrifying and humiliating but I felt no negative feelings toward him. After all, I knew how to extract myself from the situation. I *had* to prove a theorem and get the pressure off of myself. But my answer was always the same, "No sir, I cannot." It sounds cruel as I describe it now but there were no hurt feelings. It was as much comical as cruel. Someone had to be the worst student in class and it was me....

'On the last class meeting before Christmas, Dr. Moore spent the entire hour stating theorems and definitions. He really loaded us up with work to do. I took my notes and went to be with my parents for Christmas. [They] had moved that year from San Angelo to Big Lake, Texas, a small oil town 80 miles to the west.... I was as isolated and undistracted as I could possibly be for the task before me. I believe that I began as usual to make good copies of the latest proofs that were given in class. This time, as I read the proof of Theorem 25 that my classmate had presented, I somehow managed to understand the argument. I saw how he had made use of a previous theorem and the definitions which were involved. I was astonished! I was understanding one of the proofs for the first time. I went to the proof of Theorem 24 and I managed to understand that one too. I remember thinking that perhaps I could prove some of those theorems myself. I worked my way back through the theorem sequence and proved almost all of the theorems on my own. I worked all day every day through Christmas doing this. Finally, with a few days left before heading back to Austin, I was ready to look at the work that Dr. Moore had given us to do during the break.

'I wish that I had the ability to describe the exhilaration that I felt to the core of my existence when I put together a proof of Theorem 26. I marched forward proving the next theorem and the next. I am not saying it was easy; it was exhausting work. I would get up early in the morning and work all day....

'The time came to go back to Austin and a great fear came over me as I visualized the inevitable scene that would unfold in the classroom. I was ready and confident with proofs of several of the upcoming theorems but I was scared to death. I had never "been to the board". When the time came, Dr. Moore looked at me and asked the usual "Mister Young, can you prove Theorem 26?" I am pretty sure that my exact answer was "Yes sir, I can." I had learned that he did not like for anyone to say "Yes, I *think* I can." The fear that I felt at that moment is indescribable but I want to repeat that I was *confident* that my proof was correct. The fact that I *knew* I had it, the fact that I was confident in spite of the awful fear of the moment is a tribute to the Moore method of teaching. I went to the board and presented my proof.

'My friends who have had courses from Dr. Moore are always surprised when I tell them what happened next. He never called upon the same person who had just presented a proof. It was just not done. He would turn the pressure on someone else. But I distinctly remember that he looked around the room as if trying to decide whom to call upon. His eyes fell upon me again and he asked "Mister Young, can you prove Theorem 27?" I said that I could and went to the board again. When I finished I sat down again and again he looked around pretending to decide and said "Mister Young, can you prove Theorem 28?" This impish routine continued for an hour and started again at the next class meeting. In fact, no one else was called upon for the remainder of the term. I proved all the theorems that we had time for.... I will always remember that Christmas in Big Lake and the feeling of triumph that resulted — a measure of success for the Moore teaching method and success for myself which propelled me toward a career in mathematics.... I will always treasure the influence that he had upon me.'[16]

---

[16] Sam W. Young completed his PhD in analysis under H.S. Wall, and went on to a lifelong career in mathematics, eventually as Associate Professor Emeritus at Auburn University. The above quotations are from his essay, 'Christmas in Big Lake' for The Legacy of R.L. Moore Project, March 1998.

In the increasingly frenetic era of the 1960s and beyond, however, it would soon become something of a luxury in many establishments to embrace a teaching method that, at a higher level, was best suited to smaller classes. But it was certainly not impossible. In recalling Moore students who went on to instigate modified Moore Method teaching in their own careers, Wilder commented: 'Some of us, especially during periods of high enrollments, have had to cope with classes of as many as 30 students or more. I can report from experience that even with a class this large, the method can be used. Inevitably, a few, sometimes only two or three students, would star in the production.'

Wilder, by then, was among a dozen or so ex-Moore students in major universities across America who were already using his method by the early 1960s, not to the exception of all other forms of tuition, of course, and often in a modified version, but the general tenor of Moore remained central to their efforts. The network of communication, as mentioned in earlier chapters, had also grown into a sort of brotherhood and a code developed between them. It might be utilized if, for example, one of them wanted to make known to one of his colleagues in another university the availability of potentially creative material. The 'pons asinorum', as Wilder put it, of Moore's original axiom system was Theorem 15[17] and if one of Moore's graduates wished to place a student for further work under the tutelage of another of Moore's students at a different institution, he would include in his recommendation the statement, 'He proved Theorem 15'. The doors would open instantly.

Theorem 15 had been kicked around in Moore's classes almost since the beginning of his career. It may be recalled that when he was teaching at Penn he first stated the theorem in December 1913 and J.R. Kline presented a proof the following April. Another member of the class left the room rather than listen to Kline's proof, and when Moore saw him 12 years later, he admitted he still hadn't resolved it. Moore referred to Theorem 15 in the notes he prepared in advance of the making of the

---

[17] In the 1932 edition of Moore's book, this is Theorem 1 of Chapter II. Immediately after a statement of Axiom 2, which introduces a condition of local connectivity, Moore states Theorem 1: If A and B are distinct points of a connected domain D there exists a simple continuous arc from A to B that lies wholly in D. [In the presence of Axioms 0 and 1 introduced in Chapter I, domain is equivalent to open set in current standard terminology. For this equivalence, the first three conditions of Axiom 1 are not superfluous. A simple continuous arc from A to B is a closed, compact, connected point set containing A and B which is disconnected by the omission of any one of its points except A and B.]

MAA film on his method, *Challenge in the Classroom,* in 1966. He provided an up-to-date summary of some of the most discussed, and controversial, aspects of his graduate courses, and in his own words, we are able to hear his explanations for utilizing them:

'I am still giving a course bearing the same title but it is based on quite a different set of axioms and a theorem corresponding to the old Theorem 15 now comes much later in the treatment. The number of this course is 688. Axiom 0 states that every region is a point set and Axiom 1 now states that there exists a sequence satisfying certain conditions numbered 1, 2, 3 and 4. A large body of theorems can be derived from Axioms 0 and 1 without use of the fourth of these conditions and in 688 I usually do not state the fourth one until some months have gone by. In 1958–59, when the time finally arrived to state it, it appeared that there was one member of the class, a Mr. W.,[18] who did not want to be told what it was. So the others were told but he was not.

'One day about two years later, in my 690 class, he was at the board, near the door, explaining something and I started to make a remark. He seized the doorknob, flung the door open and rushed out of the room. He thought I was about to state Condition 4. I imagine some of you are thinking that it was ridiculous for him to be so determined to remain *ignorant* concerning this condition. If so I do not agree with you at all. In the course of time he thought of quite a different condition such that if it is substituted in place of Condition 4 in the statement of Axiom 1, the resulting axiom (Axiom 1_w) is quite interesting and quite different from Axiom 1. If he had been told at the outset what Condition 4 was he would, I believe, never have thought of Axiom 1_w and never have discovered the interesting things he has proved to be true concerning it.

'Often, when I have raised some difficult questions in one of my graduate classes and days (or weeks or possibly even months) have gone by without its having been settled in that class and finally the day arrives when someone announces that he has a solution and he goes to the board to present it, some of the other members of the class (sometimes most of them) walk out and stay out until he is through. Is it a good thing for a graduate student to walk out under such circumstances? I think that often it is but *sometimes* it is not and in the case of some students it is very hard for me to decide whether or not to discourage it. I do not believe I ever said anything to discourage it in the case of Mr. W.

---

[18] John M. Worrell, Jr.

I don't believe there was a single instance where he walked out when it would have been better for him had he stayed in....

'Some years ago I had in 688 a student (Mr. S)[19] of outstanding ability who, I think seldom, if ever, had any *occasion* to walk out. Seldom, if ever, did any other student present, in this class, a proof of a theorem that S had not already succeeded in proving. One of the other members of the class said that being in a class with S was like being in a lecture course (with S as a lecturer) and he did not like lecture courses so he did not want to take 689 (a sequel to 688) the next long session because S would be taking it then. He asked whether if he and another student[20] should stay away from The University of Texas the whole of the next long session they could count on my giving 689 again in the long session immediately following the next one. I indicated that I expected to do so and they both got positions, stayed away from Texas one session and returned later according to plan. I think that under the circumstances their decision to stay away was a good one. They were both good students and went on to receive PhD degrees.'

Moore's method had long been observed by other universities, and his former students who were themselves proponents of it were producing a new breed of teachers in their own mold. From time to time, papers and articles were published by former students that pointedly highlighted the writers' opinions that in spite of the pressure of modern times, the Moore Method was being adapted to the requirements of modern conditions. In a 1965 article discussing the quantity of mathematics taught, and on the issues of teaching calculus to engineers, for example, former Moore student Ed Moise, then at Harvard, pointed out that Moore's teaching had been effective, to say the least, and it was his belief that all teaching would profit if the basis of his method were better understood. It seemed more likely that the success of the method depended on special conditions and he made an interesting conjecture that teaching in this style was not merely a matter of not lecturing. It made great demands of both mathematical and psychological depth. Second, it required the teacher to dominate the environment his students lived in. Moore had done this at Texas, but Moise doubted that he or

---

[19] Moore was probably referring at this point to John Slye, who gained his PhD in mathematics in 1953 with a thesis entitled *Flat Spaces for Which the Jordan Curve Theorem Holds True.*

[20] W.S. Mahavier and S. Armentrout.

anybody else could have done it at Harvard, Princeton or Chicago well enough to persuade students not to talk to each other or well enough to persuade other professors to avoid telling them things that would help in their problems. The method might also require that the subject matter be logically primitive and fairly isolated from that of other courses. 'In spite of all these reservations, however,' Moise concluded, 'I believe Moore's work proves something of broad significance ... that sheer knowledge does not play the crucial role in mathematical development that most people suppose. The amount of knowledge that a small class can acquire, struggling at every stage to produce its own proofs, is quite small. The resulting ignorance ought to be a hopeless handicap but in fact it isn't. The only way I can see to resolve this paradox is to conclude that mathematics is capable of being learned as an activity and that knowledge which is acquired in this way has a power which is out of all proportion to its quantity.'[21]

It was an interesting thought in the year of 1965, when it was published. Ironically, at that time, Moore's position at the University was again under pressure.

---

[21] Edwin E. Moise, Activity and Motivation in Mathematics, *Amer. Math. Monthly*, 72 (1965), 407–412, quote from p. 409.

19

The Final Years
(1965–1969)

The administration had succeeded in merging Pure and Applied, but Moore refused to recognize their cohabitation. He went on as before. A time of total dedication to his students: between 1953, when he was forced into modified service, and 1969, Moore added a further 24 PhD students, ending up with 50. Building to the high drama of Moore's departure in 1969.

R.L. Moore in class (1969)

From the beginning of the 1960s, the isolation of the third floor as a teaching unit intensified as successive departmental chairmen and administrators tried to unify what in effect was still two mathematics departments, together in one building, but operating separately and often at loggerheads. Furthermore, the aspect that the administrators thought they had resolved years earlier, that of curtailing the working hours of Moore, had barely altered. He had accepted modified service, but was still teaching fifteen hours a week and five or six courses a semester. Ettlinger began referring to it as 'mortified' service, a term which Joe Frantz said Moore fully agreed with. 'No one in the administration,' Frantz wrote,[1] 'not even the tough President Logan Wilson, had the guts to tell the old man to remain home. If he was losing his touch with graduate students, it wasn't apparent. He just kept on teaching into his eighties [and] if a student had the stamina, Moore was accessible.'

Thus, there was no let up whatsoever in either his output or his resistance to change. In some respects Moore and Ettlinger (combined with the exceedingly prodigious contribution of H.S. Wall) seemed to be gathering pace in almost defiant manner. It is therefore worth bearing in mind the following summary that extends to the point when, by 1969, all three of them had left The University of Texas. Between 1953, when he was forced into modified service, and 1969, Moore added a further 24 names to his list of PhD students, only two fewer than in his previous 45 years as a teacher and ending up with a total of 50. There were numerous distinguished scholars among them, distributed throughout the land in positions of influence and authority. The figures in this period reflected Moore's cessation of publishing his own research and his dedication to producing research mathematicians and teachers. Ettlinger's record was also impressive with his lifetime supervision of 105 Masters and 22 doctoral students. Wall's figures were equally remarkable, considering he did not arrive at Texas until 1946: 80 Master's candidates and 57 doctoral candidates, in addition to the 25 Master's and 5 doctoral candidates he supervised at Northwestern.

The figures also reflect the fact that by the early 1960s, these three teachers were still undertaking the bulk of Pure Mathematics courses, assisted principally by Renke Lubben, who had remained in situ since graduating under Moore in 1923, and Ralph Lane, who had gained a

---

[1] Joe Frantz, *The Forty-Acre Follies*, p. 121.

Master's degree from Northwestern University in 1943 and followed Wall to UT for his doctorate in 1948. He joined the faculty in 1950 and became an associate professor in 1952. His untimely death in December 13, 1962, in Ames, Iowa, where he was a visiting professor at Iowa State University, left Pure Mathematics once again down to four members, with Wall, 60 in 1962, Lubben 64 and the two very senior citizens, Moore, 80, and Ettlinger, 73, assisted by their ever-ready supply of graduate teachers.

To some extent, this influenced the policy of the administrators toward mathematics, coupled with the need to accommodate the rapidly increasing numbers of students seeking mathematics courses in association with a major in other subjects. The result was the build-up of the faculty of Applied Mathematics while leaving Pure Mathematics in the hands of those outlined above. Undergraduate students were therefore increasingly steered toward enrollment in second floor courses, which meant that those who wished to take courses from Moore really had to start with him, otherwise they were disqualified because they had too much knowledge. Underhand sabotage was not unheard of in the second floor by teachers who were aware of his stipulation and thus made sure that any transfer students, particularly at advanced level, were fully exposed to concepts that Moore might introduce in his own fashion. This, of course, resulted in unpleasantness as students were turned away or excluded from his classes. On the second floor, a number of the faculty positively and openly discouraged their students from getting involved with the third floor. There were several who did not, including Professor H.V. Craig. In fact, he devised his own 'method' for his graduate courses in vector and tensor analysis. His technique was to have a volunteer student go to the board for virtually the whole class and work through examples while Craig himself was seated at a student desk. Albert Lewis recalled:

'It somehow became known amongst his students that Craig was one of the few second floor professors "approved" by the third floor. That is, third floor students who needed or wanted additional courses beyond third floor offerings could attend Craig's courses without opprobrium. One factor was that even though many of Craig's students were from physics (needing the tensor theory used in general relativity theory), he never stinted on requiring an understanding of the mathematically rigorous conditions underlying the theorems used in physical applications. This underpinning was also reflected in the textbook he authored. Also, I gather that Craig tended to side with the third floor in resisting some of the later reforms like doctoral qualifying exams. Texas came late to introducing these, largely because of the resistance of R.L. Moore and like-minded faculty who believed the only criterion needed for getting a doctorate was the demonstrated ability to do publishable research. I know that if Craig had not retired as soon as he did and I had stuck with him for my doctoral degree, he would have opposed my having to take such qualifying exams and all the courses leading to them. These exams, by the way, were defended on the grounds that a student should not get a doctorate who was ignorant of a certain "core" knowledge from each of the main branches of advanced mathematics.'[2]

There was also another unfortunate aspect arising from this general atmosphere of 'them and us'. Each was critical of the other, but the net result rested more heavily on Moore and his associates, given the build-up of myth and legend that surrounded them. Stories, anecdotes and unkind jokes abounded and while on the one hand they tended to put off some students interested in Moore's classes, others were intrigued enough to sign up. In spite of the gulf between the two floors, the department functioned under a single managerial structure, although eventually only Wall was young enough and senior enough on the third floor to take a seat on the departmental Budget Council. As enrollments rose and university populations increased by the thousands, federal funding became more plentiful. A national phenomenon occurred in the dash to hire more faculty, and in UT's case this meant a policy of cutting back on teaching assistants and graduate teachers and supplanting them with fewer courses taught by experienced professors in front of much larger classes. It was a trend that would continue through the 1960s and beyond, and was naturally alien to the true nature of Moore's method. Others could teach it by modification if they wished, but he had no intention of altering course in his own classroom. And he was still fit and strong enough to argue his corner, in no uncertain terms. This ability to rebuff the onset of some of the more debilitating effects of age was decidedly apparent in the 1966 MAA film, *Challenge in the Classroom*. Although by then 84 years old, Moore appeared fit, strong and healthy. His voice showed no sign of strain even during long passages in which he described the origins of his teaching method, which went right back to the days when he taught himself calculus when he was 15 years old.

---

[2] Albert Lewis to the author, September 2002.

Although he had prepared the talk with handwritten notes, already consulted in these pages, he spoke without apparent reference to them and his memory appeared faultless as he drew on incidents from years past without hesitation. He was also filmed in his classroom, seated at the side while students went to the board to give their proofs. Asked by an interviewer if he had considered writing a book on the Moore Method, he replied that he had considered it, but he was too busy, had too much to do — and anyway, he did not want to give away anything that might give his students advance knowledge of what he might require of them in the future.

The film provided more than just an insight into the Moore Method. It gently allowed the personality of the subject to unfold, and showed his reactions, facial and otherwise, to aspects of his teaching and his career, as they cropped up in the historical context. Surely, no one viewing it could have come to any other conclusion than that here was a man, firm and outspoken, whose sternness was interrupted by flashes of evident kindness and humor, dedicated to mathematics and to his students. Although he stated at the end of the interview that he 'would never be satisfied', the film stands today as a visible record of his life and his personal fitness to continue the job at that time. There were certainly fault lines in his make-up, retained from his childhood and upbringing, particularly in the uncompromising and steadfast belief in a man's right to his own viewpoint, however adversely it might fall against the sway of modern opinion. This was again demonstrated in that year of 1966, when great tragedy struck at the heart of The University of Texas. On the morning of 1 August, Charles Joseph Whitman, overcome by madness induced by depression, financial worries and family strife, murdered his wife and his mother. He then drove to a nearby hardware store and bought a .30-caliber rifle, clips, and ammunition. He went to another store to buy a 12-gauge shotgun and then meticulously assembled these two guns with a rifle and two handguns he already owned, along with 700 rounds of ammunition, a machete, food and water into a footlocker he had used during service in the Marine Corps.

Just before midday, Whitman — a former UT student — drove to the ground floor entrance of Austin's famous landmark, the University Tower, and gained a parking pass by telling the gate attendant he needed to unload equipment at the Experimental Science Building. Five minutes later he manhandled his footlocker on to a barrow and took the elevator to the twenty-seventh floor, then hauled the locker upstairs to the twenty-eighth floor, where secretary Edna Townsley was at her desk in the reception area. He shot her and hid her body. Four other people arrived, and as they came towards the office, he fired again killing two and wounding the others. At 11:45, Charles Whitman began firing indiscriminately from the top of the tower at people on the ground below. He dashed around the open-air perimeter of the tower's top deck with such speed that when police arrived they could not ascertain how many people were firing. Only when an officer in a light aircraft came close to the tower was it confirmed that there was just one man.

The shooting went on for ninety minutes, with bullets flying around those who had gone to the aid of the fallen. In all, 47 were hit, killing 16 and wounding 31. The victims included members of the faculty, students, passers-by and tourists. The dead included the unborn child of a woman who was shot in the stomach, although she herself survived. This terrible carnage was only ended when two policemen, Patrolmen Ramiro Martinez and Houston McCoy, reached the tower through underground tunnels, made their way to the top and courageously stormed the deck, with their guns blazing. Whitman fell dead from multiple bullet wounds.[3]

The University and Austin were somber places for many weeks, and what was really one of the first of too many school shootings and other massacres of unsuspecting members of the public over the coming years initiated a fierce debate on America's gun laws, or lack of them. Writing in an Austin newspaper in the immediate aftermath of those terrifying scenes at the tower, local politician David Lawrence had stated that the dangers of a combination of firearms and irresponsible users was something that might at least be reduced by legislation that would require licensing and examinations for those who owned guns. On 8 August, Moore, one of many, many Texans (and others across the country) furious at the very idea, wrote a letter, demanding to know:

'Is David Lawrence in favor of such licensing and examination?
With strong feelings of disillusionment
R.L. Moore'[4]

---

[3] The observation deck of the tower was closed for a period of two years, but then reopened. The deck was permanently closed in 1975 after a student jumped to his death, bringing the total number of suicides from the top of the tower to seven. The tower was remodelled in 1999 and opened to the public by reservation for the first time in 25 years.
[4] Copies of this exchange of correspondence are in the R.L. Moore Papers in the AAM.

Lawrence replied in the affirmative.[5]

That year of 1966 also saw a defining development that would ultimately end the teaching careers of Moore and Ettlinger and a voluntary conclusion for Wall. Toward the end of the year, the Board of Regents began informal discussions on the subject of compulsory retirement for all members of faculty upon their reaching the age of 75. These discussions were set in motion, coincidentally or otherwise, at a time of key administrative changes, which in turn had an impact on the overall situation and indeed contributed heavily towards emotionally charged events that now began to unfold, culminating in a crescendo of activity by Moore's students past and present.

Three developments had brought new faces into decision-making roles. They included the appointment of a new University president, Dr. Norman Hackerman, who took office in 1967. The renowned chemist and academic had held numerous positions at UT, including Chairman of the Department of Chemistry and Dean of Research and Sponsored Programs, prior to assuming the role of president, a post which he held until 1970. Around the same time, Professor Woodrow W. Bledsoe, a relative newcomer to UT, was appointed acting chairman of the troublesome mathematics department. A strong administrator recently arrived from industry, Bledsoe found himself reporting to an acting dean of the College of Arts and Sciences, Dr. Malcolm MacDonald of Government, who took over *ad interim* following the death of Dean J.R. Burdine until a permanent appointment was made.

Hackerman, Bledsoe and MacDonald were immediately confronted with the issue of compulsory retiring of two of their most famous professors, Moore and Ettlinger, which clearly required the most delicate handling. On 2 October 1967 Bledsoe[6] wrote to MacDonald in terms that confirmed that the Board of Regents required Moore and Ettlinger to retire in 1968. Significantly, Bledsoe requested that Moore be

---

[5] When the issue of gun licensing was debated again in the wake of the murders of Dr. Martin Luther King and Robert Kennedy in 1968, Moore once again made his feelings known, writing on 5 July 1968 to Senator John G. Tower: 'I have no sympathy for a law that would require *either* the registration of all handguns *or* the licensing of those who use them...I think there is something fundamentally wrong with anyone who favors such a law. Have things come to such a pass in some quarters that a man is not supposed to protect his own home without the permission of the Government?'

[6] W.W. Bledsoe, Acting Chairman, Department of Mathematics to H.M. MacDonald, Dean ad interim, College of Arts and Sciences, The University of Texas, 2 October 1967, in the R.L. Moore Papers in the AAM.

allowed to continue supervising those graduate students working with him who had been admitted to PhD candidacy by 31 August 1968, the deadline date set for his retirement. To this note, Dr. Hackerman had added a further suggested deadline so that this would apply only to those who would complete their degree requirements by 31 August 1969.

By then, news of the enforced retirement of professors over 75 had leaked out, and almost immediately members of the Board of Regents and other officers of the university and senior faculty were bombarded with letters, telephone calls, and telegrams from Moore students, demanding a reconsideration of this ruling, many pointing out that the loss of Moore from the faculty would damage the reputation of The University of Texas. This spontaneous reaction came from both existing and former Moore students, the latter having been invited by the former group to join them in a campaign to avert his forced retirement. They included one from R H Bing who wrote: 'Throughout the country there are teachers of varying abilities, some the students tolerate and some they like, but I know of no other place where the students have put up such a strenuous fight against the retirement of their professor. This seems to be an endorsement of the proposition that Moore is still doing a vigorous job.'[7]

Even as the campaign for a reprieve for Moore began to build, a new man moved into the hot seat to control operations, stating he had been given a mandate to modernize the Department of Mathematics at all costs. Malcolm MacDonald, the Dean *ad interim* of Arts and Sciences, had been replaced by John R. Silber, a tough, no-nonsense academic who was then serving as Chairman of the Department of Philosophy. He grew up in Texas and graduated from Trinity University in San Antonio. He studied theology, enrolled for a year at The University of Texas Law School, and then earned a PhD in philosophy from Yale. He came to UT to teach philosophy in the 1950s and later, as chairman of his department, established a reputation for bringing about changes that the Board of Regents considered beneficial to the University. His stated ambition on accepting his new post was to gain national reputations for depart-

---

[7] R H Bing, 27 November 1967, letter to R.D. Mauldin with copies to President Norman Hackerman and W. W. Bledsoe. Appears as Appendix 8 in a report *Concerning Dean John R. Silber and the Proposed Dismissal of Professor R.L. Moore*, Joseph M. Carter, H.K. Smith, Tom W. Fogwell, and Frederick A. Stiles (eds.), April 1969, in the R.L. Moore Papers in the AAM.

*Mary Ellen Rudin and Bruce Treybig at the
Spring Topology Conference, Houston, TX, 1971*

ments in his college. Some argued, of course, that the Department of Pure Mathematics had long held such a reputation. But according to Joe Frantz, Silber had committed himself to change: 'Teaching loads were heavy, especially in required courses like freshman math, and the Department of Mathematics needed an injection of new faculty to meet the demand. Silber represented the New Testament; Moore the Old. Silber thought ... that fewer sections should be taught to more students by experienced professors in huge classes.... Moore, if not looking backward, wasn't budging.'[8]

The scene was set for confrontation and Silber began building his case. Visiting scholars were brought to campus to evaluate departmental strengths and weaknesses and the potential for developing further strength. They were considered by Moore's supporters as outsiders who

---

[8] *The Forty-Acre Follies*, p. 122.

could in no way make an accurate assessment of the situation through a cursory visit. But they did make criticisms, including the most obvious ones, that the mathematics department, though formally one, functioned as two, that the department had insufficient faculty for the numbers of students expected to enroll and that there was too great a reliance on graduate students as teachers. Silber made no attempt to conceal his view that the very presence and reputation of Moore hindered the recruitment of new faculty and some of the visitors concurred. In short, an era of unrest and turmoil was foreshadowed as Silber and Moore or, more precisely, his students, went head to head. From the opening shots in what became a war of attrition, there were allegations and rumors that Dean Silber had made derogatory remarks about the standard of doctoral students supervised by Moore over the previous decade and inferred they were languishing in second-class universities. The reaction from students of that era was immediate and angry, and Dr. John Worrell expressed the feelings of most in a measured and carefully constructed analysis of the achievements of Moore's doctoral students during that time frame. In a long and detailed report, his opening remarks pointedly drew the attention of the administration of The University of Texas to the precise reason why such a report was necessary. He wrote: 'While its being prepared in response to an emergency situation and on very short notice precludes its being complete in all details, the data at hand would seem to be sufficiently massive to make the point that one of the best things The University of Texas has going for it is the record of certain alumni in mathematics whose work according to widely circulating rumors has been disparaged as one means of trying to force the retirement of Professor R.L. Moore.'[9]

Worrell, in a short space of time and with the cooperation of the University itself, compiled a comprehensive inventory of the work and publications of all Moore's students of that era, demonstrating the breadth of the subject matter, the international arena of a considerable number of publications, the academic affiliations embracing many leading universities and the industrial, military and defense connections of those students who had moved into those areas and had been engaged in major research, especially in weapons and space programs. Worrell's

---

[9] J.M. Worrell, Jr., Sandia Corporation, to Norman Hackerman, President, University of Texas, 18 January 1968, p. 2. The letter and attached report are in the R.L. Moore Papers in the AAM.

report arrived in Texas on 16 January 1968 and in that age of student protest the deluge of support for Moore seemed to have carried the day. On 30 January 1968, Chairman Bledsoe circulated a note to all members of faculty: 'I have been notified that the 75-year retirement rule that the regents passed and put into effect 1 September 1967, has been rescinded by the Board of Regents. This applies to all persons over 75. Whether persons in this department will teach next year (or later) will depend on the fitness recommendations of the Budget Council and the action of the Dean. I trust that individual cases will be considered in the near future.'[10]

Cheers from students greeted the news. It seemed that R.L. Moore had been given his reprieve and retirement was prevented yet again. His supporters claimed the reversal as a victory for academic propriety. In truth, it merely delayed the inevitable and those close to the hub knew very well that Moore's days were numbered. Within a week, Dean Silber had renewed his efforts to bring closure to the Moore situation. He called for a vote from the Budget Council as to the fitness of Ettlinger and Moore for continued employment, in accordance with the ruling of the Board of Regents, although no one had expected it would be called so quickly. The first meeting was held on 9 February 1968 and it concerned H.J. Ettlinger, then 78 years old. The council was unanimous in agreeing that he was 'physically and mentally fit' to teach for the period 1 September 1968 to 31 August 1969.[11] Two medical reports were produced and the council members confirmed he was up to speed with modern mathematics. It seemed therefore that Ettlinger was safe for another eighteen months. In regard to Moore, the budget council was split in its vote at a later meeting and subsequently voted against Moore's continuation on modified service, and called for a firm date to be set for his complete and final retirement, but in the event, their deliberations proved to be irrelevant.

At a meeting on 31 May 1968, the Board of Regents accepted the recommendations of Silber and Bledsoe, which amounted to a virtual shutdown of the Department of Pure Mathematics as it was then constituted. Silber requested that three senior members of the third floor faculty — Moore, Ettlinger and Lubben (who was just reaching his seventieth birthday) — should be retired with immediate effect. The regents

---

[10] Bledsoe to Faculty, 30 January 1968; R.L. Moore Papers in the AAM.
[11] Report of the Department of Mathematics Budget Council, signed by W.T. Guy, H.V. Craig, H.S. Wall, R.E. Greenwood.

accepted the recommendation, but set specific dates for their retirement: Ettlinger and Lubben to go by 31 August 1968 after the summer session, and for Moore to remain a further year in the interests of those students nearing completion of their doctoral studies.

Regardless of the views of any of those involved in this sorry saga, what happened next to Moore and his two colleagues must go down in the history of The University of Texas as one of the shabbiest examples in human relations, one that may have reminded Moore of the way in which his mentor, George Bruce Halsted, was sent packing almost sixty years earlier. Although rumor abounded, it was to be another fourteen days after the Board of Regents' meeting before Bledsoe finally got around to informing the three men of the decision. He did so in a curt manner, and with a final paragraph that was patronizing in its tribute to those men who had given the best part of their lives to the University, and who really wanted to go on doing so. The letter to Moore read: 'Earlier this year, the Budget Council, in a secret ballot, voted on your fitness to continue as Professor of Mathematics on modified service. The vote was split. This vote was passed on to the Dean of Arts and Sciences who ... forwarded it and his own recommendation to the Administration. Enclosed is a copy of a letter from President Hackerman to Dean Silber which shows that the Board of Regents has voted to terminate your modified service as of 1 June 1969. This is all the information I have at this time concerning this matter, but I would be happy to discuss this with you if you so desire. It is not possible to convey to you, in the limited space below, the feeling that most of us have, that you have made an extraordinary contribution to the University, the State and to Mathematics, through your brilliant career here. More appropriate recognition of your service will be forthcoming at a later date.'[12]

The 'more appropriate recognition' was already being planned before Moore received his letter. It was proposed to host a series of retirement parties for Moore, Ettlinger and Harry Vandiver who was also going. They were to be held in the summer, separately (bearing in mind that Moore and Vandiver had not uttered a sociable word to each other in at least 25 years) but in the same calendar week. Silber instructed Bledsoe to write to former students of the retirees to compile a 'medium length' testimonial for each one. Committees were formed and

---

[12] Bledsoe to Moore, 20 June 1968, from the R.L. Moore Papers in the AAM.

the whole process was already underway when a halt was called. Moore and Ettlinger made it clear, in no uncertain terms, that they did not want to retire and both stated categorically that they had no intention whatsoever of participating in retirement 'celebrations'. The whole situation seemed to be heading for an undignified, rather shoddy, finale for men of such stature. The Pure Mathematics team was, as H.S. Wall wrote in a letter to a friend, 'being put out of business'. Wall's bitterness obviously mirrored that of Moore and thus we can perhaps assume he spoke for both of them in the letter, which outlined the 'goings-on':

'Since the departments of Pure and Applied mathematics were combined, against the wishes of either, we have retained our identity by a sort of gentleman's agreement.... All this is now being destroyed and at a time when, in a certain sense, we have reached perhaps the highest point in a successful operation. By this I mean that we now have several of the most gifted students I have ever known; more than likely they would have never been known except for the efforts of our group. In pleading with the budget council, now chaired by Bledsoe, that our work should be permitted to continue, I was met by insulting statements such as the one of Bledsoe: "You have a Napoleon complex." The conspiracy to put us out of business has been well planned, indeed. The first step was to take away the offices occupied by our young instructors on the third floor of Benedict Hall. This has tended to break up the fine team spirit of the group and has had a very bad effect on the work in our classes. Next, Dr. Ettlinger and Dr. Lubben were put out of business. This has left only Dr. Moore and me to carry on with the graduate program. The treatment of Dr. Ettlinger is the foulest performance I have ever known: as far as I am concerned, anyone who had a part in this has thereby resigned from the human race! Along with these things, there has been a concerted campaign to discredit both Dr. Moore and me. Our work is "old-fashioned," "out of date," etc., etc. Professor Loomis of Harvard was quoted to me as replying to someone belittling the "old-fashioned southern mathematicians" as follows: "Those southern mathematicians have graduate students who are settling questions which we have only talked about at Harvard."

'The final step has been taken to destroy our group! For the coming summer, I no longer assign our instructors to our classes. In fact, none of our graduate students, with three exceptions, have any classes to teach in the coming summer session. Next long session, the course names, numbers etc. have all been changed and our operation will no longer exist. It has been reported to me that when Silber was told about all our students leaving Texas he replied: "Let them go! Good riddance."'[13]

As word of these developments spread, a new student revolt began to gather momentum. There was much talk of Socrates, who had taught by inductive means, questioning his listeners, and showing them the inadequacy of their answers. He had a following of young men in Athens, but was mistrusted because of his unorthodox views and his disregard of public opinion. Inevitably, he made many enemies and was brought to trial, charged with corrupting the minds of the young. Those dashing to the reference books to examine the analogy might have discovered a passage similar to the following:

'Socrates himself treats the whole matter with contempt. His defense consists in narrating the facts of his past life, which had proved that he was equally ready to defy the populace ... in the cause of right and law, and in insisting on the reality of his mission...and his determination to discharge it. The prosecutors had no desire for blood. They counted on a voluntary withdrawal of the accused from the jurisdiction before trial; the death penalty was proposed to make such a withdrawal certain. Socrates himself forced the issue by refusing at any stage to do anything involving the least shade of compromise.'[14]

Having accepted the death penalty at the end of his trial, Socrates refused the chance to escape and courageously drank the lethal dose of hemlock, stating that it was the result of decisions made by men of less clear minds. Back in the twentieth century, R.L. Moore was facing down those he saw as his persecutors, rather than prosecutors, and in an age of student revolt, Texas students enjoined the battle with what was termed locally as constructive rebellion by fighting from within the system. They became an embattled group as they strove to retain third floor philosophy and faculty. Their main weapon, like Socrates, was based upon a glorious past and the effectiveness of that group.

Beyond that, 25 of Moore's graduate students set about attempting to prove suspicions of a devious plot. Four of them, J.M. Carter, H.K. Smith, T.W. Fogwell, and F.A. Stiles, acted as coordinators of the effort and the principal target of their investigations centered on the activities

---

[13] H.S. Wall to Dr. Otis Singletary, 18 April 1969, copied to R.L. Moore. Appears as Appendix 9 in a report Concerning Dean John R. Silber and the Proposed Dismissal of Professor R.L. Moore, Joseph M. Carter, H.K. Smith, Tom W. Fogwell, and Frederick A. Stiles (eds.), April 1969, in the R.L. Moore Papers in the AAM.

[14] Socrates: *Encyclopaedia Britannica* Online, www.eb.com.

of Dean Silber. They produced letters, reports of meetings, notes of conversations, opinions of others from inside and outside the university and put the whole together in what became known as The Green Book, but officially titled *Concerning Dean John R. Silber and the Proposed Dismissal of Professor R.L. Moore*. They purported to demonstrate that Silber had for many years held strong personal animosity towards Moore, and had utilized his office to stir up that dislike and get Moore fired. Students in conversation with Silber alleged that he had used such terms as 'I'll kick his butt out' and cast doubt on the abilities of Moore students over the past few years. The students accused him of unprofessional and unethical behavior in order to mislead the Board of Regents into supporting his view that Moore's service should be brought to an end.

Silber strongly denied these accusations, although he admitted he did not especially like Moore, as indeed many people did not, but that was not the motive for his recommendation to seek his retirement. The needs of the University came above what he regarded as the somewhat emotional demands of students in support of their aging teacher. Silber, in turn, pointed to the fact that on 27 March 1969 the Budget Council of the Department of Mathematics was again asked to vote on Moore's fitness to continue on modified service. The result: nine against his continuing, four in favor and one abstention. A further poll was taken among faculty of the Department of Mathematics on 1 May 1969.[15] Given that the Pure Mathematics team had only two members eligible to vote, Moore and Wall, it was something of a foregone conclusion. Of the 45 members of the departmental faculty, two were on leave in Europe and six assistants were not eligible. Of the remainder, 27 voted in favor of Moore's retirement, one voted against and nine did not return their ballot papers.

Meanwhile, as the campaign by students gathered momentum a near avalanche of letters and telegrams came pouring in from all parts of the country. But Dean Silber did not intend to turn back. Finally, as talk of a compromise solution emerged, the Board of Regents gave their approval to what some students likened to the Trial of Socrates: Chairman Frank Erwin invited a delegation of four of Moore's supporters to present their case against his dismissal at the Regents' meeting in Arlington, Texas, on 2 May 1969. The case for the defense, co-ordinat-

---

[15] Ballot forms circulated to all faculty in the Department of Mathematics, from W.W. Bledsoe, Chairman, 1 May 1969. Moore was made aware of the result of the ballot; copies of the results are in the R.L. Moore Papers in the AAM.

ed by Dr. John Worrell, was allowed 45 minutes to make their submission. Their presentation did not deal with the contents of the controversial report, *Concerning Dean John R. Silber*, but rather with the facts directly concerning Moore's eminence as a mathematician, his accomplishments as a teacher, and his continuing excellent mental and physical condition. Much of the discussion dealt with Dr. Moore's unusual and highly successful method of teaching.

The description of this man's extraordinary mathematical life, said one present, clearly came as something of a revelation to some members of the Board and, according to Chancellor H. H. Ransom, favorably impressed the Board as a whole. At the conclusion of the presentation, Dr. Hackerman cordially greeted Dr. Worrell and said: 'Don't worry. We'll work something out.'[16] But, in spite of the deep interest and surprise shown by some members of the Board, nothing was worked out. No motion was put before them, no decisions were taken and the previous sentence to banish Moore from the University, other than in the style of a visitor, still stood. There were no acceptable alternatives: if the Board changed its mind, Hackerman, Silber, Bledsoe and a good number of mathematics faculty might well have felt it necessary to resign. So now it was Hackerman who drew the short straw: either he backed Silber or sacked him with the resultant loss of a number of departmental teachers. The fact that many of Moore's students had threatened to transfer en masse to another university was troublesome, but Hackerman made it clear, he would prefer to lose 30 or so graduate students rather than half of the Mathematics Department faculty.

All avenues of compromise seemed to have been explored. There were, finally, other last-ditch possibilities being mooted, which would call for the creation of a separate (Socratic) teaching section of mathematics, or the creation of a University Teaching and Research Institute with Dr. Moore and Dr. Wall involved. It was a nice idea, but at the end of the day, in a last display of wonderful and long-awaited irascibility, Moore sent a message via F.A. Stiles to his friend, the Deputy Chancellor Dr. Charles LeMaistre: 'Dr Moore is not opposed to being made Professor Emeritus ... [but he is] very much opposed to any symposium, commemorative volume of any sort, dinner, or any other func-

---

[16] *Concerning Dean John R. Silber and the Proposed Dismissal of Professor R.L. Moore*, Joseph M. Carter, H.K. Smith, Tom W. Fogwell, and Frederick A. Stiles (eds.), April 1969, in the R.L. Moore Papers in the AAM.

tion in "honor" of his retirement. Professor Moore *is* very much opposed to the creation of any R.L. Moore Professorship, R.L. Moore Chair of Mathematics, R.L. Moore Institute, or any other recognition, both now and in the future. Professor Moore's opposition to any of the above cannot be expressed too strongly...'[17]

He also turned down an offer to join the University of Houston as Professor of Mathematics, which came via Dr. Reginald Traylor. He then also rejected out of hand a request to co-operate with Douglas Forbes from the University of Wisconsin in preparation for his PhD dissertation *The Texas System: R.L. Moore Original Edition*. The request came in a note from R H Bing, to which Moore replied: 'No — I have so many objections.... I should think you would not need to be told what they are!'[18] Moore taught his classes at The University of Texas that summer for the last time. On the last day of the session, he dismissed his final class but the students just sat there. None of them moved, as if by sitting stock still, they could somehow prevent the inevitable.[19] They didn't. Robert Lee Moore walked out of the room, and heard clapping in the background as he walked briskly away. After almost 71 years since he first came to the University, he was leaving it for the last time. He made a point, someone said, of not looking back. But he did eventually return, to honor his commitment to the only accolade he accepted, that of Professor Emeritus, and resumed his daily walks to collect his mail. His two colleagues, Ettlinger and Wall, often did the same. Wall retired rather than continue on alone in the shell of what had been Pure Mathematics, but two years later he was struck by a recurring bout of illness and died at the age of 69.

In the summer of 1972, the Department of Mathematics moved to a new building, originally called the PMA Building, to distinguish it for the study of physics, mathematics and astronomy. Before the building

---

[17] F.A. Stiles to Dr. Charles LeMaistre, 31 July 1969, copy attached to the Report of Proceedings prior to the retirement of R.L. Moore, Center for American History, The University of Texas at Austin.

[18] Exchange between R H Bing and R.L. Moore, 2 September 1969, from the R.L. Moore Papers in the AAM.

[19] During the weeks prior to this final act in the drama, and such it was, students had threatened to leave The University of Texas en masse if Moore was forced to retire. A number of Moore's students did in fact quit immediately and transferred to other universities where the Moore Method was practiced, generally by former Moore students, including Emory and Auburn Universities.

was completed, the Department of Mathematics asked for their wing to be named after R.L. Moore. The Board of Regents belatedly discovered a lasting tribute for their long and steadfast servant. They decided to name the entire building after him, which was indeed a major about-turn given that it was the largest classroom building on campus and that most such additions in modern times usually carried the name of financial benefactors. Thus, it became the Robert Lee Moore Hall.[20] Moore did not attend the dedication ceremony on 5 October 1973, pleading physical weariness and indeed his walks to the campus had become less frequent. Those involved in the dedication included Professor Bing, who in that year had returned to Texas to become Chairman of the Mathematics Department of The University of Texas.

In May 1974, Moore suffered a stroke, and a second a month later, and his doctors marveled that he survived the first, let alone the second. He had suffered brain damage, however, and subsequently was removed to a nursing home where he died on 4 October that year. His wife survived a further two years. Ettlinger, Moore's long-standing colleague in that historic era in UT Pure Mathematics, remained active for several years beyond. He survived the loss of a foot due to gangrene, and then became partially sighted due to a detached retina. But it was only in the last months of his life, when he succumbed to cancer, that his spirit began to decline prior to his death in 1986 at the age of 97.

But that wasn't the end of the story, by any means.

Regardless of the controversial aura that surrounded those long years of study and endeavor, Moore left too much of an indelible imprint to be allowed to pass into history. As already mentioned, his mathematical descendants, some 1,400 by 2002, emerged from his 50 original doctoral students and thus the Moore School grew by natural progression through their own students and students of students, to form a historically significant and influential group in American mathematics. As G.M. Reed points out:

---

[20] Ironically, Dean Silber was not there to witness this event. He had recently left the university because of his principled stand against the Board of Regents' plan to dismember the College of Arts and Science and turn it into four separate units; he stated that it was not their function to intervene in educational matters. The publicity brought him to the attention of Boston University, which was in financial and educational disarray. He was appointed president in 1971 and had considerable success in balancing the budget, while at the same time hiring distinguished new faculty, raising admission standards, expanding the campus and generally restoring the university to health. Still a controversial figure in local circles, he became Chancellor at BU in 1996.

R. L. Moore Hall on the UT campus

'Moore spaces were initially but a first stage towards R.L. Moore's successful topological characterization of the plane. However, through his extensive use of this concept in teaching "non-metric" topology to several generations of gifted students, Moore spaces have become one of the most widely studied classes of spaces in topology. Their continuing relevance has been due to the depth of positive theory shared with metric spaces, together with the wealth of non-trivial counterexamples, which differentiate them.... Of course, as anyone who cares to look through Moore's notebooks at The University of Texas will easily agree, most of Moore's own interests concerned the geometric and continua theory of Moore spaces. The work of [many of] his students reflects this legacy.'[21]

But the legacy goes much deeper, and is especially relevant in the application of the Moore Method to modern teaching. Although the drawbacks in its use in larger classes have been explored in earlier pages, there is nonetheless a very positive link to the method of inquiry-based learning recommended in a report on undergraduate education

---

[21] G.M. Reed, Set-theoretic problems in Moore spaces, published in *Open Problems in Topology*, J. van Mill and G.M. Reed (eds.), Elsevier Science, B.V. 1990, pp. 165–181.

made to the National Science Foundation.[22] Those who experienced the Moore Method would identify it as a forerunner to inquiry-led learning and, coincidentally, the NSF report arrived at a time when there was a rebirth of interest in Moore's work and teaching. This culminated in the formation of The Legacy of R.L. Moore Project sponsored by the Educational Advancement Foundation, under the leadership of former Moore student Harry Lucas Jr, in close collaboration with other former students. The formation of this enterprising organization was underpinned by the simple premise that 'every student has the capability for creative and critical thinking and the method pioneered by Moore has proven to be very conducive to recognizing, nurturing, and developing this ability'. The Project included the setting up of an archive of Oral History and a comprehensive library of original papers on Moore and his work, and both include many famous mathematical contributors.

Annual conferences and study groups followed, attracting widespread attention and support from diverse and important groups within the mechanism of higher learning across the United States of America and Europe.[23] The Project has subsequently linked into major mathematical events, including the Joint Meetings of the AMS and MAA. Moore-related sessions at the 2000 New Orleans Joint Meetings, for example, created such interest that the audience spilled out into the corridors. The movement is growing, with interest and activity arising in many quarters. Through these activities, the work of one of America's great mathematical figures lives on.

---

[22] Report to the NSF, entitled *Shaping the Future,* published 1996.
[23] The aims and goals of The Legacy of R.L. Moore Project are set out on the website, www.discovery.utexas.edu along with contact details and a list of study materials and online papers.

Appendix 1

The Moore Genealogy Project

R.L. Moore's interest in genetic inheritance was apparent from the numerous articles he had marked on the subject in *Science*, a journal of the American Association for the Advancement of Science, found among his papers. Around the age of sixty, he also became a regular customer of the Goodspeed Book Store in Boston where over a period of several years he purchased by mail order a succession of books on genealogy which he used to launch himself into the compilation of a family tree, apparently inspired by a more modest version sent to him by a relative. He set about the task with the same vigor and systematic intensity that he applied to solving some piece of mathematical mystery.

The project took him on a long journey of discovery that would occupy many hours of his life for a dozen years or more, naturally involving numerous exchanges with relatives, searches of public records and published material across the United States and onwards into local and national historical records. In his extensive research R.L. Moore discovered that on his mother's side there was a direct collateral line to the Taylor family of Virginia, which put him only a generation from Zachary Taylor, the twelfth president of the United States (1849–1850). Although he does not list it in his own charts, there was an important secondary aspect to this connection that may not have been apparent to him at the time, because it was overshadowed by other events.

Zachary Taylor was the father-in-law of Jefferson Davis, the first and only president of the Confederate States of America. In one branch, therefore, Moore had two particularly strong figures in American history and it is worth taking a moment to expand upon these connections in that they lead us on a circular tour, arriving ultimately at Moore's birth in what became his beloved state of Texas.

Taylor's parents, Richard Taylor and Mary Strother, migrated to Kentucky from Virginia shortly after Zachary, the third of their nine children, was born. After a childhood spent on a farm on the Kentucky frontier, Zachary Taylor enlisted in the army in 1806 as a career officer. In 1810 he married Margaret Mackall Smith, with whom he had six children and served in the army for almost 40 years, advancing to the rank of major general. He commanded troops in the field in the War of 1812 and thereafter all major conflicts in the first half of the century. The brief but tragic Black Hawk War in 1832, in which the army cleared Illinois of Sauk and Fox Indians, was especially noteworthy. Among the young officers under his command were Abraham Lincoln, a captain of volunteers, and Jefferson Davis, an army lieutenant, who were of course destined to wage their own war, against each other, three decades later. Davis was, at the time, serving under Zachary Taylor and the association led to his meeting Taylor's daughter, Sarah. They married in 1835.

Taylor himself marched on in a successful military career and after the annexation of Texas in 1845, President James K. Polk ordered him to take an army of 4,000 men to the Rio Grande, whereupon he was engaged by Mexican troops in a skirmish that marked the beginning of the Mexican-American War. Subsequently, at the Battle of Buena Vista, Taylor's vastly outnumbered troops won a brilliant victory while General Winfield Scott captured Mexico City itself on 14 September 1847. In the Treaty of Guadalupe Hidalgo, signed the following year, Mexico gave up its claim to Texas and also ceded an area now in the states of New Mexico, Utah, Nevada, Arizona, California, and western Colorado.

Taylor returned a national hero and was immediately taken up by Whig politicians as a presidential candidate, and won. It was to be a brief administration, beset with problems not least of which was the controversy of the extension of slavery into the Mexican territories and his personal opposition to the creation of these new slave states. In December 1849, he also called for immediate statehood for California when a new constitution prohibited slavery. Southerners in Congress fought bitterly against the proposal, fearing repercussions in the mounting tension over such issues. Zachary Taylor did not, however, live to witness the eventual outcome. He died sixteen months after taking office. It was not without some irony, therefore, that he missed the developments that ensued.

Jefferson Davis had also entered politics. In the year of his marriage to Sarah Taylor, he resigned his commission and became a planter near

Vicksburg, Mississippi, on land given to him by his brother. Within three months, his bride was dead from malarial fever and, heartbroken, Davis remained in seclusion for seven years, working his plantation and reading constitutional law. He was elected to the House of Representatives in 1845, and, in the same year, married Varina Howell, a Natchez aristocrat who was 18 years his junior.

He resigned his seat in Congress after eighteen months to serve in the war with Mexico as a colonel commanding the First Mississippi volunteers, once again alongside his former father-in-law, Zachary Taylor, at Buena Vista. Severely wounded, Davis returned to politics and battled with his former father-in-law over the issue of the slave states in the former Mexican territories. He argued that the Southern case for slavery was on firm moral foundations, first because it gave them a better life than in Africa: 'Our slaves are happy and contented. Why, I have no more fear of them than I do our cattle.' Davis continued his upward thrust in politics, and became Secretary of War under President Franklin Pierce in 1853, just six years away from the calamitous developments that inexorably drew the states into Civil War.

Davis had initially opposed secession but gradually it became unstoppable and, from the Deep South's point of view, inevitable. On 18 February 1862, he was elected President of the Confederacy. Even at that late stage, he sent peace emissaries to Washington, DC, but President Lincoln refused to see them and instead sent armed ships to Charleston, SC, to re-supply the beleaguered Union garrison at Fort Sumter. Davis appointed Robert E. Lee as head of the Confederate Army, and ordered the bombardment of the fort on 12 April. So began the American Civil War in which Zachary Taylor's son Robert, a direct relation of Robert Lee Moore, supported his brother-in-law Jefferson Davis and served as a lieutenant general in the Confederate Army.

Whether Moore had established the detail of these historic connections or not, the unraveling of these side branches to his family tree must have given him great pleasure. He also often made great play on the names Robert Lee, and was proud of the statue of his hero, erected on the campus of The University of Texas in 1924. For sheer volume and effort, Moore's genealogy research was impressive and there were further revelations yet to come. Tracking his paternal side led to a distant connection to Grover Cleveland, 22nd and 24th president of the United States. Cleveland, a Democrat from humble beginnings, was known for his honesty and integrity, rising to President on an anti-corruption ticket.

His second term, however, became so imbued with Conservative pro-business connections that he became the first and only President to be repudiated by his own party. He had supported the bi-metallic standard of gold and silver to expand the nation's money supply that led to the famous Cross of Gold speech by his opponent William Jennings Bryan. Cleveland retired to Princeton, New Jersey, where he became a lecturer in public affairs at the university and a trustee from 1901–1908, coincidentally overlapping the period when Robert Lee Moore was himself at Princeton but, presumably, neither realized the connection at that time.

Yet more colorful connections were to be revealed in what became a star-studded grand finale to Moore's genealogical opus, containing a list of European ancestors. In it, he purported to show — and it is now impossible to verify — that his distinguished antecedents included three kings, and not just any old European kings, as he explained to his sister Eleanor in a letter on 13 July 1949:

Dear Sister,

Would you be surprised to hear that we are descended from King Alfred the Great of England, from King Henry I of France and from Donald Bane, King of Scots. And if someone told you it was so would you be very interested and ask for details?

Your affectionate brother,

Robert[1]

Eleanor replied that she would be very interested and Moore then sent her a copy of the proof of his theory, which took them back into the first millennium. He also sent one to his older brother Jennings, by then 79, along with several letters explaining his findings. Jennings, however, seemed mesmerized by the whole thing. He wrote back querying a number of what were apparently Jewish names in the list, such as Moses Lyman, Isaac Cowles (from their paternal grandmother's side) and Aaron Cleveland, a relative of Grover. Moore explained to his brother, over several exchanges of letters, that in the seventeenth and eighteenth centuries such Christian names were common among families of New England, noting in increasingly exasperated tones that one of their number, Jeremiah Cowles, was in fact a Congregational minister. Jennings was still not satisfied and in a further exchange he reported to his brother that he had been advised by a friend that there was some connection

---

[1] R.L. Moore Papers in the AAM.

between the names of Moore and Isaacs. R.L. Moore replied immediately, clearly upset by the inference and frustrated by his brother's unfortunate reaction to the chart. He dismissed this allegation in one sentence on a single sheet of paper: 'If anybody *tells* you that Isaacs and Moore are two words with the same meaning, tell him that I say he is a damned liar![2]

---

[2] R.L. Moore to Jennings Moore, 17 July 1949; R.L. Moore Papers in the AAM.

Appendix 2

The PhD Students of R. L. Moore

(written and compiled by Dr. Ben Fitzpatrick, Jr.)

John R. Kline (1891–1955), PhD, University of Pennsylvania, 1916. A.B. (1912) Muhlenburg; A.M. (1914) Pennsylvania. Thesis title: *Double elliptic geometry in terms of point and order*, 19 pages in journal article.

Kline was Professor of Mathematics at Penn from 1920 (the year Moore returned to Texas) until his death in 1955. He was Chair of the Department from 1933 to 1954. He was a Guggenheim Fellow (Göttingen) in 1926–1927. He was Associate Editor of the *Transactions of the American Mathematical Society*, the *Bulletin of the American Mathematical Society*, and the *American Journal of Mathematics* at various times and served on the editorial board of the AMS Colloquium Publications. He was Associate Secretary of the AMS from 1933 to 1936 and Secretary from 1936 to 1950. He directed the PhD theses of 13 students, including National Research Fellows W.L. Ayres, H.M. Gehman, N.E. Rutt, and Leo Zippin. Several of his students went to Austin in their Fellowship years.

He also directed the PhD theses of W.W. Dudley and W.W.S. Claytor, the second and third African-American mathematicians to earn their PhDs. On his recommendation, Claytor went to the University of Michigan on a post-doctoral fellowship, where he worked with R.L. Wilder. After this, he was offered the opportunity for further study at Princeton but declined it in order to accept a position at Howard University, where he had a distinguished career. Kline was the thesis director of record for Lida K. Barrett, but she began her work with R.L. Moore at Texas; the thesis problems were suggested by Kline after her move to the University of Pennsylvania and essentially solved under his direction. After Kline suffered a heart attack the work was completed under the direction of R.D. Anderson, who had joined the faculty at Pennsylvania in 1948. Kline's best-known student was Leo Zippin, who wrote with Deane Montgomery the classic monograph *Topological Transformation Groups*. The work described in this treatise, together with work done by Andrew Gleason, provides a solution to the first, and famous, part of Hilbert's Fifth Problem. Throughout his career, Kline maintained regular contact with R.L. Moore; for example, he was instrumental in getting Mary-Elizabeth Hamstrom to go to Austin for her graduate studies. Kline wrote only one joint paper, and it was with R.L. Moore. Title: "On the most general closed and bounded plane point set through which it is possible to pass an arc."

George H. Hallett (1895–1985), PhD, University of Pennsylvania, 1918. A.B. (1915) Haverford College; A.M. (1916) Harvard. Thesis title: *Linear order in three-dimensional Euclidean and double elliptical spaces.*

Hallett did not enter academia; instead, he had a long career in public service, beginning with a post as Secretary of the Proportional Representation League. In 1937 he wrote *Proportional Representation: The Key to Democracy*. He taught courses in government at several colleges in New York City: Brooklyn, Hunter, NYU, and CCNY. He was given special awards from LaGuardia Memorial Association (1963), New York City Club (1964), and Citizens Union (1947, 1969). During the Second World War he was active in the Committee for Civilian Defense. In his book *Creative Teaching: The Heritage of R.L. Moore*, D. R. Traylor quotes Hallett as saying, "... I think such success as I've had in the field of government probably has a good deal to do with [Moore's teaching] — because they don't catch me up very often in theories of logic in bills, or different parts of bills, that don't hang together."

Anna M. Mullikin (1893–1975), PhD, University of Pennsylvania, 1922. A.B. (1915) Goucher College; A.M. (1919) U. Pennsylvania. Thesis title: *Certain theorems relating to plane connected point sets.*

When Moore went to Texas in 1920, Mullikin went there, as well, to continue her work with him. Her thesis, *Certain theorems relating to plane connected point sets*, received considerable attention. It overlapped somewhat with some of Janiszewski's work (which had been published in Polish). Mullikin went on to a long career as a high school teacher in Pennsylvania; Mary-Elizabeth Hamstrom was one of her students. During the Second World War she was active in Civilian Defense. In 1952, Goucher College, where she received a B.A. in 1915, honored her with an alumnae achievement citation.

Raymond L. Wilder (1896–1982), PhD, University of Texas, 1923. B.Phil. (1920) Brown; M.Sc. (1921) Brown. Thesis title: *Concerning continuous curves*, (62 pages). Signers: R.L. Moore, H.J. Ettlinger, A.A. Bennett, H.Y. Benedict, M.B. Porter.

Wilder was a member of the National Academy of Sciences. He served as President of the American Mathematical Society in 1955–1956 and as President of the Mathematical Association of America in 1965–1966. The bulk of his academic career was at the University of Michigan, where he directed twenty-five PhD students. In his honor the University of Michigan established the Raymond L. Wilder Professorship of Mathematics. Before going to Michigan he taught at Ohio State for two years, and after retiring from Michigan he taught at UC Santa Barbara. He was the AMS Colloquium Lecturer in 1943 (*The Topology of Manifolds* was published in the Colloquium series), and the Gibbs Lecturer in 1969. He wrote *Introduction to the Foundations of Mathematics, The Evolution of Mathematical Concepts; an Elementary Study,* and *Mathematics as a Cultural System.* He was a recipient of the Distinguished Service Award of the MAA and the Lester R. Ford Award. He has observed that he went from Brown to Texas to study actuarial mathematics. He recalled that Moore did not at first want him as a student ('I was a Yankee') and did not really accept him as a member of the class until he had proved the arcwise connectedness theorem. Wilder was one of the developers of algebraic topology. He maintained close contacts with Moore and also with Lefschetz at Princeton. Wilder was largely responsible for the recognition and development of the abilities of Norman Steenrod. (See the extensive correspondence between them at the Archives of American Mathematics at the Center for American History at The University of Texas at Austin.) He also had influence on E.G. Begle and mathematics education reform.

Renke G. Lubben (1898–1980), PhD, University of Texas, 1925. B.A. (1921) Texas. Thesis title: *The double-elliptic case of the Lie-Riemann-Helmholtz-Hilbert problem of the foundations of geometry*, (102 pages). Signers: R.L. Moore, M.B. Porter, H.J. Ettlinger, G. Watts Cunningham, Edward L. Dodd, A. P. Brogan, Albert A. Bennett.

Lubben's PhD thesis gave the solution to the then last remaining problem in the foundations of geometry. He was an independent discoverer of maximal compactifications of completely regular spaces, although priority in publication is assigned to Stone and Čech. He was a National Research Fellow (Göttingen) in 1926–1927. He served on the Texas mathematics faculty until his retirement.

Gordon T. Whyburn (1904–1969), PhD, University of Texas, 1927. A.B. (1925) Texas; M.A. (Chemistry) (1926), Texas. Thesis title: *Concerning continua in the plane*, (57 pages). Signers: R.L. Moore, M.B. Porter, H.J. Ettlinger, Edward L. Dodd, J.R. Bailey, H.L. Lochte, Arnold Romberg.

Whyburn was a member of the National Academy of Sciences. He was President of the American Mathematical Society in 1953–1954. After leaving Texas, he taught at Johns Hopkins, but his principal academic appointment was at the University of Virginia, where he and E.J. McShane were brought in to develop a graduate program in mathematics. He served as Chair of the department from 1934 until 1966. He was the AMS Colloquium Lecturer in 1940; his Colloquium book *Analytic Topology* resulted. He received the Chauvenet Prize of the MAA. He directed thirty-two PhD students, many of whom had distinguished academic careers. Whyburn studied chemistry as an undergraduate and in graduate school, obtaining an M.A. in Chemistry before deciding to concentrate on mathematics.

John H. Roberts (1906–1997), PhD, University of Texas, 1929. A.B. (1927) Texas. Thesis title: *Concerning non-dense plane continua*, (47 pages). Signers: R.L. Moore, M.B. Porter, H.S. Vandiver, H.J. Ettlinger, Edward L. Dodd, J.W. Calhoun, A.E. Cooper.

On leaving Texas, Roberts went to the University of Pennsylvania for a year, where he worked with J.R. Kline. He went to Duke University in 1931, and he remained there until his retirement. During the Second World War he served in the United States Navy. He directed twenty-four PhD students, a number of whom are prominent mathematicians. He was Director of Graduate Studies in the Department from 1948 until 1960 and managing editor of the *Duke Mathematical Journal* from 1951 until 1960. He also served as Chair of the Department. His research centered at first on continua, then on dimension theory.

Clark M. Cleveland (1892–1969), PhD, University of Texas, 1930. B.S.in C.E. (1917) Mississippi. Thesis title: *On the existence of acyclic* *curves satisfying certain conditions with respect to a given continuous curve*, Signers: R.L. Moore, M.B. Porter, Edward L. Dodd, H.J. Ettlinger, J.M. Kuehner, Arnold Romberg.

Cleveland joined the Department of Applied Mathematics and Astronomy at Texas and remained there throughout his academic career. He became Chair of the Department of Mathematics in 1953 when the Department of Applied Mathematics and Astronomy and the Department of Pure Mathematics were merged.

Joe L. Dorroh (1904–1989), PhD, University of Texas, 1930. B.A. (1926) Texas; M.A. (1927) Texas. Thesis title: *Some metric properties of descriptive planes*, (38 pages). Signers: R.L. Moore, H.S. Vandiver, M.B. Porter, H.J. Ettlinger, J.M. Kuehner, S. Leroy Broun.

Dorroh was a National Research Fellow at Cal Tech in 1930–1931 and at Princeton in 1931–1932. He taught at LSU from 1942 to 1946 and at Illinois Tech in 1946–1947. He then went to Texas A&I until his retirement in 1966. He was Chair there from 1952 until his retirement.

Charles W. Vickery (1906–1982), PhD, University of Texas, 1932. B.A. (1928) Texas. Thesis title: *Spaces in which there exist uncountable convergent sequences of points*, (37 pages). Signers: R.L. Moore, M.B. Porter, H.S. Vandiver, Edward L. Dodd, E.T. Mitchell, A.P. Brogan.

He worked as a statistician and economist for the State of Texas and the U.S. Government. Following the Second World War he taught at LSU but returned to work for the government and in the aircraft industry. He published in *Econometrica*, the *Bulletin of the American Mathematical Society*, and the *American Mathematical Monthly*. He was a Fellow of the Royal Statistical Society.

Edmund C. Klipple (1906–1992), PhD, University of Texas, 1932. B.A. (1926) Texas. Thesis title: *Spaces in which there exist contiguous points*, (42 pages). Signers: R.L. Moore, H.S. Vandiver, M.B. Porter, Edward L. Dodd, H.J. Ettlinger, Homer V. Craig.

Klipple joined the Texas A&M mathematics faculty in 1935, and he stayed there until his retirement in 1971. He was Chair of the department for many years; he resigned as Chair in 1966 when asked by the Dean to rank his faculty in order, 1–40. In 1968 he was given a Faculty Distinguished Achievement Award. He nurtured a good many students who later became well known mathematicians, including Peter Lax, Efraim Armendariz, and William T. Guy.

Robert E. Basye (1908–2000), PhD, University of Texas, 1933. B.A. (1929) Missouri; M.A. (1931) Princeton. Thesis title: *Simply connected sets*, (26 pages). Signers: R.L. Moore, M.B. Porter, H.S. Vandiver, P.M. Batchelder, H.J. Ettlinger, R.V. Haskell.

Basye's principal academic appointment was at Texas A&M University, from 1940 until his retirement in 1968. After retirement, he devoted his full time to rose research, becoming a renowned genetic hybridizer and grower. Among his many achievements was Basye's Purple Rose. He served on active duty in the U.S. Naval Reserve during the Second World War.

F. Burton Jones (1910–1999), PhD, University of Texas, 1935. B.A. (Chemistry) (1932) Texas. Thesis title: *Concerning R.L. Moore's Axiom 5-1*, (80 pages). Signers: R.L. Moore, M.B. Porter, H.J. Ettlinger, Edward L. Dodd, E.P. Tchoch, W.A. Felsing.

As was the case with Whyburn before him, Jones was a chemistry student who changed to work with Moore. He stayed on the Texas faculty until 1950, serving as chair. He later served as Chair at the University of North Carolina (Chapel Hill) and at the University of California, Riverside. During the war years he worked for the U.S. Navy in Cambridge, on methods for locating and identifying submarines. He directed fifteen PhD students, many of whom became well known mathematicians. At Texas he helped develop a number of students; in fact, Mary Ellen Rudin has described him as the mathematician who had the greatest influence on her development. After leaving Texas he continued to provide counsel and support and guidance to younger mathematicians trained in the Texas tradition. He worked in continua and in abstract spaces. He originated the famous normal Moore space problem. In 1975 he received a Fulbright-Hays Fellowship to visit Canterbury University in Christchurch, New Zealand. He was twice a fellow at the Institute for Advanced Study and spent two summers in Europe on the National Academy of Sciences Exchange Program.

Robert L. Swain (1913–1962), PhD, University of Texas, 1941. B.A. (1934) Reed College. Thesis title: *I. Proper and reductive transformations. II. Continua obtained from sequences of simple chains of point sets. III. Distance axioms in Moore spaces. IV. Linear metric space. V. A space in which there may exist uncountable convergent sequences of points*, (101 pages). Signers: R.L. Moore, H.S. Vandiver, M.B. Porter, P.M. Batchelder, Edward L. Dodd, H.J. Ettlinger.

Swain's major academic appointments were at the University of Wisconsin (Madison), Teacher's College at New Paltz, and Rutgers University. In 1955–1956 he held a Ford Foundation Faculty Fellowship.

Robert H. Sorgenfrey (1915–1996), PhD, University of Texas, 1941. B.A. (1937) UCLA. Thesis title: *Concerning triodic continua*, (56 pages). Signers: R.L. Moore, F. Burton Jones, H.S. Vandiver, P.M. Batchelder, Edward L. Dodd, H.J. Ettlinger.

W.M. Whyburn, who was Department Head at UCLA at the time of Sorgenfrey's undergraduate work, arranged for him to study with Moore. Sorgenfrey was on the faculty at UCLA from 1942 until his retirement in January of 1979. In 1963 he received the UCLA Distinguished Teaching Award, the first mathematician to do so. He is known for his work in general topology, especially in the production of counterexamples. He directed four PhD students. After retirement he wrote several successful high school mathematics textbooks.

Harlan C. Miller (1896–1981), PhD, University of Texas, 1941. B.A. (1916) Wellesley; M.A. (1930) Columbia. Thesis title: *On compact unicoherent continua*, (72 pages). Signers: R.L. Moore, R.G. Lubben, H.S. Vandiver, P.M. Batchelder, Edward L. Dodd, H.J. Ettlinger.

Prior to her graduate program, Miller taught for a number of years at the Hockaday school in Dallas. After her PhD, she taught at Winthrop College and North Texas for one year each before joining the faculty at Texas Woman's University, where she spent the rest of her academic career. She was active in University administration there, serving as Director of Mathematics. She helped direct Lida K. Barrett toward studying with R.L. Moore. She advised J.R. Boyd in his graduate program. He later developed the Moore-style mathematics program at Guilford College. Texas Woman's University has an annual Harlan Miller Lecture Series.

Gail S. Young (1915–1999), PhD, University of Texas, 1942. B.A. (1939) Texas. Thesis title: *Concerning the outer boundaries of certain connected domains*, (70 pages). Signers: R.L. Moore, F. Burton Jones, R.G. Lubben, Edward L. Dodd, H.S. Vandiver, N. Coburn, H.J. Ettlinger.

Young held appointments at Purdue, Michigan, Tulane, Rochester, Case-Western Reserve, Wyoming, and Columbia. He was Chair at at least two of these. He was a President of the Mathematical Association of America. He won the Distinguished Service Award of the MAA in 1987. He worked with the School Mathematics Study Group and with the Committee on the Undergraduate Program in Mathematics. He directed fourteen PhD students, with one of whom (John Hocking) he wrote the successful textbook *Topology*. Another was Beauregard Stubblefield, who has engagingly described in an interview conducted by Albert C. Lewis for the Center for American History how he became aware of Moore's principles of teaching through R.L. Wilder, Gail Young, and Ed Moise, three of Moore's students at Michigan at the time of Stubblefield's graduate program.

R H Bing (1914–1986), PhD, University of Texas, 1945. B.S. (1935) Southwest Texas; M.Ed. (1938) Texas. Thesis title: *Concerning simple plane webs*, (34 pages). Signers: R.L. Moore, Edwin Ford Beckenbach, H.J. Ettlinger, H.S. Vandiver, P. M. Batchelder, C.T. Gray, J.G. Umstattd, Hob Gray.

Bing's principal academic appointments were at Wisconsin (1943–1973) and Texas (1973–1978). He was a member of the National Academy of Sciences. He was President of the American Mathematical Society and of the Mathematical Association of America. His work was centered at first on continuum theory, then on 3-manifolds. He is also well known for the Bing metrization theorem. He was the Colloquium Lecturer in 1970, resulting in his groundbreaking Colloquium book *Topology of 3-manifolds*. The American Mathematical Society published, in two volumes, *The Collected Papers of R H Bing*. He directed thirty-eight PhD students, many of whom developed substantial reputations. His background was that of a high school teacher and football coach. When F.B. Jones was asked whether R.L. Moore at first did not recognize Bing's talent, Jones replied, with a twinkle in his eye, "In later years Moore didn't remember it that way." Bing served as Chair of the Wisconsin and of the Texas Mathematics Departments. He was responsible for the MAA's film *Challenge in the Classroom*, which was about Moore's teaching methods.

Edwin E. Moise (1919–1998), PhD, University of Texas, 1947. B.A. (1940) Tulane. Thesis title: *An indecomposable continuum which is homeomorphic to each of its nondegenerate subcontinua*, (24 pages). Signers: R.L. Moore, H.S. Wall, F. Burton Jones, H.J. Ettlinger.

Moise's dissertation involved the pseudo-arc, a term he coined. It was used to solve an old problem of Knaster. He held academic appointments at Michigan, Harvard, and Queen's College, CUNY. It was at Michigan that he began his most important work on 3-manifolds, culminating in his proof, completed at the Institute for Advanced Study, that every 3-manifold can be triangulated. He went to Harvard as James B. Conant Professor of Mathematics and Education. He was a Vice-President of the American Mathematical Society and President of the Mathematical Association of America. He wrote a number of successful textbooks, and a treatise on *Geometric Topology in Dimensions 2 and 3*. He directed three PhD students. In his last years he devoted his attention to 19th century English poetry. During the Second World War he served in the U.S. Navy as a Japanese translator.

Richard D. Anderson (1922–), PhD, University of Texas, 1948. B.A. (1941) Minnesota. Thesis title: *Concerning upper semi-continuous collections of continua*, (22 pages). Signers: R.L. Moore, F. Burton Jones, H.S. Wall, H.J. Ettlinger, H.S. Vandiver.

Anderson was recruited by Moore to do graduate work in the Fall of 1941. His graduate program was interrupted by a tour of duty in the U.S. Navy, where he served at sea. His principal academic appointments were at Pennsylvania and LSU. As a number of Moore's other students, including Bing, Jones, Moise, and Burgess, he held appointments at the Institute for Advanced Study in Princeton. His work at first centered around the geometric topology of continua. He subsequently was largely responsible, along with his students, for developing infinite-dimensional topology. He directed ten PhD students at LSU and, as noted earlier, contributed to the direction of Lida Barrett at Pennsylvania. A number of his students have had distinguished careers. He served as Vice-President of the American Mathematical Society and as President of the Mathematical Association of America. He received the Distinguished Service Award of the MAA. He has in more recent years devoted his major efforts to reform in mathematics education, more generally, in science education. He is currently Senior Consultant to the NSF sponsored Louisiana Systems Initiatives Program.

Mary Ellen (Estill) Rudin (1924–), PhD, University of Texas, 1949. B.A. (1944) Texas. Thesis title: *Concerning abstract spaces*, (27 pages). Signers: R.L. Moore, H.J. Ettlinger, H.S. Wall, F. Burton Jones, E.F. Mitchell, David L. Miller.

Rudin's principal academic appointments were at Duke, where she met Walter Rudin, Rochester, and Wisconsin, from which she retired as

Grace Chisolm Young Professor of Mathematics. She was a Vice-President of the American Mathematical Society and has been very active in AMS affairs and committee work. Her research has been in set-theoretic topology, especially using axiomatic set theory. She has, as C.E. Aull has noted, ushered in the Rudin Era in general topology. She has directed sixteen PhD students and is largely responsible for directing the PhD research of several others, including Judy Roitman and William Fleissner at UC Berkeley.

Note: In their article 'By their fruits shall ye know them: some remarks on the interaction of general topology with other areas of mathematics', appearing in *History of Topology*, edited by I.M. James, Elsevier, 1999, Teun Koetsier and Jan van Mill write: "In that period general topology rather unexpectedly succeeded in solving several difficult problems outside its own area of research, in functional analysis and in geometric and algebraic topology.... There were in that period at least two major developments in general topology that revolutionized the field: the creations of *infinite-dimensional topology* and *set theoretic topology*. It was mainly due to the efforts of Dick Anderson and Mary Ellen Rudin that these fields have played such a dominant role in general topology ever since."

It is interesting to note that Anderson and Rudin comprised a two-person class under Moore in the immediate post-war years. Note that of the preceding five students, four became Presidents of the MAA and four became vice-president of the AMS. One might wonder whether this is duplicated by any other successive group of five students by any one thesis advisor.

Cecil E. Burgess (1920–), PhD, University of Texas, 1951. B.S. (1941) West Texas. Thesis title: *Concerning continua and their complementary domains in the plane*, (38 pages). Signers: R.L. Moore, H.S. Wall, H.S. Vandiver, H.J. Ettlinger.

Burgess's graduate program was also interrupted by service in the U.S. Navy. After leaving Texas, he went to the University of Utah, where he remained throughout his career, except for leaves, which he usually spent working with R H Bing. Most of his work, and that of his students, has been centered on Bing-style topology. He directed ten PhD students, and some of them are quite prominent. He served for a number of years as Chair of the Department.

B.J. Ball (1925–1996), PhD, University of Texas, 1952. B.A. (1948) Texas. Thesis title: *Concerning continuous and equicontinuous collec-*

tions of arcs, (36 pages). Signers: R.L. Moore, H.S. Wall, R.G. Lubben, H.J. Ettlinger, Homer V. Craig.

Ball entered the Navy before getting his BA. On his return in 1946 he moved into Moore's graduate program. His major academic appointments were at Virginia and Georgia. He served as Chair at Georgia for a number of years. His work was in continuum theory, general topology, and, in later years, shape theory. He directed eight PhD students and contributed to the direction of many others.

Eldon Dyer (1929–1993), PhD, University of Texas, 1952. B.A. (1947) Texas; B.S. (1947) Texas. Thesis title: *Certain conditions under which the sum of the elements of a continuous collection of continua is an arc*, (14 pages). Signers: R.L. Moore, H.S. Wall, R.G. Lubben, H.J. Ettlinger, Homer V. Craig.

Dyer's academic appointments included Georgia, Johns Hopkins, Chicago, Rice, and CUNY, from which he retired as Distinguished Professor in 1991. He chaired the Department of Mathematics at CUNY, Center for Graduate Studies 1967–1970. He is best known for his work in algebraic topology and for his six PhD students, among whom is Robion Kirby. As were Wilder and Bing, he was a consulting editor for the *Encyclopedia Britannica*. He held a Sloan Fellowship in 1960–1962 and an NSF Post-Doctoral Fellowship in 1955–1956. On two occasions he was a visiting member of the Institute for Advanced Study in Princeton. He served as Editor of the *Proceedings of the American Mathematical Society* 1960–1965 and as Associate Editor of the *Transactions of the American Mathematical Society*.

Mary-Elizabeth Hamstrom (1927–), PhD, University of Texas, 1952. B.A. (1948) Pennsylvania. Thesis title: *Concerning webs in the plane*, (35 pages). Signers: R.L. Moore, H.S. Wall, R.G. Lubben, H.J. Ettlinger, Homer V. Craig.

Hamstrom's principal academic appointment was at the University of Illinois. Her research was mostly in geometric topology. She directed nine PhD students. Before going to Austin Hamstrom had been taught by two of Moore's earlier students: Anna Mullikin in high school and J.R. Kline at Penn. As were many of the students in Austin in the 1940s and early 1950s, she was strongly influenced by F.B. Jones.

John M. Slye (1923–), PhD, University of Texas, 1953. B.S. (1945) California Institute of Technology. Thesis title: *Flat spaces for which the Jordan Curve Theorem holds true*, (19 pages). Signers: R.L. Moore,

H.S. Wall, H.J. Ettlinger, D.S. Hughes, R.N. Little, Jr.

Slye's academic appointments were at the Universities of Minnesota and Houston. His work was in geometric topology. He directed two PhD students. During the Second World War he served in the U.S. Navy.

John T. Mohat (1924–1993), PhD, University of Texas, 1955. B.A. (1950) Texas Western. Thesis title: *Concerning spirals in the plane*, (76 pages). Signers: R.L. Moore, H.S.Wall, H.J. Ettlinger, Homer V. Craig.

Mohat spent his academic career at the University of North Texas. He served in the U.S. Army during the Second World War.

Bennie J. Pearson (1929–), PhD, University of Texas, 1955. B.A. (1950) Texas. Thesis title: *A connected point set in the plane that spirals down on each of its points*, (18 pages). Signers: R.L. Moore, H.S.Wall, H.J. Ettlinger, Homer V. Craig.

Pearson's major academic appointment was at the University of Missouri-Kansas City. He served as Chair of the Department for six years. He directed three PhD students.

Steve Armentrout (1930–), PhD, University of Texas, 1956. B.A. (1951) Texas. Thesis title: *On spirals in the plane*, (34 pages). Signers: R.L. Moore, H.J. Ettlinger, D.S. Hughes, C.W. Horton, H.S. Wall.

Armentrout taught at the University of Iowa for a number of years and then at Penn State. He has been active in AMS committee work and served as Treasurer of the AMS. He has worked in geometric topology and differential topology. He has directed twelve PhD students, many of whom are very active.

William S. Mahavier (1930–), PhD, University of Texas, 1957. B.S. (1951) Texas. Thesis title: *A theorem on spirals in the plane*, (26 pages). Signers: R.L. Moore, H.S. Wall, H.J. Ettlinger, C.W. Horton.

Mahavier was a physics major; in fact, his only degree in mathematics was the PhD. Mahavier's academic appointments included Illinois Institute of Technology, University of Tennessee, and Emory University. His work has centered on continuum theory. He has directed eight PhD students.

L. Bruce Treybig (1931–), PhD, University of Texas, 1958. B.S. (1953) Texas. Thesis title: *Concerning locally peripherally separable spaces*, (21 pages). Signers: R.L. Moore, H.J. Ettlinger, F.A. Matsen, Homer V. Craig.

Treybig has taught at Tulane University and at Texas A&M University. His work has been primarily in general topology and continuum theory. He has directed seven PhD students.

James N. Younglove (1927–), PhD, University of Texas, 1958. B.A. (1951) Texas. Thesis title: *Concerning dense metric subspaces of certain non-metric spaces*, (22 pages). Signers: R.L. Moore, R.G. Lubben, R.N. Little, H.S. Wall.

After graduating from high school, Younglove served two years in the U.S. Navy before entering Texas as a freshman. Younglove's academic appointments were at University of Missouri (Columbia), and at University of Houston, where he served as Chair for a number of years. His work was primarily in general topology, especially metrication theory. He directed one PhD student.

George W. Henderson (1936–), PhD, University of Texas, 1959. B.A. (1958) Texas. Thesis title: *Proof that every compact continuum which is topologically equivalent to each of its nondegenerate subcontinua is an arc*, (19 pages). Signers: R.L. Moore, H.P. Hanson, R.E. Lane, H.S. Wall.

Henderson's thesis was a proof that every decomposable hereditarily equivalent continuum is an arc. (The word "decomposable" was inadvertently omitted from title pages of final copies of his thesis.) He taught at the University of North Carolina, University of Virginia, Rutgers, and the University of Wisconsin, Milwaukee, where he directed one PhD student.

John M. Worrell (1933–), PhD, University of Texas, 1961. B.A. (1954) Texas; M.D. (1957) Texas. Thesis title: *Concerning scattered point sets*, (41 pages). Signers: R.L. Moore, H.S. Wall, H.J. Ettlinger.

Before entering graduate school, Worrell obtained an M.D. degree. After he received his PhD in mathematics, he worked at Sandia for a number of years, on his own research and on problems of interest to the space program. His mathematical work has centered on general topology, often in collaboration with Howard Wicke. He later taught at Ohio University and is now in private (medical) practice. While at Ohio University he created and developed, with the assistance of George M. Reed, the Institute for Medicine and Mathematics. He held an NSF post-doctoral fellowship in 1961–1962.

Howard Cook (1933–), PhD, University of Texas, 1962. B.S. (1956) Clemson. Thesis title: *On the most general closed and bounded plane* *point set through which it is possible to pass a pseudo-arc*, (34 pages). Signers: R.L. Moore, H.J. Ettlinger, R.E. Lane, H.S. Wall.

Cook's academic appointments have been at Auburn, North Carolina, Georgia, Tasmania, and Houston, mostly at Houston. In his thesis he characterized those compact sets in the plane that can be embedded in pseudo-arcs, an analogue of the Moore-Kline characterization of those that are subsets of arcs. His work has been in continuum theory and in general topology (Moore spaces). He has directed five PhD students.

James L. Cornette (1935–), PhD, University of Texas, 1962. B.S. (1956) West Texas; M.A. (1959) Texas. Thesis title: *Continuumwise accessibility*, (52 pages). Signers: R.L. Moore, H.J. Ettlinger, R.E. Lane, H.S. Wall.

Cornette's principal position has been at Iowa State University. His earlier work was in continuum theory; he has turned to biomathematics in more recent years. He has directed seven PhD students. He is currently University Professor of Mathematics and Director of the Center for Bioinfomatics and Biological Statistics at Iowa State. In 1985 he began a collaborative program of research with three other scientists, which has resulted in twenty-one journal articles, ten review and expository articles, and three patents.

Dennis K. Reed (1933–1986), PhD, University of Texas, 1965. B.S. (1959) Texas. Thesis title: *Concerning upper semi-continuous collections of finite point sets*, (41 pages). Signers: R.L. Moore, H.S. Wall, Homer V. Craig, Patrick L. Odell.

Reed's academic career was at the University of Utah. He won the University's Distinguished Teaching Award in 1973.

Harvy L. Baker (1938–), PhD, University of Texas, 1965. B.A. (1960) Texas. Thesis title: *Complete amonotonic collections*, (35 pages). Signers: R.L. Moore, Homer V. Craig, R.G. Lubben, H.S. Wall.

Baker has taught at the University of Nebraska, where he directed two PhD students, and at The University of Texas at Arlington.

Blanche Joanne (Monger) Baker (1934–), PhD, University of Texas, 1965. B.A. (1956) Lamar State; M.A. (1958) Texas. Thesis title: *Concerning uncountable collections of triods*, (45 pages). Signers: R.L. Moore, R.G. Lubben, Homer V. Craig, H.S. Wall.

Baker has taught at the University of Nebraska and at Lamar University.

Roy D. Davis (1938–), PhD, University of Texas, 1966. B.A. (1961) Texas; M.A. (1964) Texas. Thesis title: *Concerning the sides from which certain sequences of arcs converge to a compact irreducible continuum*, (35 pages). Signers: R.L. Moore, H.J. Ettlinger, R.G. Lubben, H.S. Wall.

After leaving Texas, Davis worked in the aerospace industry in Southern California.

Jack W. Rogers (1943–), PhD, University of Texas, 1966. B.A. (1963) Texas; M.A. (1965) Texas. Thesis title: *A space whose regions are the simple domains of another space*, (34 pages). Signers: R.L. Moore, H.J. Ettlinger, R.G. Lubben, W.E. Millett, H.S. Wall.

Rogers has taught at Emory University and at Auburn University, where he is currently Professor of Mathematics and Director of the Auburn University Honors College. His early work was in continuum theory; he later changed to applied mathematics and computational linear algebra. He has directed three PhD students. Rogers first encountered Moore as a 10th grade high school student. He took the summer geometry course that year.

Martin D. Secker (1927–), PhD, University of Texas, 1966. B.A. (1949) North Texas; M.A. (1950) North Texas; M.A. (1964) Texas. Thesis title: *Reversibly continuous bisensed transformations of an annulus into itself*, (28 pages). Signers: R.L. Moore, H.J. Ettlinger, R.G. Lubben, H.S. Wall.

After leaving Austin, Secker taught first at Iowa State University, then at Branson School, a private college preparatory school in California, and at the College of Marin.

David E. Cook (1935–), PhD, University of Texas, 1966. B.A. (1958) Texas; M.A. (1960) Texas. Thesis title: *Concerning compact point sets with noncompact closures*, (39 pages). Signers: R.L. Moore, Homer V. Craig, W.T. Guy, Jr., Harold P Hanson, R.G. Lubben.

Cook's main academic appointment was at the University of Mississippi, where he directed three PhD students.

John W. Hinrichsen (1940–), PhD, University of Texas, 1967. B.A. (1961) Texas; M.A. (1964) Texas. Thesis title: *Certain web-like continua*, (26 pages). Signers: R.L. Moore, W.T. Guy, Jr., R.G. Lubben, H.S. Wall, W.E. Millett. Hinrichsen's academic career was spent at Auburn University. His research has been in continuum theory.

Joel L. O'Connor (1942–), PhD, University of Texas, 1967. B.A. (1962) Texas. Thesis title: *Holes in two-dimensional space*, (51 pages). Signers: R.L. Moore, Homer V. Craig, Robert E. Greenwood, W.E. Millett, H.S.Wall.

O'Connor taught at the University of Florida and then went into industry as an applied mathematician. He has consulted with the NSA and with the Vanderbilt University College of Medicine and, jointly with a medical physicist, founded Clinical Database Systems. He has consulted with other private firms and with state and local governments. O'Connor first encountered Moore as a high school student in Moore's summer geometry course.

John W. Green (1943–), PhD, University of Texas, 1968. B.A. (1965) Texas; M.A. (1966) Texas. Thesis title: *Concerning the separation of certain plane-like spaces by compact dendrons*, (85 pages). Signers: R.L. Moore, Homer V. Craig, R.G. Lubben, H.S. Wall.

Green was on the University of Oklahoma faculty for fifteen years. He then obtained a PhD in mathematical statistics from Texas A&M University and afterwards taught at the University of Delaware for five years. He has since been employed by E.I. DuPont as senior research biostatistician. He has said that his success in his present position is largely due to the training he obtained in Moore's classes. He has directed four PhD students, two at Oklahoma in topology and two at Delaware in statistics.

Michael H. Proffitt (1942–), PhD, University of Texas, 1968. B.A. (1964) Texas; M.A. (1966) Texas. Thesis title: *Concerning uncountable collections of mutually exclusive compact continua*, (32 pages). Signers: R.L. Moore, Homer V. Craig, R.E. Greenwood, H.S. Wall.

After a post at SUNY New Paltz, Proffitt returned in 1972 to The University of Texas, where he was a Robert A. Welch Fellow in chemistry and later in physics. In 1980 he moved to the University of Colorado, working in atmospheric research, more specifically, measurements of ozone. His measurements identified the cause of the ozone hole and its spread into other latitudes (published as articles in *Nature* and *Science*). He has written over 100 journal publications. He retired in 2004 as Senior Scientific Officer of the World Meteorological Organization and continues his scientific work in Buenos Aires, Argentina.

Jesse A. Purifoy (1938–), PhD, University of Texas, 1969. B.A. (1963) Texas; M.A. (1965) Texas. Thesis title: *Some separation theorems*, (50 pages). Signers: R.L. Moore, P.R. Meyer, D.S. Hymann, H.S. Wall, J.R. Whiteman.

After leaving Texas, Purifoy joined the faculty at Memphis State University, where he helped start a PhD program in mathematics. While there he began consulting with the manufacture of programmable calculators and consulting with municipal bond companies and municipalities and other governmental agencies. He then moved to Houston and remained busy with computer hardware and software. He currently owns a software company, Purifoy Systems Analysis, Inc.

Robert E. Jackson (1943–), PhD, University of Texas, 1969. B.A. (1964) Texas; M.A. (1966) Texas. Thesis title: *Concerning certain plane-like domains*, (57 pages). Signers: R.L. Moore, Homer V. Craig, Robert E. Greenwood, W.T. Guy, Jr., H.S. Wall.

On leaving Texas, Jackson taught at Dickinson College, Carlisle, PA for several years and then went into industry, first with NCNB in North Carolina as a systems analyst, then with Diagnostic Laboratories, and then with BMC Software, where he is now a senior computer scientist.

Nell Elizabeth (Stevenson) Kroeger (1944–), PhD, University of Texas, 1969. B.A. (1965, in microbiology) Texas; M.A. (1968) Texas. Thesis title: *Concerning indecomposable continua and upper semi-continuous collections of nondegenerate continua*, (38 pages). Signers: R.L. Moore, Robert E. Greenwood, Homer V. Craig, W.T. Guy, Jr., H.S. Wall.

Kroeger held an academic position at SUNY Binghamton before going into the private sector, with computer software.

Moore directed very few M.A. students; two were Lucille S. Whyburn, wife of Gordon T. Whyburn, and Martin Ettlinger, son of H.J. Ettlinger. Martin Ettlinger had an illustrious career in chemistry, being a professor at Rice and then at University of Copenhagen. In an interview, Ettlinger has remarked that the intellectual atmosphere in Moore's classes was never duplicated in his experience except when he was a Junior Fellow at Harvard. Mathematicians who studied with Moore but who wrote PhD theses under others include W. L. Ayres (Kline), Lida Barrett (Kline-Anderson), Robert Williams (G.T. Whyburn), Steven Jones and Gary Richter (R H Bing), D.R. Stocks and E. Hensley (Greenwood), D.R. Traylor (Fitzpatrick), W.T. Reid, W.M. Whyburn, O.H. Hamilton, J.H. Barrett, D.H. Tucker, B. Fitzpatrick, and E.I.

Deaton (H.J. Ettlinger). Of the last group, Hamilton and Fitzpatrick subsequently worked mainly in the directions in which Moore had started them. In the 1966 film *Challenge in the Classroom*, Hamilton was the only student Moore mentioned by name. Many of H.S.Wall's PhD students were profoundly influenced by Moore. A special class of students are those who left Austin in the 1960s to study elsewhere; these include Raymond Houston and George Golightly (Houston), Michel Smith and Tom Jacob (Emory), Kenneth Van Doren, Kermit Smith, Douglas Moreman, John Bales, and Nick Williams (Auburn), and Don Fox (Riverside). A number of women who were or later became wives of students of Moore took courses from him. These include Jean Mahavier, Katherine Cook, June Treybig, Janet Rogers, and Gayle Ball.

Finally, there are those persons who did not become mathematicians at all, but were very successful in other endeavors and who attribute their success, to greater or lesser extent, to the training they got from Moore. These include James Wm. McClendon, Distinguished Scholar-in-Residence at the Fuller Theological Seminary in California, and Patricia Pound, Secretary of the Governor's Committee (for the State of Texas) for Persons with Handicaps. McClendon has written a number of books on philosophy and theology; at the time of writing he was completing a three volume systematic theology. He was long (1971–1990) a professor at the Graduate Theological Union, Berkeley, California. Also, there are Robert Boyer, Professor of Computer Science and Philosophy at The University of Texas, Harry Lucas, Jr., a successful businessman, Joel Finegold, a free-lance detergent chemist, and Margaret Ball, a successful writer. Lorene Rogers, a distinguished chemist and a President Emeritus of The University of Texas at Austin, has said that she had a severe case of mathematics anxiety, especially at the prospect of having to take calculus, until she took solid geometry and then calculus under Moore. That she was successful in calculus is evidenced by Moore's having tried to recruit her into a career in mathematics. The renowned statistician S.S. Wilks (PhD, University of Iowa, 1931), editor of the *Annals of Mathematical Statistics*, became fascinated with advanced mathematics as a student after taking a course in point set theory from Dr. Moore.[1]

Reference has been made above to several of Moore's students who did not become mathematicians and who commented very favorably on their training under him. For the sake of completeness, and for balance,

---

[1] F.F. Stephan et al., Samuel S. Wilks, *J. Amer. Statist. Assoc.* 60(1965), 939–966.

it should be noted that some of his students who did become mathematicians with successful academic careers viewed their training as a mixed blessing, and expressed their wish that they had had a wider mathematical education, specifically including the learning of algebraic methods in topology. Most of his students who directed PhD students made sure that their students did learn some algebra. At least two other mathematicians, H.S. Wall and L.E. Dickson, directed more PhD students than did Moore. It is very doubtful that anyone else directed students over a longer period of time, from 1916 to 1969. This is the more remarkable in that his first student, J.R. Kline, did not graduate until eleven years after Moore's own PhD was awarded in 1905.

Of the first half of Moore's students, more than half first encountered him in their graduate program. Of the remainder, a substantial majority took courses from him as undergraduates, and at least two studied with him, while still in high school. More precisely, of his first twenty-seven students, seventeen already had Bachelor's degrees before taking a course from him, Dyer took two courses from him the summer he received his B.A. and B.S., and Young came as a senior and started directly in Moore's graduate course. Of the other twenty-three, all save four took lower level courses either from him or from his colleagues. In looking over the PhD students' dissertation titles, we note that four of them, some of the early ones, are in geometry. Many of the later ones deal with continuum theory; not as many are in abstract spaces, and several are on spirals in the plane, a subject that has not stirred much interest outside Moore's school. The same is true of webs. Bing once remarked that if anyone wanted a reprint of the journal article based on his dissertation, he still had forty-eight of the fifty copies provided him.

Appendix 3
Publications of Robert Lee Moore

[Submissions published in solutions of problems or problems for solution.] *Amer. Math. Monthly* 8 (1901), 196; 11 (1904), 45.

(with Halsted, G.B.) The Betweenness Assumptions, *Amer. Math. Monthly* 9 (1902), 98–101.

Geometry in which the sum of the angles of every triangle is two right angles. *Trans. Amer. Math. Soc.* 8 (1907), 369–378.

Sets of metrical hypotheses for geometry. *Trans. Amer. Math. Soc.* 9 (1908), 487–512. Doctoral dissertation.

A note concerning Veblen's axioms for geometry. *Trans. Amer. Math. Soc.* 13 (1912), 74–76.

On Duhamel's theorem. *Ann. of Math.* 13 (1912), 161–166.

The linear continuum in terms of point and limit. *Ann. of Math.* 16 (1915), 123–133.

On the linear continuum. *Bull. Amer. Math. Soc.* 22 (1915), 117–122.

On a set of postulates which suffice to define a number-plane. *Trans. Amer. Math. Soc.* 16 (1915), 27–32.

Concerning a non-metrical pseudo-Archimedean axiom. *Bull. Amer. Math. Soc.* 22 (1916), 225-236.

On the foundations of plane analysis situs. *Trans. Amer. Math. Soc.* 17 (1916), 131–164; brief summary in *Proc. Nat. Acad. Sci. U.S.A.* 2 (1916), 270-272.

A theorem concerning continuous curves. *Bull. Amer. Math. Soc.* 23 (1917), 233–236.

A characterization of Jordan regions by properties having no reference to their boundaries. *Proc. Nat. Acad. Sci. U.S.A.* 4 (1918), 364–370.

Concerning a set of postulates for plane analysis situs. *Trans. Amer. Math. Soc.* 20 (1919), 169–178.

Continuous sets that have no continuous sets of condensation. *Bull. Amer. Math. Soc.* 25 (1919), 174–176.

On the most general class L of Fréchet in which the Heine-Borel-Lebesgue theorem holds true. *Proc. Nat. Acad. Sci. U.S.A.* 5 (1919), 206–210.

(with Kline, J.R.) On the most general plane closed point-set through which it is possible to pass a simple continuous arc. *Ann. Math.* 20 (1919), 218–223.

On the Lie-Riemann-Helmholtz-Hilbert problem of the foundations of geometry. *Amer. J. Math.* 41 (1919), 299–319.

[Review of] The second volume of Veblen and Young's projective geometry, *Bull. Amer. Math. Soc.* 26 (1920), 412–425.

Concerning simple continuous curves. *Trans. Amer. Math. Soc.* 21 (1920), 333–347.

Concerning certain equicontinuous systems of curves. *Trans. Amer. Math. Soc.* 22 (1921), 41–55.

Concerning connectedness im kleinen and a related property. *Fund. Math.* 3 (1922), 232–237.

Concerning continuous curves in the plane. *Math. Zeit.* 15 (1922), 254–260.

On the relation of a continuous curve to its complementary domains in space of three dimensions. *Proc. Nat. Acad. Sci. U.S.A.* 8 (1922), 33–38.

On the generation of a simple surface by means of a set of equicontinuous curves. *Fund. Math.* 4 (1923), 106–117.

Concerning the cut-points of continuous curves and of other closed and connected point sets. *Proc. Nat. Acad. Sci. U.S.A.* 9 (1923), 101–106.

An uncountable, closed, and non-dense point set each of whose complementary intervals abuts on another one at each of its ends. *Bull. Amer. Math. Soc.* 29 (1923), 49–50.

Report on continuous curves from the viewpoint of analysis situs. *Bull. Amer. Math. Soc.* 29 (1923), 289–302.

Concerning the sum of a countable number of mutually exclusive continua in the plane. *Fund. Math.* 6 (1924), 189–202.

Concerning the common boundary of two domains. *Fund. Math.* 6 (1924), 203–213.

Concerning relatively uniform convergence. *Bull. Amer. Math. Soc.* 30 (1924), 504–505.

An extension of the theorem that no countable point set is perfect. *Proc. Nat. Acad. Sci. U.S.A.* 10 (1924), 168–170.

Concerning the prime parts of certain continua which separate the plane. *Proc. Nat. Acad. Sci. U.S.A.* 10 (1924), 170–175.

Concerning upper semi-continuous collections of continua which do not separate a given continuum. *Proc. Nat. Acad. Sci. U.S.A.* 10 (1924), 356–360.

Concerning sets of segments which cover a point set in the Vitali sense. *Proc. Nat. Acad. Sci. U.S.A.* 10 (1924), 464–467.

Concerning the prime parts of a continuum. *Math. Zeit.* 22 (1925), 307–315.

A characterization of a continuous curve. *Fund. Math.* 7 (1925), 302–307.

Concerning the separation of point sets by curves. *Proc. Nat. Acad. Sci. U.S.A.* 11 (1925), 469–476.

Concerning upper semi-continuous collections of continua. *Trans. Amer. Math. Soc.* 27 (1925), 416–428.

Concerning the relation between separability and the proposition that every uncountable point set has a limit point. *Fund. Math.* 8 (1926), 189–192.

An acknowledgement, *Fund. Math.* 8 (1926), 374–375.

Conditions under which one of two given closed linear point sets may be thrown into the other one by a continuous transformation of a plane into itself. *Amer. J. Math.* 48 (1926), 67–72.

Covering theorems. *Bull. Amer. Math. Soc.* 32 (1926), 275–282.

A connected and regular point set which contains no arc. *Bull. Amer. Math. Soc.* 32 (1926), 331–332.

Concerning indecomposable continua and continua which contain no subsets that separate the plane. *Proc. Nat. Acad. Sci. U.S.A.* 12 (1926), 359–363.

Concerning paths that do not separate a given continuous curve. *Proc. Nat. Acad. Sci. U.S.A* 12 (1926), 745–753.

Some separation theorems. *Proc. Nat. Acad. Sci. U.S.A.* 13 (1927), 711–716.

Concerning triods in the plane and the junction points of plane continua. *Proc. Nat. Acad. Sci. U.S.A .* 14 (1928), 85–88.

On the separation of the plane by a continuum. *Bull. Amer. Math. Soc.* 34 (1928), 303–306.

A separation theorem. *Fund. Math.* 12 (1928), 295–297.

Concerning triodic continua in the plane. *Fund. Math.* 13 (1929), 261–263.

Concerning upper semi-continuous collections. *Monatsh. Math. Phys.* 36 (1929), 81–88.

Foundations of Point Set Theory. American Mathematical Society Colloquium Publications 13 (1932).

Concerning compact continua which contain no continuum that separates the plane. *Proc. Nat. Acad. Sci. U.S.A.* 20 (1934), 41–45.

A set of axioms for plane analysis situs. *Fund. Math.* 25 (1935), 13–28.

Fundamental theorems concerning point sets: I, Foundations of a point set theory of spaces in which some points are contiguous to others; II, Upper semi-continuous collections of the second type; III, On the structure of continua. *Rice Institute Pamphlet* 23 (1936), 1–74.

Concerning essential continua of condensation. *Trans. Amer. Math. Soc.* 42 (1937), 41–52.

Concerning accessibility. *Proc. Nat. Acad. Sci. U.S.A.* 25 (1939), 648–653.

Concerning the open subsets of a plane continuum. *Proc. Nat. Acad. Sci. U.S.A.* 26 (1940), 24–25.

Concerning separability. *Proc. Nat. Acad. Sci. U.S.A.* 28 (1942), 56–58.

Concerning intersecting continua. *Proc. Nat. Acad. Sci. U.S.A.* 28, (1942), 544–550.

Concerning a continuum and its boundary. *Proc. Nat. Acad. Sci. U.S.A.* 28 (1942), 550–555.

Concerning domains whose boundaries are compact. *Proc. Nat. Acad. Sci. U.S.A.* 28 (1942), 555–561.

Concerning webs in the plane, *Proc. Nat. Acad. Sci. U.S.A.* 29 (1943), 389–393.

Concerning continua which have dendratomic subsets. *Proc. Nat. Acad. Sci. U.S.A.* 29 (1943), 384–389.

Concerning tangents to continua in the plane. *Proc. Nat. Acad. Sci. U.S.A* 31 (1945), 67–70.

A characterization of a simple plane web. *Proc. Nat. Acad. Sci. U.S.A.* 32 (1946), 311–316.

Spirals in the plane. *Proc. Nat. Acad. Sci. U.S.A.* 39 (1953), 207–213.

Foundations of Point Set Theory. (revision) American Mathematical Society Colloquium Publications 13 (1962).

Appendix 4

Descriptions of Courses Often Taught by Moore at Texas

(as described by W.S. Mahavier)

Moore's courses varied from year to year since he would alter them so as to best serve his students. The general content described below is based on the courses in the late 1940s. At that time most students took algebra, trignometry, or analytic geometry as freshmen and then took calculus as a sophomore. In the summers Moore taught a plane geometry course and a linear measure theory course.

613 Calculus This was a reasonably standard one variable calculus course insofar as the topics covered were concerned. Moore lectured very little, mostly in the beginning to provide an intuitive feeling for limits and tangent lines. The students did most of the presentations. We derived all the usual formulas for derivatives of all the standard functions. A lot of time went to graphing functions and max-min problems, many of which were quite challenging. We covered Taylor's polynomials and series. The usual applications were covered such as volume, mass with variable density, water pressure and others. These were treated as differential equations, rather than definite integrals and an emphasis was put on the physical assumptions that were made. There were extensive problems on finding antiderivatives. The definite integral as a limit of Riemann sums was introduced. Moore's emphasis throughout the course was on the careful use of language.

624 Introduction to the Foundations of Analysis This was the first course in which the students began to prove theorems. It began with the topology of the line, using the Dedekind cut axiom as the completeness axiom. Moore introduced the concepts of limit points, closed sets, compact sets, and convergent sequences. We proved the Bolzano-Weierstrass Theorem and the Heine-Borel Theorem. Moore defined a continuous function and we developed the usual properties of functions continuous on a closed interval, including the intermediate value theorem, Rolle's theorem, and uniform continuity. The derivative was defined and the basic theorems about differentiable functions were developed. Moore's emphasis in this course was on including details in our proofs.

688–689 Introductory graduate topology courses This was the students' first contact with abstract mathematics. Many took 688 as a senior undergraduate. Moore stated the axiom from his colloquium publication, "Foundations of Point Set Theory". In modern terms the space was a regular, Hausdorff developable space. Theorems were stated and proved by the students. The axiom used was strong enough to be able to prove most of the usual theorems for a metric space, but not strong enough to imply normality. He covered the basic notions of 1st countability, regularity, conditional compactness and compactness, Baire category, connectedness, local connectivity, and arcwise connectivity. In addition various special topics would be covered such as closed mappings (as upper semi-continuous collections), irreducible continua, the Brouwer reduction theorem, indecomposable continua, and separation theorems in a euclidean plane.

690 R.L. Moore's research seminar This course was basically a continuation of 689 and could be repeated for credit, in this course Moore would ask questions and pose problems intended to lead to a dissertation. Students might be competing for a solution to a problem or they might be working on different problems. Interestingly enough we usually did not know or care if the problems we were working on were unsolved or not. Some of the special topics I recall were webs, spirals in the plane, mappings on a disc, and chainable continua.

Bibliography

Albers, D., and C. Reid, An interview with Mary Ellen Rudin, in *More Mathematical People*, D. Albers, G. Alexanderson, C. Reid (eds.), Harcourt Brace Jovanovich, Boston, 1990, pp. 283–303.

Albert, A.A., Leonard Eugene Dickson 1874–1954, *Bull. Amer. Math. Soc.* 61 (1955), 331–346.

———, *A Survey of Training and Research Potential in the Mathematical Sciences, Final Report,* February 1957; University of Chicago.

Anderson, R.D., 'I Led Three Mathematical Lives', *MER Newsletter*, (MER Forum, Fall 1998), pp. 3–11.

——— and C.E. Burgess, R H Bing: October 20, 1914–April 28, 1986, *Notices Amer. Math. Soc.* 33 (4) (1986), 595–596.

Archibald, R.C., *A Semicentennial History of the American Mathematical Society 1888–1938*, American Mathematical Society, 1938.

———, Material Concerning James Joseph Sylvester, August 1944, New York, Schuman.

——— et al., Benjamin Peirce, *Amer. Math. Monthly* 32 (1925), 1–30.

Backlund, U., and L. Persson, Moore's teaching method, *Normat* 41 (1996), 145–149.

Bernays, P., David Hilbert, in *Encyclopedia of Philosophy* III, New York, 1967, pp. 496–504.

Bing, R H, et al., (1976, January 24). Remarks at The University of Texas, Austin Mathematics Awards, Honoring the Memory of Professor Robert Lee Moore and Professor Hubert Stanley Wall.

Birkhoff, G.D., Eliakim Hastings Moore (1862–1932), *Amer. Acad. Arts and Sci.* 69 (1934), 527–528.

Bliss, G.A., Eliakim Hastings Moore, *Bull. Amer. Math. Soc.* 2nd ser. 39 (1933), 831–838.

———, The scientific work of Eliakim Hastings Moore, *Bull. Amer. Math. Soc.* 40 (1934), 501–514.

———, Oskar Bolza — In Memoriam, *Bull. Amer. Math. Soc.* 50 (1944), 478–489.

———, G.A. Bliss, Autobiographical Notes, *Amer. Math. Monthly* 59 (1952), 595–606.

——— and L.E. Dickson, *A Biographical Memoir of Eliakim Hastings Moore*, National Academy of Sciences, 1939.

Bolza, O., Heinrich Maschke: His Life and Work, *Bull. Amer. Math. Soc.* 15 (1908), 85–95.

Browder, F.E. (ed.), *The mathematical heritage of Henri Poincaré. Part 1*, American Mathematical Society, Providence, RI, 1983.

——— (ed.), *The mathematical heritage of Henri Poincaré. Part 2*, Providence, RI, 1983.

Brown, J., My experiences with the Various Texas Styles of Teaching, in the R.L. Moore Legacy Collection, Archives of American Mathematics, Center for American History, The University of Texas at Austin, (1996, April 30).

Brown, M., The mathematical work of R H Bing, *Proceedings of the 1987 Topology Conference, Birmingham, AL, 1987, Topology Proc.* 12 (1) (1987), 1–25.

Butler, L.J., George David Birkhoff, *American National Biography* 2 (Oxford, 1999), 813–814.

Chalice, D., How to teach a class by the modified Moore method, *Amer. Math. Monthly* 102 (1995), 317–321.

Clark, David, R.L. Moore and the Learning Curve, prepared for The Legacy of R.L. Moore Project, February 2001.

Cohen, D.W., A modified Moore method for teaching undergraduate mathematics, *Amer. Math. Monthly* 89 (1982), 473–474, 487–490.

Dancis, Jerome, and Neil Davidson, The Texas method and the small group discovery method (1970); the authors are both professors at the University of Maryland.

Dantzig, T., *Henri Poincaré: critic of crisis: reflections on his universe of discourse,* Scribner's, New York, 1954.

Davis, P.J., Otto E. Neugebauer: Reminiscences and Appreciation, *Amer. Math. Monthly* 101 (1994), 129–131.

Dickson, L.E., Eliakim Hastings Moore, *Science* 77 (1933), 79–80.

Eyles, J., The importance of R.L. Moore's calculus class, in the R.L. Moore Legacy Collection, Archives of American Mathematics, Center for American History, The University of Texas at Austin.

Fitzpatrick, B., Some aspects of the work and influence of R.L. Moore, in *Handbook of the History of General Topology*, C. Aull and R. Lowen (eds.), Kluwer Academic, Dordrecht, Boston, 1997, pp. 41–61.

Forbes, D.R., *The Texas System: R.L. Moore's Original Edition*, PhD thesis, University of Wisconsin, 1971.

Foster, J.A., M. Barnett, K. Van Houten, and L. Sheneman, Informal methods: teaching program derivation via the Moore method.

Frantz, J.B., *The Forty-Acre Follies*, Texas Monthly Press, Austin, 1983, pp. 111–122.

Gilman, D.C., *The Launching of a University and Other Papers*, Dodd, Mead and Co, New York, 1906.

Greenwood, R.E., Papers 1937–1993, Archives of American Mathematics, Center for American History, The University of Texas at Austin.

———, The kinship of E.H. Moore and R.L. Moore, *Historia Mathematica* 4 (1977), 153–155.

Halmos, P.R., What is teaching? *Amer. Math. Monthly* 101 (1994), 848–855.

———, How to Teach in *I Want to be a Mathematician*, Springer-Verlag, New York, 1985, pp. 253–265.

——— and E.E. Moise, The problem of learning how to teach, *Amer. Math. Monthly* 82 (1975), 466–474.

Halsted, G.B., The Betweenness Assumptions, *Amer. Math. Monthly* 9 (1902), 98–101.

———, Biography. Professor Felix Klein, *Amer. Math. Monthly* 1 (1894), 416–420.

Jackson, A., Mary Ellen Rudin, in *Profiles of Women in Mathematics: The Emmy Noether Lectures,* Association for Women in Mathematics, College Park, MD, 1984.

Jones, F. Burton, Some glimpses of the early years, in *The Work of Mary Ellen Rudin: Summer Conference on General Topology and Applications in Honor of Mary Ellen Rudin Held in Madison, Wisconsin, June 26–29, 1991,* Tall, F.D. (ed.), New York, 1993, pp. xi–xii.

———, The Beginning of Topology in the United States and the Moore School, in *Handbook of the History of General Topology*, Volume 1, Kluwer Academic Publishers, Dordrecht, Boston, 1997, pp. 97–103.

———, The Moore Method, *Amer. Math. Monthly* 84 (1997), 273–277.

———, R H Bing, *Proceedings of the 1987 Topology Conference, Birmingham, AL, 1987, Topology Proc.* 12 (1) (1987), 181–186.

——— and E.E. Floyd, Gordon T. Whyburn 1904–1969, *Bull. Amer. Math. Soc.* 77 (1971), 57–72.

Lehmer, D.H., Harry Schultz Vandiver, *Bull. Amer. Math. Soc.* 80 (1974), 817–818.

Lewis, A.C., R.L. Moore entry in the *Dictionary of Scientific Biography*, vol. 18. Charles Scribner's Sons, New York, 1990, pp. 651–653.

———, Reform and Tradition in Mathematics Education: The Example of R.L. Moore, Manuscript April 1998; Revised 1999.

———, The building of The University of Texas mathematics faculty, 1883–1938, in *A Century of Mathematics in America – Part II,* Peter Duren, (ed.), American Mathematical Society, Providence, RI, 1989, pp. 205–239.

Mac Lane, S., Jobs in the 1930s and the views of George D. Birkhoff, *Math. Intelligencer* 16 (1994), 9–10.

Mahavier, Lee, *On Three Crucial Elements of Texas-style Teaching as Shown to be Successful in the Secondary Mathematics Classroom*, paper prepared for The Legacy of R.L. Moore Project, 1999.

Mahavier, W.S., What is the Moore Method?, *Primus* 9 (2) (1999), 339–354.

Mahavier, W.T., A gentler discovery method (the modified Moore method), *College Teaching* 45 (1997), 132–135.

———, Interactive Numerical Analysis, *Creative Math Teaching* 3, 1–2.

Moise, E.E., Activity and motivation in mathematics, *Amer. Math. Monthly* 72 (1965), 407–412.

Monna, A.F., Oswald Veblen, *Math. Intelligencer* 16 (1994), 50–51.

Montgomery, D., Oswald Veblen, Obituary, *Bull. Amer. Math. Soc.* 69 (1963), 26–36.

Moore, R.L., Papers, 1889–1979, Archives of American Mathematics, Center for American History, The University of Texas at Austin.

———, Letter to Miss Hamstrom, published in *A Century of Mathematics*, American Mathematical Society, Providence, RI, 1996, pp. 295–300.

Murray, M.A.M., *Women Becoming Mathematicians*, MIT Press, Cambridge, MA, 2000.

Nemeth, L., The two Bolyais, *The New Hungarian Quarterly* 1 (1960).

Neuenschwander, E., Studies in the history of complex function theory. II. Interactions among the French school, Riemann and Weierstrass, *Bull. Amer. Math. Soc.* 5 (1981), 87–105.

Nyikos, P., F. Burton Jones's contributions to the normal Moore space problem, in *Topology Conference, Greensboro, NC, 1979*, Greensboro, NC, 1980, 27–38.

Ormes, N., *A Beginner's Guide to the Moore Method*, paper prepared for The Legacy of R.L. Moore Project, 1999.

Parker, G.E., Getting More from Moore, *Primus*, 2 (September 1992), 235–246.

Parshall, Karen Hunger, America's First School of Mathematical Research: James Joseph Sylvester at The Johns Hopkins University 1876–1883, *Arch. Hist. Exact Sci.* 38 (1988), 153–196.

———, Eliakim Hastings Moore and the Founding of a Mathematical Community in America 1892–1902, *Annals of Sciences* 41 (1984), 313–333.

———, Eliakim Hastings Moore, *American National Biography* 15 (Oxford, 1999), 748–749.

——— and D.E. Rowe, *The Emergence of the American Mathematical Research Community 1876–1900: J.J. Sylvester, Felix Klein, and E.H. Moore*, American Mathematical Society and London Mathematical Society, Providence, RI and London, 1994.

Phillips, R., Reminiscences about the 1930s, *Math. Intelligencer* 16 (3) (1994), 6–8.

Reid, C., *Hilbert/Courant*, Springer, New York, 1986.

Reid, W.T., Oskar Bolza, in *Dictionary of American Biography* Supplement Three 1941–45, Scribner, New York, 1973, pp. 86–87.

Renz, P., The Moore Method: What Discovery Learning Is and How It Works, *FOCUS: Newsletter of the Mathematical Association of America.* (August/September, 1999), 6, 8.

Rogers, J.T., Jr., F. Burton Jones (1910–1999): an appreciation, *Proceedings of the 1999 Topology and Dynamics Conference, Salt Lake City, UT, Topology Proc.* 24 (1999), 2–14.

Rowe, D.E. and J. MacCleary (eds.), *The History of Modern Mathematics*, two volumes, Academic Press, Inc., Boston, 1989.

———, David Hilbert on Poincaré, Klein, and the world of mathematics, *Math. Intelligencer* 8 (1986), 75–77.

Singh, S., R H Bing (1914–1986): a tribute, Special volume in honor of R H Bing (1914–1986), *Topology Appl.* 24 (1–3) (1986), 5–8.

———, R H Bing: A study of his life, in *R H Bing: Collected Papers* Vol. 1, American Mathematical Society, Providence, RI, 1988, pp. 3–18.

———, S. Armentrout, and R.J. Daverman (eds.), *R H Bing: Collected Papers* (2 Vols), American Mathematical Society, Providence, RI, 1988.

Smith, D.E., Heinrich Maschke, in *Dictionary of American Biography* XII, Scribner, New York, 1933, pp. 356–357.

Sneddon, N.I., Kazimierz Kuratowski Hon. F.R.S.E., in *Yearbook of the Royal Society of Edinburgh Session 1980–81,* 1982, pp. 40–47.

Starbird, M., Mary Ellen Rudin as advisor and geometer, in *The Work of Mary Ellen Rudin: Summer Conference on General Topology and Applications in Honor of Mary Ellen Rudin Held in Madison, Wisconsin, June 26–29, 1991,* Tall, F.D. (ed.), New York, 1993, pp. 114–118.

———, R H Bing's human and mathematical vitality, in *Handbook of the history of general topology, Vol. 2, San Antonio, TX, 1993,* Dordrecht, 1998, pp. 453–466.

Storr, R.J., *Harper's University: The Beginnings*, University of Chicago Press, Chicago and London, 1966.

Tall, F.D. (ed.), *The Work of Mary Ellen Rudin: Summer Conference on General Topology and Applications in Honor of Mary Ellen Rudin Held in Madison, Wisconsin, June 26–29, 1991,* New York, 1993.

Taylor, A.E., A study of Maurice Fréchet I, *Arch. Hist. Exact Sci.* 27 (1982), 233–295.

———, A study of Maurice Fréchet II, *Arch. Hist. Exact Sci.* 34 (1985), 279–380.

Toepell, M., On the origins of David Hilbert's *Grundlagen der Geometrie*, *Arch. Hist. Exact Sci.* 35 (4) (1986), 329–344.

Traylor, D.R., *Creative Teaching: The Heritage of R.L. Moore*, University of Houston, 1972.

Vandiver, H.S., Some of my recollections of George David Birkhoff, *J. Math. Anal. Appl.* 7 (1963), 271–283.

Veblen, O., and J.W. Young, *Projective Geometry*, vol. 1, Ginn and Co, Boston, 1910.

Weyl, H., Obituary: David Hilbert. 1862–1943, *Obituary Notices of Fellows of the Royal Society of London* 4 (1944), 547–553.

———, David Hilbert and his mathematical work, *Bull. Amer. Math. Soc.* 50 (1944), 612–654.

Whyburn, G.T., Dynamic topology, *Amer. Math. Monthly* 77 (1970), 556–570.

Whyburn, L.S., Student-oriented teaching — the Moore method, *Amer. Math. Monthly* 77 (1970), 351–359.

———, A visit with E.H. Moore, *The Proceedings of the 1979 Topology Conference, Topology Proc.* 4 (1) (1980), 279–283.

———, Letters from the R.L. Moore Papers, *Proceedings of the 1977 Topology Conference* I, *Topology Proc.* 2 (1) (1977), 323–338.

———, R H Bing 1949–50, *Proceedings of the 1987 Topology Conference, Birmingham, AL, 1987, Topology Proc.* 12 (1) (1987), 177–180.

Wilder, R.L., Axiomatics and the development of creative talent, in *The Axiomatic Method with Special Reference to Geometry and Physics*, L. Henken, P. Suppes, and A. Tarski (eds.), North-Holland, 1959, pp. 474–488.

———, Robert Lee Moore 1882–1974, *Bull. Amer. Math. Soc.* 82 (1976), 417–427.

———, Material and method, *Undergraduate Research in Mathematics, a Report of a Conference,* Carleton College, Northfield, Minnesota, June 16 to 23, 1961, Edited by Kenneth O. May and Seymour Schuster, pp. 9–27.

———, The mathematical work of R.L. Moore: its background, nature, and influence, *Arch. Hist. Exact Sci.* 26 (1982), 73–97.

Young, G.S., Being a student of R.L. Moore, 1938–42, in *A Century of Mathematics Meetings*, American Mathematical Society, Providence, RI, 1996, pp. 285–293.

Young, S.W., *Christmas in Big Lake*, reminiscences of his class with R.L. Moore, prepared for The Legacy of R.L. Moore Project, in the R.L. Moore Legacy Collection, Archives of American Mathematics, Center for American History, The University of Texas at Austin.

Zitarelli, David, The Origin and Early Impact of the Moore Method, *Amer. Math. Monthly* 111 (6) (2004), 465–486.

Photo Credits

Moore at age ten, p. 1. August 1893. Photographer: Webster, 239 Main St., Dallas, TX. Source: R.L. Moore Legacy Collection in the Archives of American Mathematics, Center for American History, The University of Texas at Austin [DI01195].

Moore's parents, Charles and Louisa Ann, p. 4. Ca. 1900. Photographer: Schreiber & O'Bannon, Dallas, TX. Source: R.L. Moore Legacy Collection in the AAM [DI00614].

Robert Lee as a toddler, p. 6. Undated. Photographer: Unknown, donated by Louis A. Beecherl, member of the Board of Regents of the University of Texas, 1987–1993. Source: R. L. Moore Papers in the AAM [DI01255].

Moore's feed store, p. 7. Undated. Source: R.L. Moore Legacy Collection [DI01250].

Young Master Moore, p. 13. Undated. Photographer: J. H. Webster, Dallas, TX. Source: R.L. Moore Papers in the AAM [DI01245].

Moore and unknown person in cap and gown, p. 17. Ca. 1900. Source: R.L. Moore Legacy Collection in the AAM [DI00615].

R.L. with his sister, Caroline Louisa Moore, p. 19. December 1898. Source: R.L. Moore Papers in the AAM [DI01253].

James Joseph Sylvester, p. 22. Undated. Photographer: unknown.

G. B. Halsted, p. 25. Undated. Source: R.L. Moore Papers in the AAM [DI00783].

David Hilbert, p. 35. Undated. Photographer: unknown (German postcard). Source: R.L. Moore Papers in the AAM [DI01233].

"Bobby" Moore, p. 39. Ca. 1900. Photographer: unknown, donated by Emily Cutrer. Source: R.L. Moore Papers in the AAM [DI01248].

Eliakim Hastings Moore, p. 49. Undated. Source: On EAF website. From: Archibald, R.C. (1938), *A Semicentennial History of the American Mathematical Society, 1888–1938*, American Mathematical Society, New York.

L.E. Dickson, p. 53. Undated. Photographer: Root. Source: R.L. Moore Legacy Collection in the AAM [DI01234].

Mister Robert Lee Moore, p. 57. Undated. Photographer: Schreiber & O'Bannon, Dallas, TX. Source: R. L. Moore Papers in the AAM [DI01199].

Heinrich Maschke, p. 59. Source: `www-gap.dcs.st-and.ac.uk/~history/PictDisplay/Maschke.html`.

Oskar Bolza, p. 62. Source: `www-gap.dcs.st-and.ac.uk/~history/PictDisplay/Bolza.html`.

E.H. Moore and R.L. Moore in Chicago, p. 73. Ca. 1905. Source: R.L. Moore Papers in the AAM [DI01400].

Oswald Veblen, p. 77. Source: On EAF website. From: Archibald, R.C. (1938), *A Semicentennial History of the American Mathematical Society, 1888–1938*, American Mathematical Society, New York.

R.L. Moore, relaxing at home, p. 87. Undated. Source: R.L. Moore Papers in the AAM [DI01252].

R.L. and his uncle, James Willard Moore, p. 91. Ca. 1911. Photographer: Kresge's Photo-Studio, Cleveland, OH. Source: R. L. Moore Papers in the AAM [DI01246].

Margaret MacLellan Key Moore, p. 94. November 18, 1918. Taken in Philadelphia. Source: R.L. Moore Papers in the AAM [DI01249].

John R. Kline, p. 97. Source: On EAF website. From: Pitcher, E. (1988), *A History of the Second Fifty Years, American Mathematical Society, 1939–1988*. American Mathematical Society, Providence.

Moore at his desk, p. 109. October 1930. Source: R.L. Moore Legacy Collection [DI01387].

Milton Brockett Porter, p. 113. Undated. Source: R.L. Moore Papers in the AAM [DI01389].

Harry Yandell Benedict, p. 118. Ca. 1927. Photographer: Dan E. McCaskill, University Studio, Austin, TX. Source: Prints and Photographs Collection, Center for American History, The University of Texas at Austin [DI01379].

Paul Mason Batchelder, p. 121. Undated. Source: UT Office of Public Affairs Records, Center for American History, The University of Texas at Austin [DI01394].

R. L. Moore walking down Guadalupe Street in Austin, p. 125. Ca. 1920. Photographer: Unknown, donated by Louis A. Beecherl. Source: R. L. Moore Legacy Collection in the AAM [DI01193].

Fall 1928: Back row, left to right: W. T. Reid, J. H. Roberts, C. M. Cleveland, Norman E. Rutt, and J. R. Dorrow; front row, left to right: Lucille Whyburn, G. T. Whyburn, R. L. Moore, R. G. Lubben, p. 129. Source: R. G. Lubben Papers in the AAM [DI01198].

Gordon T. Whyburn, p. 133. Source: On EAF website. From: Pitcher, E. (1988), *A History of the Second Fifty Years, American Mathematical Society, 1939–1988*. American Mathematical Society, Providence.

R. L. Moore, p. 136. October 1930. Photographer: Jenson Studio, Austin, TX. Source: R. L. Moore Legacy Collection in the AAM [DI01386].

Taken at the Mathematical Association of America/American Mathematical Society/American Assocation for the Advancement of Science meetings in Cleveland Ohio, December 1930; left to right, disregarding row: Wilfrid Wilson, J. W. Alexander, W. L. Ayres, G. T. Whyburn, R. L. Wilder, P. M. Swingle, C. N. Reynolds, W. W. Flexner, R. L. Moore, T. C. Benton, K. Menger, S. Lefschetz, p. 139. Source: R. L. Moore Papers in the AAM [DI00785].

R. L. Moore in his office, p. 143. Ca. 1930s. Photographer: R. G. Lubben. Source: R. G. Lubben Papers in the AAM [DI01197].

Raymond L. Wilder, p. 146. Ca. 1965. Source: Mathematical Association of America Records in the AAM [DI01235].

Robert E. Greenwood, p. 149. November 1974. Source: R. L. Moore Legacy Collection in the AAM [DI01396].

Spring 1931: Karl Menger, Milton B. Porter, J. H. Roberts, and R. G. Lubben, p. 153. Source: R. G. Lubben Papers in the AAM [DI01381].

R. E. Basye and E. C. Klipple, p. 158. Ca. 1930s. Photographer: R. G. Lubben. Source: R. G. Lubben Papers in the AAM [DI01393].

R. L. Moore, p. 161. Ca. 1935. Photographer: R. G. Lubben. Source: R. G. Lubben Papers in the AAM [DI01384].

Possibly taken in the Moore home. Back row, left to right: R. E. Bayse, E. C. Klipple, F. Burton Jones; front row, left to right: C. W. Vickery, R. L. Moore, R. G. Lubben, p. 164. Ca. 1935. Source: R. G. Lubben Papers in the AAM [DI01191].

R. G. Lubben, C. W. Vickery and F. C. Biesele, p. 169. Ca. 1936. Source: R. G. Lubben Papers in the AAM [DI01383].

F. Burton Jones and R. G. Lubben, p. 173. Ca. 1930s. Source: R. G. Lubben Papers in the AAM [DI01380].

Moore at his desk, p. 177. 1935. Photographer: R. G. Lubben. Source: R. G. Lubben Papers in the AAM [DI01196].

Richard D. Anderson, p. 180. Ca. 1941. Courtesy Richard D. Anderson.

Gail S. Young, p. 183. Ca. 1940s. Source: R. L. Moore Legacy Collection in the AAM [DI01404].

Robert Sorgenfrey, Robert Swain, Bernadine Sorgenfrey, Mary Ruth Coleman, Walter Coleman, Harlan Cross Miller, p. 188. June 1941. Source: R. L. Moore Legacy Collection in the AAM [DI01397].

R. L. Moore during registration at UT, p. 193. September 1939. Photographer: Dr. H. F. Kuehne. Source: R. L. Moore Papers in the AAM [DI01192].

Moore walking down Congress Avenue in Austin, p. 198. Undated. Photographer: Unknown, donated by Louis A. Beecherl. Source: R. L. Moore Legacy Collection in the AAM [DI01194].

Moore at the chalkboard, p. 207. Undated. Source: UT Texas Student Publications Inc. Photographs, Center for American History, The University of Texas at Austin [DI01309].

R H Bing, p. 209. Ca. 1960s. Source: R H Bing Papers in the AAM [DI00787].

E. E. Moise, p. 212. Ca. 1967–1968. Photographer: Unknown, donated by MAA. Source: Mathematical Association of America Records in the AAM [DI01238].

Photo Credits

Mary Ellen Estill, Lida Barrett, John Barrett and others, p. 215. Courtesy Lida Barrett.

R. L. Moore, p. 225. Ca. 1930s. Photographer: R.G. Lubben. Source: R.G. Lubben Papers in the AAM [DI01256].

Harry Schultz Vandiver, p. 228. Undated. Photographer: Walter Barnes Studio, Austin, TX. Source: Prints and Photographs Collection, Center for American History, The University of Texas at Austin [DI01391].

Hubert Stanley Wall, p. 236. Source: On EAF website. Taken from dust jacket of Wall's book *Creative Mathematics*. Credited to Walter Barnes Studio of Austin.

Moore in his office, p. 241. September 1963. Photographer: Benny Springer. Source: R.L. Moore Legacy Collection in the AAM [DI01399].

Mary-Elizabeth Hamstrom, p. 244. Undated. Courtesy Mary-Elizabeth Hamstrom.

Page 5 of the letter from R. L. Moore to Mary-Elizabeth Hamstrom, p. 247. 05/07/48. Source: R.L. Moore Legacy Collection in the AAM [DI01388].

Lida Barrett at a UT Roundup Dance, p. 251. Courtesy Lida Barrett.

Mary Ellen and Walter Rudin, p. 252. Source: Donald J. Albers, G.L. Alexanderson, and Constance Reid, *More Mathematical People*, Harcourt, Brace, Jovanovich, 1990.

Harlan Cross Miller, p. 255. Undated. Source: R.L. Moore Legacy Collection in the AAM [DI01243].

Moore in his office, p. 257. April 1966. Photographer: Paul Halmos. Source: R.L. Moore Papers in the AAM [DI01401].

Moore in his office with Michael Proffitt, p. 260. January 1970. Photographer: Paul Halmos. Source: R.L. Moore Papers in the AAM [DI00784].

A still from *Challenge in the Classroom*, p. 269. Film about the Moore Method made by the Mathematical Association of America.

Moore in his office, p. 275. September 1963. Photographer: Benny Springer. Source: R.L. Moore Legacy Collection in the AAM [DI01398].

R.L. and Margaret Moore in front of their home, p. 284. June 1954. Photographer: Fritz Key. Source: R.L. Moore Papers in the AAM [DI01385].

Moore on the UT campus, p. 293. 1969. Photographer: Homer G. Ellis. Source: R H Bing Papers in the AAM [DI00786].

H. J. Ettlinger, p. 299. Undated. Photographer: *Daily Texan*, staff photo. Source: UT Texas Student Publications Inc. Photographs in the AAM [DI01200].

Nell Elizabeth Kroeger (née Stevenson), p. 304. 1971. Photographer: Tom Ingram. Source: R.L. Moore Legacy Collection in the AAM. [DI01403]

R.L. Moore in class, p. 313. 1969. Photographer: Annette Calhoun. Source: Educational Advancement website.

Mary Ellen Rudin and Bruce Treybig, p. 321. 1971. Photographer: Tom Ingram. Source: R.L. Moore Legacy Collection in the AAM. [DI01402]

R. L. Moore Hall on the UT campus, p. 331. May 1973. Photographer: unknown. Source: Prints and Photographs Collection, Center for American History, The University of Texas at Austin. [DI01390]

Index

Adams, C. R., 169
African-American students, 12, 96, 287–290, 294, 295, 339
Albert Report, 238
Alexander, J.W., 139
Alexandroff, P.S., 302
Algebra, 23, 32, 58, 80, 84, 93, 266
Algebraic geometry, 156
Algebraic topology, 140, 191, 341
American Association for the Advancement of Science (AAAS), 31, 139, 157, 164, 178
American Association of University Professors (AAUP), 194, 204, 205
American Mathematical Society (AMS), viii–ix, 33, 53, 62, 65, 80, 87, 95, 114, 135, 144, 157, 165, 168, 172, 220, 233
Anderson, Richard D., xiii, 102, 178–181, 182, 211, 213, 215–218, 220–222, 238, 280, 304, 347, 348
Archibald, R. C., 61, 74, 75
Archives of American Mathematics (AAM), xiii
Armendariz, Efraim, 343
Armentrout, Steve, xiii, 310, 350
Auburn University, 353
Axiomatics, 60, 145–146, 305; *see also* Logic and set theory
Ayres, W.L., 81, 102, 139, 339, 355

Baker, Blanche Joanne (née Monger), xiii, 352
Baker, H.L., 352
Bales, John, 356
Ball, B.J. (Joe), 243, 244, 280, 348
Ball, Gayle, 243, 244, 356
Ball, Margaret, 356
Barrett, J.H., 215, 355

Barrett, Lida K., 102, 215, 251, 339, 345, 355
Basye, Robert E., 158, 164, 344
Batchelder, Paul Mason, 121, 122
Beckenbach, Edwin F., 150, 227
Beecherl, Louis A., 373, 375, 376
Begle, Ed, 233, 341
Benedict, Harry Yandell, 27, 118, 195
Benedict Hall, 235, 276
Benton, T.C., 139
Betweenness assumptions, 35, 36
Biesele, F. C., 169
Bing, Mary, xiii
Bing, R H, 130, 178, 208–211, 216, 218–220, 280, 290, 291, 302, 303, 320, 329, 330, 346, 355, 357
Biomathematics, 352
Biostatistics, 354
Birkhoff, George David, 53, 59, 83, 103, 110, 122, 157, 164, 169, 170
Blacks, *see* African-American students; racial segregation
Bledsoe, Woodrow W., 319–328
Bliss, Gilbert, 54, 55, 59, 64, 65, 110
Bôcher, Maxime, 119
Bolyai, János, 16, 26, 70
Bolza, Oskar, 44, 50–54, 58, 61, 62, 76, 103
Boner, C.P., 239, 276, 277
Boyd, J.R., 345
Brauer, Alfred, 231
Briles, David, 268–270
Buchanan, 183
Burgess, Cecil E., 181, 210, 280, 348
Burlington, Orville, 197
Bush, Vannevar, 171

Calculus, vii, 2, 55, 84, 127, 131, 133, 150, 257–274, 294, 310

379

Calhoun, John W., 195
Cantor, Georg, 246
Cantorean line, 113
Carter, J.M., 326
Carver, Wallace, 83
Cayley, Arthur, 23
Center for American History, 2
Challenge in the Classroom (film), vii, 115, 258, 269, 309, 316, 317
Chicago, the city, 45, 46
Chicago Mathematical Congress (1893), 51
Chittenden, E.W., 103, 112, 300
'Christmas in Big Lake,' 307
Civil War, 3, 4, 5, 45
Clarkson, L.L., 289
Claytor, William Waldron Schieffelin, 96, 339
Cleveland, Clark M., 129, 138, 342
Coble, A. B., 169
Concerning Dean John R. Silber and the Proposed Dismissal of Professor R.L. Moore ('The Green Book'), 320, 326, 327, 328
Continuum Hypothesis, 80
Cook, David E., 353
Cook, Howard, 351
Cook, Katherine, 356
Cornette, James L., 352
Courant, Richard, 168
Cowley, Don E., 270, 271
Craig, Homer V., 315
Crawley, Edwin Schofield, 97
Curtis, Otis F., 157

Dancis, Jerome, xiii, 216
Daus, Paul H., 218
Davidson, Neil, 216
Davis, J.C., 287
Davis, Roy D., 353
Deaton, E.I., 355–356
Decherd, Mary E., 37, 120
Dehn, M., 70, 87
Denjoy, Arnaud, 113
De Voto, Bernard, 194, 197, 200, 204, 205
Dewey, John, 41, 42, 48
Diaz, Joseph, 182
Dickson, Leonard Eugene, 20, 28, 29, 40, 50, 53, 54, 59, 65, 116, 357
Differential geometry, 58, 163

Dodd, E. L., 129
Dorroh, Joe L., xiii, 129, 138, 343
Dos Passos, John, 200
Dresden, Arnold, 172
Driskill Hotel, 12
Duhamel's Theorem, 95
Duke University, 342, 347
Dyer, Eldon, 250, 280, 349

Eaton, William T., xiii, 218
Edmondson, Don E., 290
Educational Advancement Foundation, xiii, 332
Eilenberg, Samuel, 192
Einstein, Albert, 163, 228
Eisenhart, Katherine (Mrs. Luther Pfahler), 25
Elder, Ralph, xiii
Emory University, 350, 353
Encyclopedia Britannica, 349
Erwin, Frank, 327
Estill, Mary Ellen
 see Rudin, Mary Ellen (née Estill)
Ettlinger, Hyman Joseph, 95, 120, 164, 227, 228, 278, 282, 285, 290, 295, 296, 299, 314, 319, 323–325, 329, 330
Ettlinger, Martin, 164, 228, 355
Eyles, J., 2

Fermat's Last Theorem, 231
Fine, Henry Burchard, 25, 82
Finite intersection property, 302
Fisher, George Egbert, 97
Fitzpatrick, Ben, xiii, 103, 227, 303, 355
Fleissner, William, 348
Flexner, W.W., 139
Fogwell, T.W., 326
Forbes, Douglas, xiii, 199, 211, 329
Foster, Mary
 see Spencer, Mary (née Foster)
Foundations of Geometry
 see *Grundlagen der Geometrie* (Hilbert, D.)
Foundations of Point Set Theory (Moore, R.L.), 126, 136, 141, 144, 154
Fox, Don, 356
Frantz, B. Joe, 3, 8, 30, 37, 105, 117, 118, 148, 196, 314
Fréchet, Maurice, 98
Fry, Thornton C., 165, 166, 169

Index

Function theory, 50, 51, 131

Gauss, C.F., 16
Gehman, Harry, 102, 339
Geometry, 23, 29, 33, 63, 81, 84, 85
    analytic, vii, 264, 365
        see also Moore, Robert Lee: and geometry
    axioms, 66, 105
    Euclidean, 26, 43, 62
    Non-Euclidean, 16, 24, 26 29, 43, 50
    summer course, 353, 354
Gleason, Andrew, 102, 340
Golightly, George, 356
Gordon, Cameron, 218
Goucher College, 341
Green Book, The, see Concerning Dean John R. Silber and the Proposed Dismissal of Professor R.L. Moore
Green, John W., ix, xiii, 273, 274, 354
Greenwood, Robert, 27, 148–150, 231, 232, 237, 238, 355
Group theory, 50, 58, 62
Grover, Blanche Bennet, 120
Grundlagen der Geometrie (Hilbert, D.), 33, 60, 61
Guy, W.T., 298, 323, 343

Hackerman, Norman, 266, 267, 319, 328
Hallet, G. H., Jr., 99, 114, 340
Hallet, George Hervey, 97, 99, 100
Halmos, Paul, 369, 377, 378
Halsted, George Bruce, 10, 12–16, 21, 24, 28, 30, 31–44, 51, 52, 67, 68, 82, 87, 88, 92, 324
        see also Moore, Robert Lee: correspondence
Hamilton, O.H., 355, 356
Hamstrom, Mary-Elizabeth, 244, 245, 249–253, 280, 340, 341, 349
Harper, William Rainey, 45–50, 92
Harvard University, 347
Heine-Borel theorem, 66, 67, 247–248
Hellinger, Ernst, 235
Hensley, Elmer Lee, 355
Henderson, George W., 295, 351
Hilbert, David, 19, 32–35, 40, 51, 60, 61
    axioms, 34–36, 87, 88, 102, 104
Hinrichsen, John W., 353
Hinstead, Ralph E., 355
Hocking, John, 191, 346

Houston, Raymond, 356
Hunt, Walker, 289
Huntington, E.V., 79

Illinois Institute for Technology, 350
Indecomposable continua, 221, 295
Institute for Advanced Study in Princeton, 74, 163, 221, 347
Invariant theory, 24, 51
Iowa State University, 352, 353

Jackson, Robert E., 355
Jacob, Tom, 356
Janiszewski, Zygmund, 340
Johnson, Raymond, 290, 291
Jones, F. Burton, 103, 104, 150–155, 163, 164, 173, 174, 182, 188, 189, 216, 239, 242, 303, 344, 349
Jones, Steven, 355
Jordan (simple closed) curve, 75, 79, 105, 112

Kelley, John L., 155
Kelley, 'Spider,' 85
Kilgore, H.M., 171
Kirby, Robion, 349
Klein, Felix, 49, 50, 51, 62
Kline, John R., 96, 97, 101, 102, 112–114, 130, 221, 339, 342, 349
Kline Sphere Problem, 219
Klipple, Edmund, 138, 158, 164, 343
Knaster, Bronislaw, 138, 346
Kroeger, Nell Elizabeth (née Stevenson), 304, 355
Kuratowski, C., 134, 303

Lamar University, 352
Lane, Ralph, 314, 351–352
Lax, Peter, 235, 343
Lefschetz, Solomon, 139, 140, 155, 156, 163, 191, 342
Legacy of R.L. Moore Project, xiii, 103, 163, 272, 332
LeMaistre, Charles, 328, 329
Lennes, N. J., 63, 67, 78, 79
Levi-Civita, T., 168
Lewis, Albert C., xiii, 42, 315
Lindemann, L. A., 171
Lobachevsky, Nikolai, 26, 27
Logic and set theory, 25, 54, 58, 79, 80, 98, 100, 105, 110, 137, 138, 173, 297, 298

see also Axiomatics
Loomis, Lynn, 325
Los Alamos National Laboratory, 272
Louisiana State University, 347
Lovelace, Randy, 301
Lowell, James Russell, 21
Lubben, Renke G., 129, 135, 153, 164, 169, 173, 190, 314, 315, 323, 325, 341
Lucas, Harry, Jr., xiii, 103, 332, 356

McAuley, Louis, 155
McClendon, James W., 356
MacDonald, Malcolm, 319, 320
McShane, E.J., 342
Mahavier, Jean, xiv, 356
Mahavier, Lee, xiii
Mahavier, Ted, xiv
Mahavier, William S., 263, 265, 310, 350
Malcolmson, Waldemar, 2, 8, 9, 10, 11
Marconi, Guglielmo, 171
Marshall High School (Texas), 40, 41
Mars Viking Project, 301–302
Maschke, Heinrich, 51, 52, 54, 58, 59, 61, 76, 103
Mathematical Association of America (MAA), 139, 178, 208, 220, 233
Mathematical Reviews, 154, 169, 170
Mauldin, R.D., 320
Mayes, Vivienne M., 290
Mazurkiewicz, Stefan, 295
Measure theory course, 297–298, 365
Menger, Karl, 139, 153
Miller, Harlan C., 188, 242, 255, 345
Mohat, John T., x, 350
Moise, Edwin E., 178, 181, 210–214, 220, 221, 233, 280, 288, 295, 310, 311, 346
Monger, Blanche Joanne
   see Baker, Blanche Joanne (née Monger)
Montgomery, Deane, 74, 82, 102, 340
Moore, Charles Jonathan (RLM's father), 4, 5, 8, 85
Moore, Eliakim Hastings, 20, 28, 29, 35, 36, 40, 44, 48, 49, 50–61, 66, 67, 74, 76, 79, 88, 92, 116, 117, 141
Moore, James Willard (RLM's uncle), 91
Moore, Louisa Anne (RLM's mother), 3, 4, 5, 8
Moore, Margaret MacLellan (RLM's wife), 93, 94, 284, 330

Moore Genealogy Project, 333–337
Moore Method, vii, viii, ix, 2, 56, 110, 132, 181, 188, 285
  characteristics of, 100, 127, 147
  criticisms of, 126–127, 150, 214, 232, 250, 271
  curriculum, 365–366
  experienced by RLM's students, 99, 151–153, 181–191, 211–214, 234–235, 250–257, 273–274, 305–307
  and grades, 213
  influence on education, 216, 233, 302, 331–332
  origin, 98, 107, 258–259, 317
  used and modified by others, 114–115, 154, 155, 188, 216, 232–233, 235–236, 288, 303, 308, 310–311
  see also calculus
Moore, Robert Lee
  algebra, 266, 366
  and American Mathematical Society, 55, 74, 104, 110, 136, 139
    presidency, viii, 52, 155, 156, 167, 178, 195
    Visiting Lectureship, viii, 74, 139
  and analysis, 60
  ancestry, 3–8, 175, 333–337
  and anti-Semitism, 163, 164, 214
  appearance, viii, 85, 269, 271, 294, 316
  attitudes and beliefs, ix, x, 17, 42, 54, 59, 76, 77, 84, 89, 115, 130, 132, 157–159, 162, 167, 170–172, 179, 187, 201–204, 217, 232–234, 237, 260, 295, 296, 304, 317
  boxing, interest in, 85, 86
  and calculus, 127, 131, 150, 258–268, 365
  Challenge in the Classroom (film), vii, 115, 258, 269, 309, 316, 317
  character, 150, 179, 202, 259
  correspondence
    with G.D. Birkhoff, 139, 140, 157
    with G.B. Halsted, 10, 17, 38, 40–44, 55, 68, 69, 71, 77, 82, 104, 105
    with Mary-Elizabeth Hamstrom, 245, 246–249
    with J.R. Kline, 101, 221
    with Mary Spencer (née Foster), 255
    with Oswald Veblen, 79, 86, 95

Index

death, 330
diary, 12, 15, 17, 31, 32, 40, 41, 44, 65, 76, 77, 78
and differential invariants, 61
dissertation: *Sets of Metrical Hypotheses for Geometry*, 75
and driving, 213
education, 14, 32
   elementary, 2, 8–10, 55, 233
   influence on, 54, 233, 286, 294, 331
     see also Moore Method
   University of Chicago, 44, 45, 46, 47, 50, 52, 54
   University of Texas, 12–17, 115
and genealogy, 3, 5, 333–337
and geometry, 16, 26, 29, 60, 61, 75
   see also Moore, Robert Lee: and topology
and guns, 229, 318, 319
Halsted, G. B., relationship with, 15, 16, 17, 20, 21
Lefschetz, Solomon, relationship with, 191, 192
Legacy Project, xiii, 103, 272, 332
M.A. students, 232, 356, 357
malaria, effect of, 76, 77, 84
marriage, 94
Marshall High School (Texas), teaching post at, 40, 41
Moore, E. H., relationship with, 54, 55, 56
National Academy of Sciences, election to, 190
PhD students, i, viii, ix, 144, 145, 178, 231, 232, 314, 330, 339–357
politics, 162, 165, 166, 171, 187, 194, 198, 199
productivity, 84, 95, 104, 114, 126, 144, 174, 178, 231, 285
publications, 52, 61, 89,
   articles and papers, 87, 89, 95, 104, 110, 112–114, 126, 128, 144, 174, 175
   Foundations of Point Set Theory, 126, 136, 141, 144, 154
   On the Foundations of Plane Analysis Situs, 105, 110
race prejudice, 244, 287–290
   see also Moore, Robert Lee: and anti-Semitism
and religion, 42, 43
Report on Continuous Curves from the Viewpoint of Analysis Situs (Moore, R.L.), 128
reputation, 12, 65, 105, 130, 259
research, viii, 2, 20, 51, 55, 110, 111, 126, 231, 366
retirement, 276–285, 296, 319–324, 329
Science Mobilization Bill, opposition to, 171, 172
sense of humor, 224, 273, 317
Socrates, compared with, 146, 305, 326
and students
   female, 242, 243
   recruitment of, 128–130, 147–148, 150, 183, 259, 305
   relationship with, 41, 52, 100, 102, 103, 136, 149, 150, 179, 215, 216, 218, 234, 300
   see also Moore, Robert Lee: M.A. students; Moore, Robert Lee: PhD students
teaching posts
   Marshall High School (Texas), 40, 41
   University of Tennessee, Knoxville, 76, 80, 82, 84
   Princeton University, 82, 83, 84
   Northwestern University, Evanston, 91, 92, 93, 95
   University of Pennsylvania, 96, 100, 114
   University of Texas at Austin, 12, 66, 98, 115, 117
teaching style, vii, 2, 10, 54, 55, 99, 100, 106, 111, 115, 116, 137, 185, 186, 217, 250, 257–274, 299, 305
   see also Moore Method
thesis, 67, 78, 86, 87, 88
and topology, 79, 98, 103, 104, 110, 111–113, 144, 151, 366
training scientists, ix, 153, 174, 259, 265, 268, 272, 273, 295, 300–302, 356
Vandiver, Harry Schultz, relationship with, 226–231
Veblen, Oswald
   correspondence with, 79, 86, 95
   relationship with, 61, 62, 74, 75
Wall, H. S., relationship with, 236, 237

women, attitude toward, 242
Moore School, 110, 163, 178, 300, 330
Moore spaces, 153, 154, 331, 344, 352
Moreman, Douglas, 356
Morse, Marston, 169
Mullikin, Anna M., 106, 107, 114, 184, 242, 340, 349

National Academy of Sciences, ix, 126, 140, 190, 217
National Aeronautics and Space Administration (NASA), 301
National Science Foundation, viii, 238, 300, 332
Neuberger, John, xiv, 235, 272
Neugebauer, Otto, 168, 170
Newton, Hubert Anson, 48, 49
Normal Moore space problem, 344
Northwestern University, 92, 93, 95, 235

Ochoa, James, xiv
O'Connor, Joel L., 354
O'Daniel, W. Lee 'Pappy', 195–197
Office of Naval Research, 295
Ohio University, 351
On the Foundations of Plane Analysis Situs (Moore, R.L.), 105, 110
Osgood, William Fogg, 119
Osserman, Robert, 295

Painter, T.S., 204, 276
Parshall, Karen Hunger, 21, 52, 53, 58
Pearson, Bennie J., 246–248, 350
Peirce, Benjamin, 23, 24, 25
Peirce, Charles, 25, 41
Pennsylvania State University, University Park, 350
Poincaré, Henri, 68, 70, 145
Point set theory, 105, 112, 113, 131, 137, 153, 173
Polish School, 128, 138
Pons asinorum, 308
Porcelli, Pasquale, 235
Porter, Milton Brockett, 27, 113, 118, 127, 153
Pound, Patricia, 356
Princeton University, 21, 64, 74, 82–85, 92, 122, 156, 163
Proffitt, Michael H., 260, 354
Proportional Representation League, 340
Pseudo arcs, 221, 295, 346, 352

Purifoy, Jesse A., 354
Putnam, T.M., 28

R.L. Moore Oral History Project, 229, 263, 287
Racial segregation, 242, 286–289
Rainey, Homer P., 194, 196, 200–204
Ransom, H.H., 328
Reed, Coke, xiv
Reed, Dennis K., 352
Reed, G.M. (Mike), xiv, 303, 330, 331, 351
Reid, W.T., 129, 218, 355
Report on Continuous Curves from the Viewpoint of Analysis Situs (Moore, R.L.), 128
Reynolds, C.N., 139
Richter, Gary, 355
Riesz, Frigyes, 98, 113
Robbins, R.B., 166, 167
Roberts, John H., 129, 138, 153, 154, 342
Rockefeller, John D., 46, 47
Rogers, Jack W., 353
Rogers, Janet, 356
Rogers, Lorene, 356
Roitman, Judy, 348
Rowe, David E., 23, 58
Rudin, Mary Ellen (née Estill), 189, 210–212, 215, 242, 252–254, 272, 280, 303, 321, 344, 347
Rudin, Walter, 252–254
Rudolph, Frederick, 47
Rutt, N.E., 102, 129, 339

Sandia, space program, 301, 302
Schmitt, Cooper D., 80, 81, 82
Schumaker, Carol, xiv
Schur, Friedrich, 36, 61
Schwartz, Hermann, 50
Schwatt, Isaac J., 97
Secker, Martin D., 353
Sierpiński, Wacław, 128, 134
Silber, John R., 320–330
Singh, Sukhijit (Suji), 218
Slye, John M., 280, 310, 349
Smith, Kermit, 356
Smith, Michel, 218
Socrates, 326, 327
Sorgenfrey, Robert H., 188, 345
Souslin's problem, 254
Spencer, Mary (née Foster), 254, 255
Spirals in the plane, 357

Splawa-Neyman, J., 298
Springer, Julius, 168
Starbird, Michael, 218
Statistics, 343, 354, 356
Steenrod, Norman, 155, 176, 192, 341
Stevenson, Nell Elizabeth
   see Kroeger, Nell Elizabeth (née Stevenson)
Stewart, A. N., 289
Stiles, F. A., 326
Stocks, D. R., 355
Stone-Čech compactification, 342
Stone, Ormond, 48
Stone, Wilson, 268
Stubblefield, Beauregard, 191, 287, 288, 346
Swain, Robert L., 188, 344
Sweatt, Herman M., 289
Swingle, P. M., 139
Sylvester, James Joseph, 21, 22, 23, 41, 104

Tamarkin, J. D., 168, 170
Taylor, Thomas Ulvan, 27, 31
Texas A&M University, 343, 344, 351
Texas Method, 98, 236
   see also Moore Method
Texas Woman's University, 345
Theorems, 56, 188
   proving, 2, 99, 114, 121, 187, 190, 217, 235, 258, 271, 307
Topology, 58, 67, 128, 153, 173, 331
   'Moore spaces', 153, 154, 300, 331
Traylor, D. R., 41, 59, 105, 329, 355
Treybig, June, 356
Treybig, L. Bruce, 321, 350
Trigonometry, 271
Tucker, D. H., 355

'Un-American Activities', 196
University of California, Santa Barbara, 341
University of Chicago, 40, 44–58, 62, 74, 92, 156
   Department of Mathematics, 29, 48, 50, 58
   foundation, 45, 47, 48
   Mathematics Club, 56, 65
University of Houston, 350, 351
University of Illinois, Urbana-Champaign, 349

University of Michigan, Ann Arbor, 341, 347
University of Minnesota, Twin Cities, 350
University of Mississippi, 353
University of Missouri, Columbia, 351
University of Missouri, Kansas City, 350
University of Nebraska, Lincoln, 352
University of North Texas, 350
University of Pennsylvania, 96, 100, 102, 105, 114, 117, 118, 129, 194, 195, 221
University of Tennessee, Knoxville, 76, 80, 82, 84
University of Texas at Austin, 11, 26, 36, 37, 38, 40, 85, 102, 276, 277, 317, 318, 329, 346
   mathematics departments, 29, 118, 119, 120, 127, 137, 148, 154, 226, 227, 230, 238, 239, 276, 296, 314, 316, 323, 329, 330
   student life, 12–14
University of Utah, 348, 352
University of Wisconsin, Madison, 346, 347
University of Wisconsin, Milwaukee, 351
Upper semi-continuous collections, 175, 221, 300
Urysohn, Pavel, 112

Vandiver, Frank, 228–230, 324
Vandiver, Harry Schultz, 64, 65, 120, 122, 123, 182, 205, 226–231
Van Doren, Kenneth, 356
Veblen, Oswald, 53, 59–95, 103–105, 110, 112, 163, 170
Vickery, Charles W., 138, 164, 169, 343
Vitali coverings, 297, 300
Vivian, Roxana Hayward, 96

Wall, Hubert Stanley, 120, 235, 236, 272, 285, 295, 296, 314, 325, 326, 329, 356, 357
Webb, Walter Prescott, 229
Webs, 357
Weierstrass, Karl, 49, 50, 52
Weyl, Hermann, 145
Whyburn, Gordon T., 128, 133, 134, 137, 138, 139, 156, 169, 178, 342
Whyburn, Lucille, 129, 134, 148, 149, 355
Whyburn, William M., 134, 345, 355
Wicke, Howard, 300, 302, 351

Wilder, Raymond L., 33, 79, 80, 89, 96, 104, 111, 128–132, 139, 144, 146, 153, 154, 174, 178, 288, 305, 308, 339, 341
Wilks, S. S., 356
Williams, Nick, 356
Williams, Robert, 355
Wilson, Thomas Woodrow, 82, 83
Wilson, Wilfrid, 139
Woodard, Dudley Weldon, 96
Worrell, John, xiv, 259, 266–268, 272, 297–304, 309–310, 322, 328, 351

Young, Gail S., 179, 181–191, 221, 288, 345
Young, Sam W., 305–307
Younglove, James N., 351

Zentralblatt für Mathematik und ihre Grenzgebiete, 167–169
Zippin, Leo, 102, 340
Zitarelli, David E., 45, 372
Zoretti, L., 113

About the Author

British author **John Parker** has been a journalist and writer all his working life. He went straight from Kettering Grammar School to join *Northampton Chronicle and Echo* as a trainee reporter and remained in local newspapers in the UK until securing a position with the *Nassau Daily Tribune* in the Bahamas.

He later worked for *Life* magazine, New York, before returning to the UK to join the *Daily Mirror*, then one of Britain's foremost and highly respected daily newspapers as a sub-editor. Hometown beckoned again when at 30, he was appointed editor of *Northamptonshire Evening Telegraph,* the country's youngest evening newspaper editor at that time.

Towards the 1980s, he returned to Fleet Street, rejoining the Mirror Group to become night editor of the *Daily Mirror* and later deputy editor of the *Sunday Mirror*. Along with a number of other Mirror stalwarts, he resigned during the reign of Robert Maxwell to concentrate wholly on his writing. To date, he has published 30 books in hardback which have appeared in 64 editions in the UK and more than 40 international editions.